AF389312

Sustainable Catalytic Conversion of Biomass for the Production of Biofuels and Bioproducts

Sustainable Catalytic Conversion of Biomass for the Production of Biofuels and Bioproducts

Special Issue Editors

Gabriel Morales
Jose Iglesias
Juan A. Melero

MDPI • Basel • Beijing • Wuhan • Barcelona • Belgrade • Manchester • Tokyo • Cluj • Tianjin

Special Issue Editors

Gabriel Morales	Jose Iglesias	Juan A. Melero
Universidad Rey Juan Carlos	Universidad Rey Juan Carlos	Universidad Rey Juan Carlos
Spain	Spain	Spain

Editorial Office
MDPI
St. Alban-Anlage 66
4052 Basel, Switzerland

This is a reprint of articles from the Special Issue published online in the open access journal *Catalysts* (ISSN 2073-4344) (available at: https://www.mdpi.com/journal/catalysts/special_issues/Catalytic_Conversion_Biomass_Biofuels_and_Bioproducts).

For citation purposes, cite each article independently as indicated on the article page online and as indicated below:

LastName, A.A.; LastName, B.B.; LastName, C.C. Article Title. *Journal Name* **Year**, *Article Number*, Page Range.

ISBN 978-3-03936-433-6 (Hbk)
ISBN 978-3-03936-434-3 (PDF)

Contents

About the Special Issue Editors

Gabriel Morales (Ph.D., Associate Professor) was born in Madrid (Spain) in 1977, and completed his MEng in Chemical Engineering at Universidad Complutense de Madrid in 2000. He completed his Ph.D. in 2005 (Universidad Rey Juan Carlos), for which he worked on the synthesis and characterization of sulfonic acid-functionalized mesostructured materials. He has held several teaching positions in the Department of Chemical and Environmental Technology, gaining a permanent position in 2008. He was a visiting researcher at the Department of Chemical Engineering in the University of California at Santa Barbara (USA) in 2004 and 2006, under the supervision of Prof. Brad F. Chmelka; and at the Department of Applied Chemistry, University of Strathclyde (Glasgow, UK), under the supervision of Prof. Peter Cormack. During his research career, his main research has been focused on the synthesis, characterization and catalytic application of organically-functionalized mesostructured materials; the immobilization of enzymes on silica mesoporous supports for different applications; the valorization of glycerol to obtain oxygenated additives for fuel blends; the production of biodiesel over heterogeneous acid catalysts; and more recently, the catalytic up-grading of cellulosic biomass. His current research interests focus on the application of catalysis to achieve a more sustainable and bio-based economy, through the biorefinery concept.

Jose Iglesias (Ph.D., Associate Professor) was born in Benavente (Spain) in 1976. He completed his MEng at Universidad Complutense, Madrid (1999, UCM), and his PhD in Chemical Engineering at Universidad Rey Juan Carlos (2005, URJC), working on the development of new heterogeneous catalysts for asymmetric epoxidation. He has undertaken several research positions: in the group of Professor Brian F.G. Johnson at the University of Cambridge (2003), working on the synthesis of new heterogeneous catalysts for green technology, in the group of Professor S. David Jackson at the University of Glasgow (2006), working on the heterogenization of BINAPO-Ir catalytic systems, and in the CDFA group at IRCELYON (2017), working on the development of new cascade synthesis strategies for biomass valorization. Currently, he is an associate professor of Chemical Engineering at the Department of Chemical, Energy and Mechanics Technology, URJC. His main research interests are focused on the rational design of heterogeneous catalytic systems for green and fine chemistry, lately applied to the chemical valorization of biomass.

Juan A. Melero (Ph.D., Full Professor) was born in Madrid (Spain) in 1970. He studied Chemistry at Universidad Complutense, Madrid (1988–1993), and completed his Ph.D. in 1998, for which he worked on the synthesis and applications of zeolitic materials for redox and acid-catalyzed reactions. He undertook postdoctoral research in the group of Prof. G. Stucky and B. F Chmelka at California University in Santa Barbara for one year, working on the synthesis and applications of organically modified mesostructured materials. Currently, he is a full-time professor of chemical engineering at Universidad Rey Juan Carlos (Madrid). His research is focused on the synthesis and characterization of porous solids with enhanced properties, and their catalytic application in fine chemistry, environmental catalysis and biomass valorization, towards bio-products and biofuels, being a leader-researcher of several active projects related to these topics.

 *catalysts*

Editorial

Sustainable Catalytic Conversion of Biomass for the Production of Biofuels and Bioproducts

Gabriel Morales *, Jose Iglesias * and Juan A. Melero *

Chemical and Environmental Engineering Group, Escuela Superior de Ciencias Experimentales y Tecnología (ESCET), Universidad Rey Juan Carlos, 28933 Móstoles, Madrid, Spain
* Correspondence: gabriel.morales@urjc.es (G.M.); jose.iglesias@urjc.es (J.I.); juan.melero@urjc.es (J.A.M.)

Received: 18 May 2020; Accepted: 20 May 2020; Published: 22 May 2020

Biomass, in its many forms—oils and fats, lignocellulose, algae, etc.—is widely contemplated as a potential alternative to dwindling fossil fuel reserves, as its enormous and widespread availability allows for its consideration as a source for biofuels and a wide range of bio-based chemicals. The search for new sustainable and efficient alternatives to fossil feedstock is gaining increasing relevance within the chemical industry, wherein the role of catalysis is often critical for the development of clean and sustainable processes, with a high efficiency and atom economy [1]. In transportation, the penetration of biofuels has reached so far just modest figures because of a number of limitations associated to the so-called first-generation biofuels (mainly ethanol from sugars containing plants or cereal crops, and biodiesel from edible vegetable oils). Accordingly, the development of new types of biofuels has been identified as an urgent need. Likewise, the term "advanced biofuels" has spread to refer to biofuels that comply with sustainability criteria and do not compete with the food market. They should be produced preferentially from forestry and agriculture residues, non-food energy crops, industrial wastes and other non-conventional sources, such as microalgae and microorganisms. This scheme is expected to exhibit many benefits, as it would allow combining the large-scale production of biofuels, accompanied by a significant reduction of greenhouse gases emissions, with an efficient management and valorisation of different types of biomass wastes. The biorefinery concept has been envisaged to integrate within a single facility the upgrading of different raw materials (biomass and wastes) into transportation fuels, chemicals and energy. Noticeably, a relatively small number of platform chemicals has been identified and proposed as key intermediates to the final target products.

The main routes for the conversion and valorisation of biomass are sorted according to its chemical nature. For lipids and oils, the main routes are transesterification, catalytic cracking and hydrotreatment to produce different types of biofuels within the gasoline and diesel range. Transesterification remains the main catalytic route for the production of fatty acid alkyl esters as a replacement for diesel fuel [2]. Catalytic cracking and hydrocracking present the advantage of leading to higher liquid fuels yield. Numerous examples can be found in the literature about the use of heterogeneous catalysts for both types of transformations. A key aspect of these catalysts is their ability to remove oxygen contained in the oleaginous feedstock by decarboxylation, decarbonylation, dehydration and hydrodeoxygenation reactions, usually requiring the incorporation of metal species to an acidic solid support [3].

The major route currently in use for the conversion of sugar carbohydrates is the combination of enzymatic saccharification and fermentation treatments to produce alcohols, mainly ethanol. In contrast, when using lignocellulosic biomass, a large variety of alternative processes are possible, such as gasification (usually combined with Fisher-Tropsch synthesis), pyrolysis, liquefaction, hydrolysis and aqueous phase reforming. Moreover, aqueous sugars may be catalytically upgraded through chemical strategies, usually in the presence of catalysts, to produce platform molecules and bio-based chemicals [4]. The development of efficient processes for the valorisation of lignocellulosic biomass is currently of great relevance, since it would open up the possibility of generating valuable bio-products from a great variety of abundant and renewable sources, such as agriculture, pruning and forestry residues, agroindustrial waste or even dedicated non-edible energy crops. For the conversion of lignocellulose, both thermocatalytic and chemocatalytic transformations are envisioned. The former involves the conversion of lignocellulose through a fast pyrolysis process into bio-oil, which is subsequently catalytically upgraded through distinct types of treatments, such as pyrolysis, esterification, ketonization, aldol condensation and hydrodeoxygenation. The chemocatalytic routes mostly start from the sugars obtained by lignocellulose hydrolysis, which can be further transformed by isomerization, dehydration and hydrogenation into different intermediate compounds, usually called platform molecules (e.g., 5-HMF, furfural, levulinic acid, sorbitol, etc.). In turn, such platforms can be converted into a large variety of final products with relevant applications in the formulation of advanced biofuels and/or the commercially valuable chemicals [5,6].

To summarize, there is a wide range of biomass sources, conversion routes and products, where the use of catalysis is of major importance to provide clean and sustainable processes with a high efficiency and atom economy. Although large efforts are worldwide invested in the implementation of heterogeneous catalysts for biomass transformation, results are still less straightforward than initially thought. This is because biomass presents a number of particular features that difficult the development of feasible valorisation processes: (i) biomass typically consists of complex and bulky molecules with multiple functionalities, the processing of which is not an easy task, usually requiring pre-treatments to facilitate the contact catalyst-biomass; (ii) high water content, and consequently hydrothermal conditions, in many cases with non-neutral pH, which may have severe detrimental effects on the catalysts; (iii) aside of carbon and hydrogen, biomass contains significant amounts of heteroatoms like oxygen (reaching values close to 50 wt% in many biomass components), nitrogen and phosphorus, which typically need to be partially or totally removed to provide the target products; and (iv) formation of carbonaceous deposits through non-desired transformations, due to the complex composition of biomass, making it highly reactive because of the presence of a multitude of components with different functional groups. The development of new catalytic processes, able to perform multifunctional cascade transformations and aiming to process intensification principles, is one of the keys to face these challenges. Classic heterogeneous catalysts need to be tailored for biomass valorisation, especially in terms of enhanced accessibility, tight control of the acidic features, generation of basic sites, surface polarity and the preparation of multifunctional materials by incorporation of metal phases [7].

This Special Issue gathers works at the cutting edge of investigation in application of catalysis for the sustainable conversion of biomass into biofuels and bio-based chemicals. Lignocellulosic biomass attracts most of the attention, with up to eight manuscripts related to this topic. Thus, Huiling Li et al. present a study on the catalytic hydrolysis of lignocellulose over magnetic solid acid in an efficient pathway, where a bamboo-derived carbonaceous magnetic solid acid catalyst was synthesized by $FeCl_3$ impregnation, followed by carbonization and $-SO_3H$ group functionalization [8]. Xueru Sheng et al. show the use of waste-seashell-derived CaO catalysts as high-performance solid base catalysts for biomass-derived cyclopentanone self-condensation, which is an important reaction in bio-jet fuel or perfume precursor synthesis [9]. Michèle Besson et al. present the comparative selective catalytic C-O bond hydrogenolysis of C5-C6 polyols, sugars, and their mixtures for the production of valuable C6 and C5 deoxygenated products over ReO_x-Rh/ZrO_2 catalysts. They show that C-O bond cleavage occurs significantly via multiple consecutive deoxygenation steps, leading to the formation of linear deoxygenated C6 or C5 polyols [10]. Elísabet Pires et al. show the application of sulfonated hydrothermal carbons from cellulose and glucose as catalysts for glycerol ketalization. In their work, the sulfonated hydrothermal carbons were also coated on a graphite microfiber felt (SHTC@GF), and they report the first results on the continuous flow production of solketal [11]. The group of Dr. Anders Riisager contributes with an efficient and selective catalytic method for the aerobic oxidation of lignin and lignin model compounds to aromatics. Particularly, their work relates to the oxidative cleavage catalysed by Ru/Al_2O_3 of a representative lignin model compound, guaiacyl glycerol-β-guaiacyl ether [12]. Karen Wilson et al. present the acetic acid ketonization catalysed by Ga/HSM-5 for the upgrading of biomass pyrolysis vapours. Pyrolysis bio-oils contain significant amounts of carboxylic acids, which limit their utility as biofuels. Ketonisation of carboxylic acids within biomass pyrolysis vapours is a potential route to upgrade the energy content and stability of the resulting bio-oil condensate, but requires active, selective and coke-resistant solid acid catalysts. In their work, Ga-doped HZSM-5 exhibited good stability for over five h on-stream acetic acid ketonization [13]. The group of Dr. Robert A. Dagle analyses the viability of using a $Zn_xZr_yO_z$ mixed oxide catalyst for the direct production of C4 olefins from the aqueous phase derived from three different bio-oils. Complete conversion of the carboxylic acids was achieved over said catalyst for all the feedstocks investigated, being the main reaction product isobutene (>30% selectivity) [14]. Pedro Maireles-Torres et al. have contributed with a work about the selective conversion of glucose into HMF using L-type zeolites with different morphologies, where they also show that the addition of $CaCl_2$ has a positive influence on the catalytic performance [15].

Within the field of biodiesel and oleaginous feedstock, Ching-Chang Chen et al. present the use of bauxite as a low-cost solid base catalyst precursor for the production of biodiesel via transesterification. Bauxite is economic, contains a high percentage of Si and Al species, and can replace expensive commercial materials [16]. Likewise, Mark Crocker et al. have demonstrated the promotional effect of Cu, Fe and Pt on the performance of Ni/Al_2O_3 in the deoxygenation of used cooking oil to fuel-like hydrocarbons [17].

Finally, Xiaohui Zhang et al. make an evaluation on the methane production potential of wood waste pre-treated with NaOH and co-digested with pig manure, to provide an effective method to apply wood waste in anaerobic digestion [18]. Last, but not least, Dr. Zhou & Dr. Hu provide a comprehensive review analysing the catalytic thermochemical conversion of algae and upgrading of algal oil for the production of high-grade liquid fuel [19].

In conclusion, we personally feel that the present Special Issue *"Sustainable Catalytic Conversion of Biomass for the Production of Biofuels and Bioproducts"* is of great interest and relevance, as it covers many of the new aspects of catalysis applied to biomass, and significantly contribute to further development in this field. We are honoured to have been Guest Editors for this issue, and would like to thank all the contributors and reviewers for providing us with their valuable manuscripts and comments. We are also grateful to Keith Hohn, the Editor-in-Chief, and all the staff of the Catalysts Editorial Office.

Conflicts of Interest: The authors declare no conflict of interest.

References

1. Aricò, F.; Moreno-Tost, R. Editorial Overview: Editorial: Toward new sustainable development goals. *Curr. Opin. Green Sustain. Chem.* **2020**, A1–A3. [CrossRef]
2. Thangaraj, B.; Solomon, P.R.; Muniyandi, B.; Ranganathan, S.; Lin, L. Catalysis in biodiesel production—A review. *Clean Energy* **2019**, *3*, 2–23. [CrossRef]
3. Kim, S.; Kwon, E.E.; Kim, Y.T.; Jung, S.; Kim, H.J.; Huber, G.W.; Lee, J. Recent advances in hydrodeoxygenation of biomass-derived oxygenates over heterogeneous catalysts. *Green Chem.* **2019**, *21*, 3715–3743. [CrossRef]
4. Mika, L.T.; Cséfalvay, E.; Németh, Á. Catalytic Conversion of Carbohydrates to Initial Platform Chemicals: Chemistry and Sustainability. *Chem. Rev.* **2018**, *118*, 505–613. [CrossRef] [PubMed]
5. Costa, F.F.; de Oliveira, D.T.; Brito, Y.P.; da Rocha Filho, G.N.; Alvarado, C.G.; Balu, A.M.; Luque, R.; Nascimento, L.A.S. Do Lignocellulosics to biofuels: An overview of recent and relevant advances. *Curr. Opin. Green Sustain. Chem.* **2020**, *24*, 21–25. [CrossRef]
6. Arias, P.L.; Cecilia, J.A.; Gandarias, I.; Iglesias, J.; López Granados, M.; Mariscal, R.; Morales, G.; Moreno-Tost, R.; Maireles-Torres, P. Oxidation of lignocellulosic platform molecules to value-added chemicals using heterogeneous catalytic technologies. *Catal. Sci. Technol.* **2020**, *10*, 2721–2757. [CrossRef]
7. Serrano, D.P.; Melero, J.A.; Morales, G.; Iglesias, J.; Pizarro, P. Progress in the design of zeolite catalysts for biomass conversion into biofuels and bio-based chemicals. *Catal. Rev. Sci. Eng.* **2018**, *60*, 1–70. [CrossRef]
8. Zhu, Y.; Huang, J.; Sun, S.; Wu, A.; Li, H. Effect of dilute acid and alkali pretreatments on the catalytic performance of bamboo-derived carbonaceous magnetic solid acid. *Catalysts* **2019**, *9*, 245. [CrossRef]

9. Sheng, X.; Xu, Q.; Wang, X.; Li, N.; Jia, H.; Shi, H.; Niu, M.; Zhang, J.; Ping, Q.W. Waste seashells as a highly active catalyst for cyclopentanone self-aldol condensation. *Catalysts* **2019**, *9*, 661. [CrossRef]

10. Mounguengui-Dialo, M.; Sadier, A.; Noly, E.; Perez, D.D.S.; Pinel, C.; Perret, N.; Besson, M. C-O bond hydrogenolysis of aqueous mixtures of sugar polyols and sugars over ReOx-Rh/ZrO2 catalyst: Application to an hemicelluloses extracted liquor. *Catalysts* **2019**, *9*, 740. [CrossRef]

11. Fernández, P.; Fraile, J.M.; García-Bordejé, E.; Pires, E. Sulfonated hydrothermal carbons from cellulose and glucose as catalysts for glycerol ketalization. *Catalysts* **2019**, *9*, 804. [CrossRef]

12. Melián-Rodríguez, M.; Saravanamurugan, S.; Meier, S.; Kegnæs, S.; Riisager, A. Ru-catalyzed oxidative cleavage of guaiacyl glycerol-β-guaiacyl ether-a representative β-O-4 lignin model compound. *Catalysts* **2019**, *9*, 832. [CrossRef]

13. Jahangiri, H.; Osatiashtiani, A.; Ouadi, M.; Hornung, A.; Lee, A.F.; Wilson, K. Ga/HZSM-5 catalysed acetic acid ketonisation for upgrading of biomass pyrolysis vapours. *Catalysts* **2019**, *9*, 841. [CrossRef]

14. Davidson, S.D.; Lopez-Ruiz, J.A.; Flake, M.; Cooper, A.R.; Elkasabi, Y.; Morgano, M.T.; Dagle, V.L.; Albrecht, K.O.; Dagle, R.A. Cleanup and conversion of biomass liquefaction aqueous phase to C3–C5 olefins over ZnxZryOz catalyst. *Catalysts* **2019**, *9*, 923. [CrossRef]

15. Ginés-Molina, M.J.; Ahmad, N.H.; Mérida-Morales, S.; García-Sancho, C.; Mintova, S.; Ng, E.P.; Maireles-Torres, P. Selective conversion of glucose to 5- hydroxymethylfurfural by using L-type zeolites with different morphologies. *Catalysts* **2019**, *9*, 1073. [CrossRef]

16. Dai, Y.M.; Hsieh, C.H.; Lin, J.H.; Chen, F.H.; Chen, C.C. Biodiesel production using bauxite in low-cost solid base catalyst precursors. *Catalysts* **2019**, *9*, 1064. [CrossRef]

17. Silva, G.C.R.; Qian, D.; Pace, R.; Heintz, O.; Caboche, G.; Santillan-Jimenez, E.; Crocker, M. Promotional effect of Cu, Fe and Pt on the performance of Ni/Al2O3 in the deoxygenation of used cooking oil to fuel-like hydrocarbons. *Catalysts* **2020**, *10*, 91. [CrossRef]

18. Li, R.; Tan, W.; Zhao, X.; Dang, Q.; Song, Q.; Xi, B.; Zhang, X. Evaluation on the methane production potential of wood waste pretreated with NaOH and Co-digested with pig manure. *Catalysts* **2019**, *9*, 539. [CrossRef]

19. Zhou, Y.; Hu, C. Catalytic thermochemical conversion of algae and upgrading of algal oil for the production of high-grade liquid fuel: A review. *Catalysts* **2020**, *10*, 145. [CrossRef]

 *catalysts*

Article

Promotional Effect of Cu, Fe and Pt on the Performance of Ni/Al₂O₃ in the Deoxygenation of Used Cooking Oil to Fuel-Like Hydrocarbons

Gisele C. R. Silva [1], Dali Qian [2], Robert Pace [2,3], Olivier Heintz [4], Gilles Caboche [4], Eduardo Santillan-Jimenez [2,3,4] and Mark Crocker [2,3,*]

[1]   Campus Centro Oeste Dona Lindu, Universidade Federal de São João del Rei, Rua São Sebastião 400 Divinópolis, MG 35501-029, Brazil; giselec@ufsj.edu.br
[2]   Center for Applied Energy Research, University of Kentucky, 2540 Research Park Drive, Lexington, KY 40511, USA; dali.qian@uky.edu (D.Q.); robert.pace@uky.edu (R.P.); esant3@uky.edu (E.S.-J.)
[3]   Department of Chemistry, University of Kentucky, Lexington, KY 40506, USA
[4]   Laboratoire Interdisciplinaire Carnot de Bourgogne (LICB), UMR CNRS 6303, Université de Bourgogne Franche-Comté, 9 avenue Alain Savary, 21078 Dijon CEDEX, France; olivier.heintz@u-bourgogne.fr (O.H.); gilles.caboche@u-bourgogne.fr (G.C.)
*   Correspondence: mark.crocker@uky.edu; Tel.: +1-859-257-0295

Received: 20 December 2019; Accepted: 3 January 2020; Published: 7 January 2020

Abstract: Inexpensive Ni-based catalysts can afford comparable performance to costly precious metal formulations in the conversion of fat, oil, or greases (FOG) to fuel-like hydrocarbons via decarboxylation/decarbonylation (deCO$_x$). While the addition of certain metals has been observed to promote Ni-based deCO$_x$ catalysts, the steady-state performance of bimetallic formulations must be ascertained using industrially relevant feeds and reaction conditions in order to make meaningful comparisons. In the present work, used cooking oil (UCO) was upgraded to renewable diesel via deCO$_x$ over Ni/Al₂O₃ promoted with Cu, Fe, or Pt in a fixed-bed reactor at 375 °C using a weight hourly space velocity (WHSV) of 1 h^{-1}. Although all catalysts fully deoxygenated the feed to hydrocarbons throughout the entire 76 h duration of these experiments, the cracking activity (and the evolution thereof) was distinct for each formulation. Indeed, that of the Ni-Cu catalyst was low and relatively stable, that of the Ni-Fe formulation was initially high but progressively dropped to become negligible, and that of the Ni-Pt catalyst started as moderate, varied considerably, and finished high. Analysis of the spent catalysts suggests that the evolution of the cracking activity can be mainly ascribed to changes in the composition of the metal particles.

Keywords: used cooking oil; deoxygenation; decarboxylation; decarbonylation; nickel; copper; iron; platinum; hydrocarbons

---

1. Introduction

Interest in renewable energy sources has increased considerably, mainly due to concerns related to the climate change caused by the atmospheric accumulation of greenhouse gases resulting from fossil fuel use [1]. A promising alternative to the fossil fuels used in the transportation sector is the production of biofuels—e.g., biodiesel, green diesel, and biokerosene—from renewable feedstocks, including vegetable oils and animal fats [2,3]. Moreover, to improve the economics of these biofuels and avoid disrupting the food supply, attention has shifted to low-cost inedible feedstocks, including used cooking oil (UCO), which is also known as yellow grease (YG) [4–6]. This particular waste stream is both abundant and inexpensive (~$463/ton), with ca. one million tons being produced annually in the U.S. alone [7].

Of the biofuels mentioned above, biodiesel is the name given to the fatty acid methyl esters (FAMEs) resulting from the catalytic transesterification of fat, oil, or greases (FOG) with methanol. Biodiesel has several advantages over petroleum-based diesel, such as being biodegradable and producing less harmful gas emissions and particulate matter upon combustion [8,9]; however, it also displays several drawbacks, including poor cold flow properties as well as relatively low thermal and oxidative stability, mostly arising from its oxygen content [10]. The catalytic deoxygenation of FOG to fuel-like hydrocarbons offers advantages over biodiesel in terms of fuel quality and feedstock flexibility. Indeed, deoxygenation processes are able to handle FOG feeds with significantly higher free fatty acid concentrations relative to those typically required for biodiesel synthesis [11].

As shown in Scheme 1, FOG deoxygenation can proceed via hydrodeoxygenation (HDO) and decarboxylation/decarbonylation (deCO$_x$). In HDO, oxygen is removed as H_2O, and the alkanes produced have the same number of carbon atoms as the corresponding fatty acid chains comprising the FOG. In decarboxylation, oxygen is removed in the form of CO_2, and in decarbonylation, oxygen is removed as H_2O and CO. In both cases, the resulting alkanes have one carbon atom less than the corresponding fatty acid bound in the triglyceride. Depending on the catalyst and the experimental conditions employed, the CO and CO_2 produced in the gas phase may react with hydrogen to form CH_4. In fact, because of the number of confounding reactions, which include not only methanation but also the water gas shift (WGS) and the Boudouard reaction, these routes cannot be easily distinguished from each other based solely on the amounts of H_2O, CO_2, and CO produced [12,13].

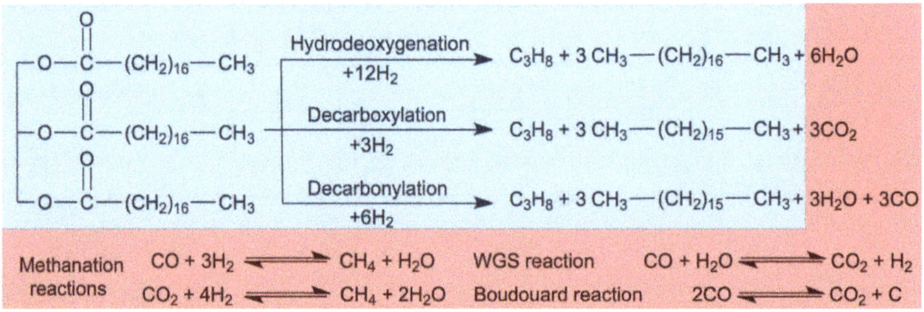

Scheme 1. Deoxygenation routes for tristearin as a model compound representing triglycerides (blue shading) and concomitant reactions confounding oxygen-bearing deoxygenation products (red shading).

In recent years, the production of fuel-like hydrocarbons via deCO$_x$ has been intensively investigated as a way to avoid the large amounts and pressures of hydrogen, as well as the problematic sulfide catalysts required by HDO, since the hydrogen requirements of deCO$_x$ are lower and these reactions proceed over simple supported metal catalysts [14]. Although the majority of deCO$_x$ studies have focused on Pd and Pt catalysts, the high price of these metals has spurred the search for alternatives. Saliently, inexpensive Ni-based catalysts can provide comparable results to Pd and Pt formulations in the deCO$_x$ of FOG to hydrocarbons [15,16].

Since the high activity of Ni in C–C hydrogenolysis can decrease the carbon yield and the hydrogen efficiency of deCO$_x$ processes, the incorporation of a second metal has been investigated as a means to modify the electronic and geometric properties of Ni to ultimately improve its activity and selectivity [17]. Indeed, Ni/Al$_2$O$_3$ promoted with Cu or Pt can afford near quantitative diesel yields in the conversion of both model and realistic lipid feeds to fuel-like hydrocarbons [13,18]. The promotion effect displayed by these bimetallic catalysts is in large part attributed to the ability of Cu and Pt to facilitate NiO reduction at relatively low temperatures since metallic Ni sites constitute the active site for the deCO$_x$ reaction. Moreover, Pt addition also curbs the adsorption of CO on the catalyst surface, helping to avoid catalyst inhibition by any CO evolved via decarbonylation and the catalyst coking resulting from the disproportionation of CO via the Boudouard reaction [13].

Supported Ni catalysts promoted with Fe have also afforded promising results in the conversion of model and realistic lipid feeds to hydrocarbons [17], the promoting effect of Fe being attributed to the synergy between nickel sites possessing the ability to activate hydrogen and iron sites with strong oxophilicity [17,19]. Indeed, since Fe has a higher oxygen affinity than Ni, oxygen vacancies within iron oxide species can facilitate the adsorption and subsequent activation of oxygenates. Specifically, H_2 activated through its facile dissociative adsorption on Ni sites can spill over to neighboring Fe sites onto which the oxygen atoms of C=O groups are adsorbed, subsequent hydrogenation leading to deoxygenation products. In addition, the formation of Ni-Fe alloys with Fe-rich surfaces disrupts the adjacency of Ni atoms, a geometric effect known to suppress C–C hydrogenolysis, which requires Ni ensembles. Cu is also known to decrease the C–C hydrogenolysis activity of Ni through the same geometric effect [20].

In order to develop practicable catalytic $deCO_x$ technology for the conversion of FOG to fuel-like hydrocarbons, it is necessary to study the most promising catalysts using industrially relevant feeds and reaction conditions. In addition, in order to make meaningful comparisons between the performance of different catalysts, measurements must be made at a steady state. Against this backdrop, the present work investigated the conversion of UCO to renewable diesel via $deCO_x$ over supported Ni catalysts promoted with Cu, Fe or Pt. The performance of these formulations was tested in a fixed bed reactor using industrially-relevant reaction conditions for 76 h of time on stream (TOS), as previous work has shown that catalysts of this type require >48 h of TOS to attain a steady state [11]. In addition, the analysis of the fresh, spent, and regenerated catalysts was undertaken in an effort to understand the distinct performance displayed by these formulations.

2. Results and Discussion

2.1. Catalytic Deoxygenation of UCO Over Ni/Al$_2$O$_3$ Promoted with Cu, Fe, and Pt

The composition of the UCO employed in this work is shown in Table A1 within Appendix A. The feed is mostly triolein (~95%) with a small amount (~5%) of oleic acid. This feed was upgraded in a fixed bed reactor using a WHSV of 1 h^{-1} and a reaction temperature of 375 °C (see Section 3.3) in order to investigate and compare the relative effect of Cu, Fe and Pt promotion on the performance of Ni/Al$_2$O$_3$ in the conversion of UCO to diesel-like hydrocarbons. The results of the gas chromatography-mass spectrometry (GC-MS) analysis of the liquid products collected at representative times on stream are summarized in Figure 1, and are presented in more detail in Appendix A (Tables A2–A4), while the gaseous products are shown in Figure 2. In addition, a blank (sans catalyst) run was performed using an identical set of conditions in order to assess the extent of thermal (as opposed to catalytic) contributions to UCO conversion and diesel yield. The GC-MS analysis of the liquid products obtained in this blank run (see Table A5 in Appendix A) revealed the vast majority (>79%) of the products to be fatty acids and monolein stemming from the thermal conversion of triolein. In addition, the amount of hydrocarbons obtained was <21%, and olefins represented the vast majority of hydrocarbon products irrespective of TOS, which is unsurprising in the absence of a hydrogenation catalyst. Thus, it can be concluded that under the experimental conditions employed, thermal contributions to the conversion of UCO to diesel-like hydrocarbons are relatively minor.

Remarkably, complete deoxygenation occurs over all catalysts tested (see Figure 1 and Tables A2–A4), the concentration of diesel-like (C10–C20) hydrocarbons in the reaction products being >82% irrespective of both catalyst and TOS, heavier (C21–C35) hydrocarbons comprising the remainder of the product mixtures. Whereas heavier hydrocarbons stem from the deoxygenation of long-chain ester intermediates [21], diesel-like hydrocarbons are the result of the deoxygenation of the triglycerides and the fatty acids constituting the UCO feed (as well as of the cracking of heavier hydrocarbons) [11,18].

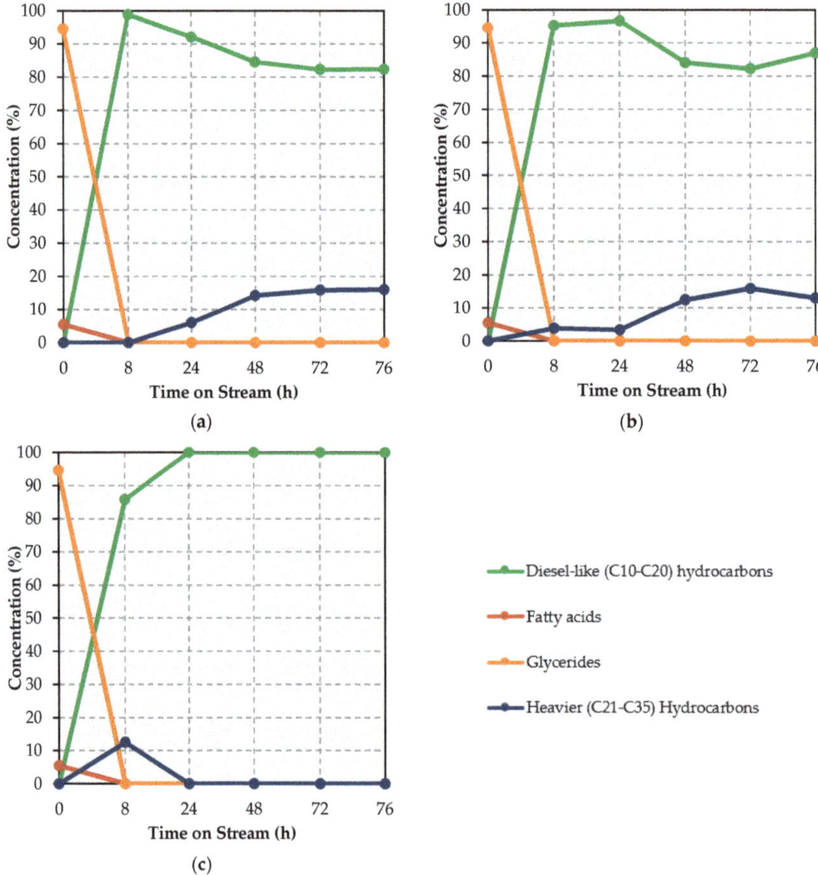

Figure 1. GC-MS analysis of the liquid products resulting from the deoxygenation of UCO over 20%Ni–5%Cu/Al$_2$O$_3$ (**a**), 20%Ni–5%Fe/Al$_2$O$_3$ (**b**), 20%Ni–0.5%Pt/Al$_2$O$_3$ (**c**).

It is also worth noting that the Ni-Pt catalyst afforded liquid product mixtures comprised solely of diesel-like hydrocarbons after 24 h on stream. In contrast, the Ni-Cu and Ni-Fe catalysts showed lower diesel amounts along with a higher yield of heavier (C21–C35) hydrocarbons (particularly at ≥48 h on stream) in their liquid products, which suggests that Ni-Pt disfavors the production of long-chain ester intermediates and/or favors cracking reactions. A closer look at the individual components of the liquid products in the diesel range, namely, C10, C11, C14, C15, and C16 (C12 and C13 could not be determined due to the interference of the reaction solvent) provides valuable insights vis-à-vis the changes in selectivity that take place as the reaction progresses over each catalyst (see Tables A2–A4). While the vast majority of the feed comprised triolein and oleic acid, the most abundant product obtained over all the catalysts is heptadecane (C17), suggesting that the reaction proceeds mainly via deCO$_x$ as opposed to HDO, which would afford octadecane (C18). Nevertheless, C18 is also produced in significant amounts, which indicates that HDO also occurs. Whereas the amount of C17 and C18 produced over the Ni-Cu and Ni-Fe catalysts remains relatively stable throughout, the corresponding values drop considerably beyond 28 h on stream over the Ni-Pt formulation. Tellingly, the amount of lighter (C10–C14) diesel-like hydrocarbons—which is the result of (and, thus, a proxy for) cracking activity—remains low and fairly stable, drops significantly, and increases considerably with TOS over the Ni-Cu, Ni-Fe and Ni-Pt catalysts, respectively.

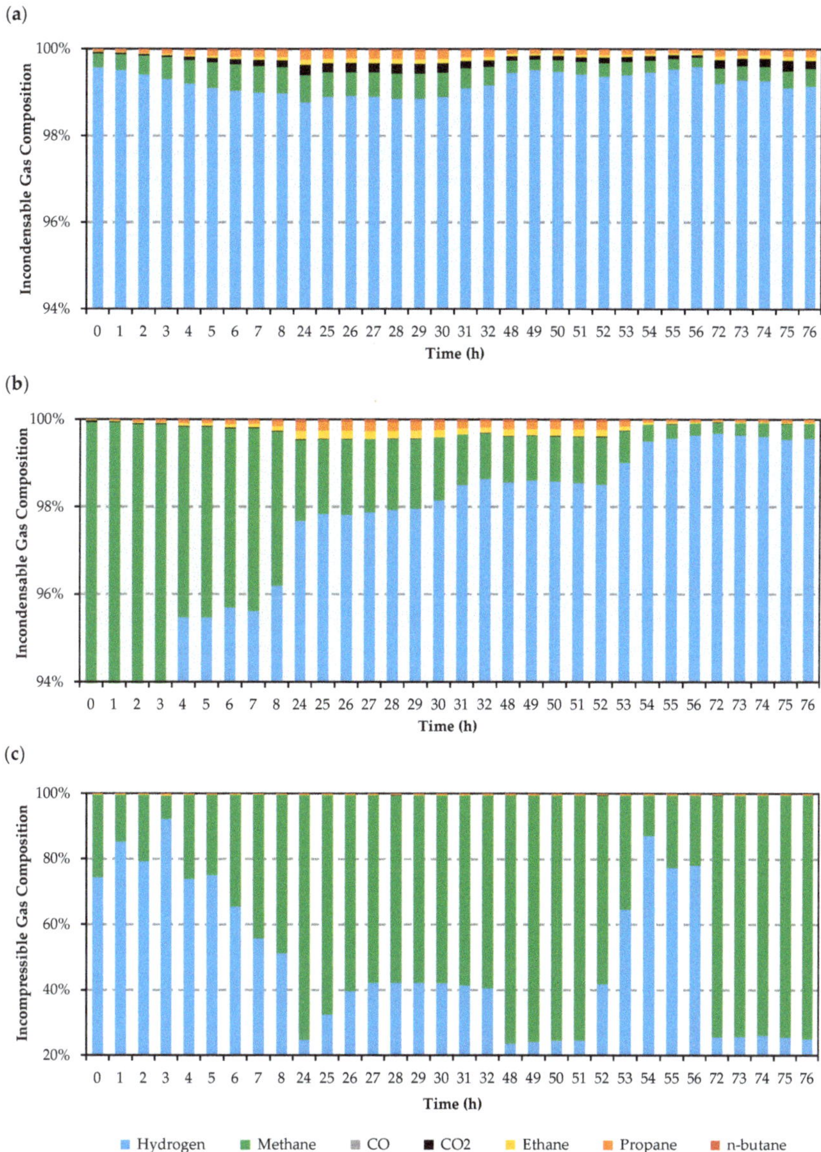

Figure 2. GC analysis of the gaseous products from deoxygenation of UCO at different catalysts for 20%Ni–5%Cu/Al$_2$O$_3$ (**a**), 20%Ni–5%Fe/Al$_2$O$_3$ (**b**), 20%Ni–0.5%Pt/Al$_2$O$_3$ (**c**).

Differences in the evolution of the incondensable gas products observed over each catalyst (see Figure 2) provide insights that are both consistent and complementary to those drawn from the composition of the liquid product mixtures. Briefly, whereas H$_2$ represents the reaction atmosphere, CO and CO$_2$ are produced from glycerides and fatty acids via decarbonylation and decarboxylation, respectively. Butane, propane, and ethane are produced through the internal chain cracking of glycerides, fatty acids, and long-chain hydrocarbons, while propane can also stem from the triglyceride backbone and its cracking can afford additional ethane and methane. Lastly, methane is also produced from the methanation of CO_x as well as from the cracking of glycerides, fatty acids, and long-chain hydrocarbons via terminal carbon loss, the main chain shortening mechanism according to a previous report [22]. With this in mind, the first thing worth noting is the small amount of CO_x detected. Indeed, the amount of CO_x is practically negligible in the gaseous products evolved over both the Fe- and Pt-promoted catalysts (see Figure 2b,c), indicating that the entirety of these gases is converted to methane and/or remain adsorbed on the surface of these formulations. In contrast, a small amount of CO_2 is detected in the gaseous products evolved over the Cu-promoted catalyst (see Figure 2a), particularly after a brief induction period observed in the first hours of the experiment. Parenthetically, this induction period has been observed in previous work and attributed to the accumulation of CO_2 on the catalyst surface as alumina-bound carbonates [22]. While CO_2 eventually breaks through and is detected in the gaseous products, CO remains undetected, likely indicating its full conversion to methane, to CO_2, and coke via the Boudouard reaction, or its strong and irreversible adsorption on the surface of the Ni-Cu catalyst (see Section 2.3).

The amount of ethane, propane, and butane in the gaseous products is also telling. Over both the Cu- and Fe-promoted catalysts, the amount of these gases detected at the beginning of the reaction is practically negligible, gradually increasing over the first 24 h on stream (see Figure 2a,b). Beyond this point, the amount of these gases evolved over the Cu-promoted catalyst remains relatively stable for the remainder of the experiment, whereas it becomes negligible once again towards the end of the run over the Fe-promoted formulation. In contrast, the amount of C2–C4 gaseous products evolved over the Pt-promoted catalyst is both higher and constant throughout the entirety of the run (see Figure 2c). Nevertheless, the amount of ethane, propane, and butane is always smaller than that of methane irrespective of both catalyst and TOS; the amount of methane detected being particularly informative. Whereas the quantity of methane evolved over the Ni-Cu catalyst is both small ($< 0.6\%$) and stable (see Figure 2a), the latter only applies to the end of the experiment involving the Ni-Fe formulation (see Figure 2b). At the beginning of the run performed over the Fe-promoted catalysts, the amount of methane in the gaseous products varies from 71% at t = 0 h to 64% at t = 3 h (results not shown), before dropping precipitously to 4% at t = 4 h and then more gradually to reach stability around 0.4% at t = 54 h. In stark contrast, the amount of methane in the gaseous products evolved over the Ni-Pt catalyst varies widely and can be as high as ~75% at t = 24, 48, and 72 h (see Figure 2c).

All of these trends indicate that while the Ni-Cu and Ni-Fe catalysts retain their deoxygenation activity within the time period investigated, cracking activity either remains constant or declines with TOS, consistent with results reported in other studies [11,17,23]. Although the Ni-Pt catalyst also retains its deoxygenation activity throughout the entire experiment, cracking reactions remain prevalent during the entirety of the run. In short, a comparison of the results obtained with 20% Ni–5% Cu/Al$_2$O$_3$, 20% Ni–5% Fe/Al$_2$O$_3$, and 20% Ni–0.5% Pt/Al$_2$O$_3$ catalysts suggests that Cu- and Fe-promoted catalysts are preferable to Ni-Pt formulations. Indeed, the latter is rendered disadvantageous by its higher price and cracking activity, which would reduce the cost and carbon efficiency of a process designed to convert UCO to diesel-like hydrocarbons.

2.2. Characterization of Fresh and Spent Catalysts

The textural properties of the catalysts used in this study are compiled in Table 1. The surface area, pore volume, and pore size of all catalysts fall in very narrow ranges, which is consistent with their total metal loadings and particle sizes (*vide infra*) and the fact that all catalysts were prepared using the same alumina support. These results indicate that the effects of differences in these properties on catalyst performance should be minimal.

Figure 3 includes the X-ray diffractograms of the catalysts employed in this study. Since diffractograms were acquired using the fresh catalysts in their oxidized form—catalysts were subjected to XRD after the calcination in air constituting the final step of their preparation (see Section 3.1)—the fact that all Ni detected is present as NiO is unsurprising. Indeed, the three diffractograms display several peaks (at 37.2°, 43.3°, 62.9°, 75.4°, and 79.4°) assigned to NiO [24].

The fact that diffraction peaks attributed to Fe_3O_4 and to Ni-Fe alloy phases [17,19,25,26] are absent from the diffractogram corresponding to 20% Ni–5% Fe/Al_2O_3 (Figure 3a) is unsurprising since these phases would only be expected in reduced (as opposed to oxidized) catalysts. Peaks associated with Fe_2O_3 are also absent from this diffractogram, indicating that Fe is highly dispersed. As previously reported [18], the fact that peaks at 35.5° and 38.7° corresponding to a CuO phase [27] are not observed in the diffractogram of 20% Ni–5% Cu/Al_2O_3 (Figure 3b) can be similarly attributed to the high dispersion of the Cu phase [24,28]. Likewise, no distinct Pt-related features (peaks or peak shifts) can be observed in the diffractogram corresponding to 20% Ni-0.5% Pt/Al_2O_3 (Figure 3c).

Table 1. Textural properties of the catalysts studied.

Catalyst	BET Surface Area (m^2/g)	Pore vol. (cm^3/g)	Avg. Pore Diameter (nm)
20% Ni–5% Cu/Al_2O_3	129	0.28	8.8
20% Ni–5% Fe/Al_2O_3	145	0.29	6.7
20% Ni-0.5% Pt/Al_2O_3	134	0.29	6.9

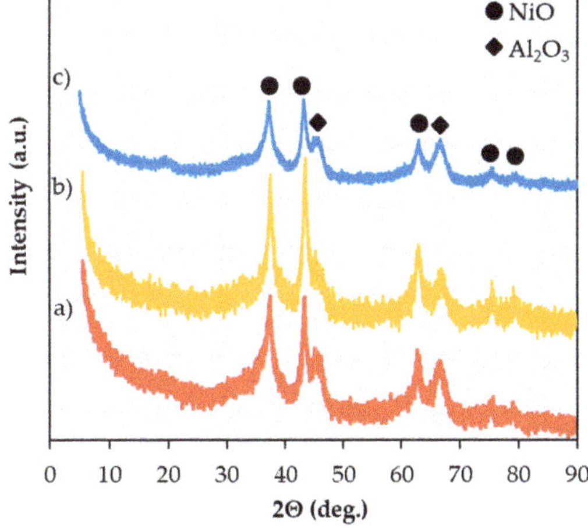

Figure 3. X-ray diffraction patterns for 20%Ni–5%Fe/Al_2O_3 (**a**), 20%Ni–5%Cu/Al_2O_3 (**b**), 20%Ni–0.5%Pt/Al_2O_3 (**c**).

The temperature-programmed reduction (TPR) profiles shown in Figure 4 clearly illustrate that the three catalysts employed in this study display very different reduction behavior. As discussed in a previous report [18], the TPR profile for 20% Ni–5% Cu/Al_2O_3 shows four distinct reduction events: (1) a sharp peak at 180 °C attributed to the reduction of copper oxide [24,29]; (2) a broader but well-defined peak with a maximum at 360 °C assigned to the reduction of a NiO-CuO phase [30]; (3) a shoulder with a local maximum at 460 °C signaling the reduction of NiO [31]; and (4) a weak and broad signal around 690 °C indicating the reduction of nickel aluminate ($NiAl_2O_4$) [32]. The TPR profile for 20% Ni–5% Fe/Al_2O_3 also shows four (but less distinct) reduction events, namely: (1) a small signal with a maximum at 235 °C corresponding to large (10–50 nm) NiO ensembles (*vide infra*); (2,3) a very large and broad peak ranging from 260 to 675 °C with a maximum at 350 °C, commingling the reduction of nickel and iron oxides (leading to the formation of a Ni-Fe alloy) [17,19]; and (4) a high-temperature tail of the latter peak, assigned to $NiAl_2O_4$ reduction. Lastly, as discussed in a recent report [13] the TPR profile for 20% Ni-0.5% Pt/Al_2O_3 also displays several reduction events, including (1) a small and broad peak between 300 and 350 °C attributed both to the reduction of surface Pt and of large NiO particles in close proximity to Pt [13,33]; (2) an intense and well-defined signal with a maximum at 460 °C assigned to the Pt-assisted reduction of smaller NiO particles [13]; and (3) a broad peak above 500 °C with a high temperature (>700 °C) shoulder attributed to the reduction of NiO and NiAl$_2$O$_4$, respectively.

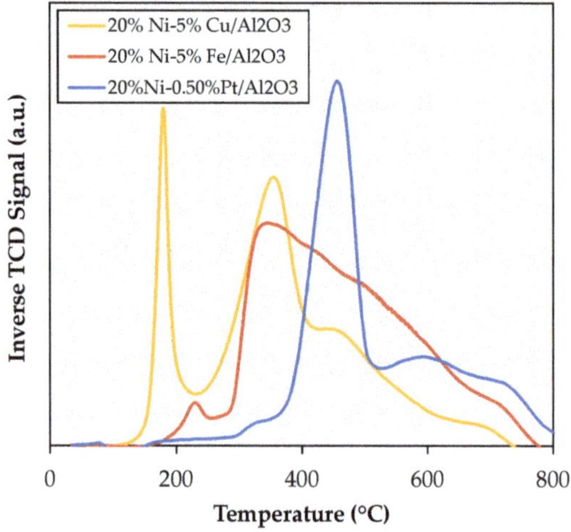

Figure 4. TPR profiles of the catalysts studied.

Given that Ni-based formulations used in the deoxygenation of FOG to fuel-like hydrocarbons are known to be particularly susceptible to coking [1], the spent catalysts were subjected to thermogravimetric analysis (TGA) in air, the resulting profiles being shown in Figure 5.

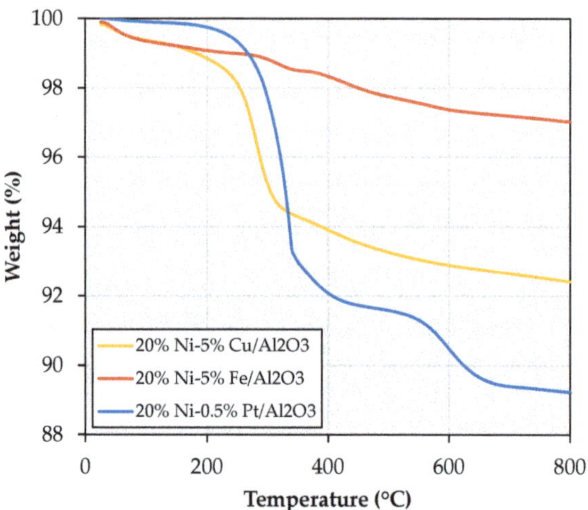

Figure 5. TGA of the spent catalysts recovered from the deoxygenation experiments.

The TGA profiles indicate that the total mass loss displayed by the spent catalysts follows the trend Ni-Fe (3.0%) < Ni-Cu (7.6%) < Ni-Pt (10.8%). In addition, the temperature at which mass loss takes place is also noteworthy, since mass loss events <400 °C can be attributed to strongly adsorbed reactants, intermediates and products (or soft coke) and mass loss events >400 °C can be assigned to more recalcitrant carbonaceous deposits (graphitic or hard coke). Tellingly, albeit the majority of the weight loss displayed by all spent catalyst takes place below 400 °C, the Ni-Pt formulation also shows distinct and significant weight loss above this temperature. The increased coking observed on the Ni-Pt catalyst is consistent with its considerably higher cracking activity, which is evinced by the copious amounts of methane produced by this formulation (see Section 2.1). In turn, the fact that both the Ni-Cu and the Ni-Fe catalysts display lower amounts of carbonaceous deposits is in agreement with the known ability of both Cu and Fe to curb the hydrogenolysis activity of Ni via geometric effects (see Section 1).

Table 2 shows the surface concentration (in at.%) of the elements detected via x-ray photoelectron spectroscopy (XPS) in the catalysts after (i) 76 h of TOS, followed by washing with dodecane and drying (spent); (ii) their subsequent calcination for 5 h at 450 °C under air (calcined); and (iii) their successive reduction for 3 h at 400 °C under H_2 (re-reduced). All XPS spectra can be found in Appendix B (Figures A1–A8). Carbon in the samples can be divided into inorganic (carbide) and organic (coke) carbon, the inorganic carbon being mostly associated with the SiC used as a diluent in the upgrading experiments (see Section 3.3) and being indicative of the relative amount of sample components—catalyst or diluent—analyzed during XPS measurements. The amount of inorganic carbon can considerably impact the interpretation of the data in Table 2. This is also the case for the amount of organic carbon (calculated by subtracting the inorganic from the total carbon) associated with coke deposits and whose Ni-Pt > Ni-Cu > Ni-Fe trend in the spent catalysts is in agreement with the results of TGA (*vide supra*). The fact that both spent Ni-Fe and Ni-Pt catalysts display a very similar surface concentration of Si and inorganic carbon indicates that a similar fraction of SiC diluent and catalyst is being analyzed, which in turn confirms that organic carbon (coke) deposits are much more abundant on the Pt- than on the Fe-promoted formulation. Similarly, while both the Si and inorganic carbon concentration of the spent Ni-Cu sample is lower, suggesting that a lower amount of C is contributed by SiC, the fact that the concentration of C_{Tot} (and, thus, of C_{Org}) in this formulation is higher than on Ni-Fe indicates that coke deposits on the surface of Ni-Cu are intermediate to those on Ni-Pt and Ni-Fe, which is also consistent with TGA data.

Table 2. Surface concentration (at.%) of elements detected via XPS.

Sample	C_{Tot}	C_{Inorg} [a]	C_{Org} [b]	Si [c]	Al	Ni (Ni^0) [d]	Cu	Fe
20% Ni–5% Cu/Al$_2$O$_3$—Spent	31.7	3.8	27.9	3.9	18.6	2.4 (0.38)	0.95	–
20% Ni–5% Fe/Al$_2$O$_3$—Spent	16.5	6.1	10.4	5.6	13.3	9.4 (3.10)	–	2.70
20% Ni-0.5% Pt/Al$_2$O$_3$—Spent	52.6	6.6	46.0	5.9	11.2	1.6 (0.29)	–	–
20% Ni–5% Cu/Al$_2$O$_3$—Calcined	4.2	2.3	1.9	4.2	25.0	4.6 (0.32)	3.20	–
20% Ni–5% Fe/Al$_2$O$_3$—Calcined	8.4	5.4	3.0	8.2	13.5	9.6 (0.13)	–	4.15
20% Ni-0.5% Pt/Al$_2$O$_3$—Calcined	15.2	11.4	3.8	9.5	14.3	9.3 (0.22)	–	–
20% Ni–5% Cu/Al$_2$O$_3$—Re-reduced	4.9	ND	4.9	1.5	30.1	3.5 (2.10)	0.90	–
20% Ni–5% Fe/Al$_2$O$_3$—Re-reduced	8.2	5.6	2.6	7.7	13.2	11.0 (3.0)	–	4.70
20% Ni-0.5% Pt/Al$_2$O$_3$—Re-reduced	45.7	33.0	12.7	1.3	12.9	3.5 (2.00)	–	–

[a] $C_{Tot} \times$ Carbide %Area; [b] C_{Tot}-C_{Inorg}; [c] This amount should be equal to the amount of C contained in SiC; [d] Ni $\times$ Ni^0 %Area.

Notably, the amount of Ni in general and Ni^0 in particular, is considerably higher on the surface of the spent Ni-Fe catalyst than on the other two formulations, although the observed trend (Ni-Fe >> Ni-Cu > Ni-Pt) likely stems—at least in part—from the relative amount of coke deposits on the surface of these catalysts. Considering the promoter metals, it is worth noting that a much higher amount of Fe in the Ni-Fe catalyst (relative to Cu in the Ni-Cu formulation) is detected at the surface. While this cannot be attributed to the amount of SiC artificially depressing the amount of Cu detected (since the Si and inorganic carbon concentration is higher in the Ni-Fe catalyst) or to the slightly lower atomic weight (A_r) of Fe—the A_r difference is too small—the higher amount of coke deposits on the surface of the Ni-Cu catalyst could partially explain the relatively low Cu concentration, as with the case of Ni. Lastly, while the entirety of Cu is present as Cu$^+$, which is indicative of Cu$_2$O as copper does not form a carbide phase, a small amount (ca. 13%) of Fe is in the metallic state, the remainder being present in oxidic form (see Figures A7 and A8 in Appendix B).

Upon calcination, the amount of C_{Tot} is significantly reduced mainly due to the combustion of coke deposits. Consistent with the more graphitic—and thus recalcitrant—nature of the coke on this formulation as indicated by TGA results (see Figure 5), the amount of residual coke after calcination is highest for the Ni-Pt catalyst. Although the amount of Ni in the calcined catalysts is lowest for 20% Ni–5% Cu/Al$_2$O$_3$, the amount of Ni^0 follows the trend Ni-Cu > Ni-Pt > Ni-Fe, which is indicative of the resistance of surface Ni in each catalyst to oxidation. Regarding the metallic promoters within the calcined catalyst, which increase in concentration due to the removal of coke, Cu is present as a mixture of Cu$_2$O and CuO, the vast majority of Fe also being present in the oxidized form (only ~3% being present as Fe^0).

Changes observed upon the reduction of the calcined catalysts are also informative. The similar amounts of Si, Ni, and Ni^0 on the re-reduced Ni-Cu and Ni-Pt formulations and the changes in these values relative to those displayed by their calcined counterparts indicate that (i) the region of the samples analyzed are catalyst-rich and SiC-poor; and (ii) sintering takes place during reduction based on the loss of surface Ni relative to the calcined catalysts. Changes in the Ni/Al ratio—from 0.18 to 0.12 and from 0.65 to 0.27 for calcined to re-reduced Ni-Cu and Ni-Pt, respectively—also suggests that sintering takes place during reduction. Moreover, the amount of C_{Tot} and C_{Inorg} on the Ni-Pt catalyst is striking and suggests the formation of a considerable amount of metallic carbide(s). Unfortunately, the presence of the latter could not be conclusively confirmed since Pt could not be observed (due to the small amount of Pt and the overlapping of the Pt4f and the Al2p XPS regions) and a distinct nickel carbide signal is not resolved. It is also noteworthy that the surface concentration of Ni (and to a lesser degree that of Ni^0) on the re-reduced Ni-Fe catalyst is significantly higher than that of its Ni-Cu and Ni-Pt counterparts, particularly taking into account the considerably larger amount of SiC being analyzed alongside Ni-Fe. Finally, it is interesting to note that the surface concentration of Cu is lower on the re-reduced Ni-Cu catalyst than on its spent and calcined counterparts—particularly taking into account the lower amount of SiC analyzed alongside the re-reduced material and the higher amount of coke on the spent formulation—which may indicate the alloying of Cu with Ni. Regarding the oxidation state of the promoter metals after reduction, while the entirety of Cu is present in the metallic form, only ~9% of iron is present as Fe^0, the remainder being present as Fe^{2+} or Fe^{3+}. Since it has been reported that Fe$_2$O$_3$ undergoes reduction to Fe$_3$O$_4$ in the ~300–400 °C range, while Fe$_3$O$_4$ is reduced between ~400 and 500 °C [34], it is unsurprising that after reduction at 400 °C most of the Fe detected by means of XPS is oxidic.

In short, XPS results indicate that the trend related to the amount of organic (coke) deposits on the surface of spent catalysts (Ni-Pt >> Ni-Cu > Ni-Fe) explains the relative amounts of Ni and Ni^0—as well as of promoter metals—on the surface of spent formulations (Ni-Fe >> Ni-Cu > Ni-Pt). Moreover, the trends related to the amount of Ni^0 on the surface of calcined and re-reduced catalysts (Ni-Cu > Ni-Pt > Ni-Fe and Ni-Fe > Ni-Cu $\approx$ Ni-Pt, respectively) indicate that Ni displays distinct redox behavior within each formulation. Indeed, Ni on the surface of 20% Ni–5% Fe/Al$_2$O$_3$ is easier to oxidize and reduce than Ni on the surface of 20% Ni–5% Cu/Al$_2$O$_3$ or 20% Ni–0.5% Pt/Al$_2$O$_3$. Finally, XPS results evince that while the Ni-Cu and Ni-Pt catalysts experience metal particle sintering during re-reduction, Ni-Cu and Ni-Fe alloys form within the Cu- and Fe-promoted catalysts and metallic carbides may form within the Pt-promoted formulation.

The analysis of the fresh and spent catalysts via transmission electron microscopy-energy dispersive X-ray spectroscopy (TEM-EDS) also afforded significant insights. In the case of the Cu-promoted catalyst, TEM results indicate that the particle size distribution—which is narrow and centered around 4 nm particles in the fresh catalyst—is both broader and centered around larger particles in the spent formulation (see Figure 6a), signaling particle sintering. The TEM-EDS results in Figure 6b reveal that the metal particles in the fresh catalyst display a composition that is close to that of the bulk formulation (80% Ni-20% Cu considering only the metallic phase), albeit Ni-rich particles containing 85–95% Ni are also observed. Notably, the spent catalyst comprises particles slightly more enriched in Cu relative to those in the fresh formulation, all particles in the spent catalyst containing between 65 and 80% Ni. This indicates that particles not only grow in size but also become Cu-rich during the reaction, which is also in line with previously reported results [11]. This conclusion is clearly illustrated by the TEM micrographs and the TEM-EDS elemental maps included in Figure A9; the elemental maps also showing that Ni and Cu are present in close association on both the fresh and spent catalysts. The Cu map of the spent Ni-Cu catalyst shown in Figure A9 also provides an example of the Cu-hollow space not observed in the fresh formulation, as Ni-Cu particles likely undergo Cu-hollowing through a mechanism based on the Kirkendall effect [35]. These observations are consistent with the widely reported bulk and surface enrichment of Ni-Cu nanoparticles with Cu [11,36,37].

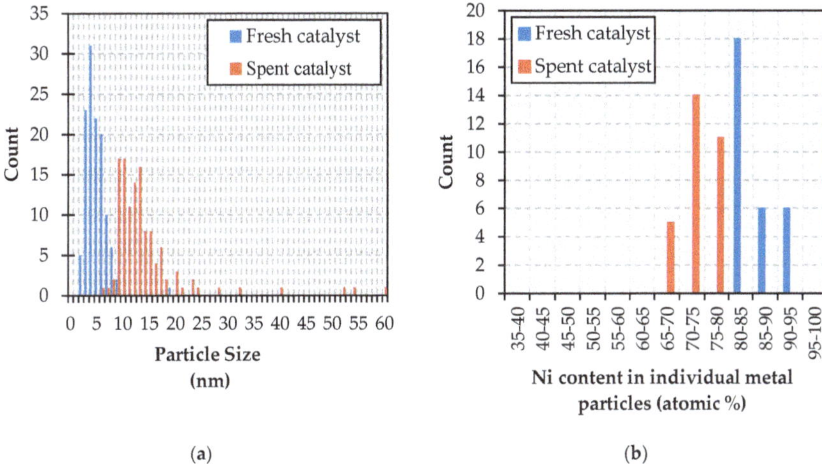

Figure 6. Particle size distribution (**a**) and metallic composition (**b**) histograms for the Ni-Cu catalyst.

In the case of the Fe-promoted catalyst, TEM results indicate that a similar particle size change takes place during reaction. Indeed, the particle size distribution—which is narrow in the fresh catalyst as the vast majority of particles range from 3 to 7 nm—is both broader and shifted to larger (8–30 nm) particles in the spent formulation (see Figure 7a). However, the composition of the metal particles does not change much during the reaction in contrast with the Cu-promoted catalyst; the only change observed being the disappearance of Ni-rich (90–100% Ni) particles according to the TEM-EDS results in Figure 7b. These observations are illustrated by the TEM micrographs and the TEM-EDS elemental maps in Figure A10, which show the degree of association between Ni and Fe in both the fresh and the spent catalyst. The Fe maps included in Figure A10 also evince Fe-hollow spaces on both the fresh and spent formulation, albeit the latter displays more of these spaces. Analogous to the case of the Ni-Cu catalyst, Fe-hollowing may occur through a mechanism based on the Kirkendall effect, which has also been reported for Ni-Fe bimetallic formulations [38,39].

In the case of the Pt-promoted catalyst, TEM results indicate a similar particle size change as those experienced during the reaction by the other formulations. Specifically, whereas the fresh catalyst shows a fairly narrow particle size distribution with the vast majority of particles falling within the 5–10 nm range, the spent formulations show a much broader distribution with particles as small as 3 nm and as large as 28 nm (see Figure 8a). Nevertheless, changes in the composition of metal particles within the Pt-promoted catalyst are noteworthy (see Figure 8b). Indeed, the fresh catalyst shows a significant amount of Pt-rich particles—relative to the bulk formulation (97.6 wt.% Ni-2.4 wt.% Pt or 99.3 at. % Ni-0.7 at.% Pt)—whereas the vast majority of metal particles in the spent catalyst show a composition very close to that of the bulk. The TEM micrographs and the TEM-EDS elemental maps in Figure A11 support these conclusions, showing both the increase in metal particle size and the closer association of Ni and Pt in the spent catalysts than in the fresh state.

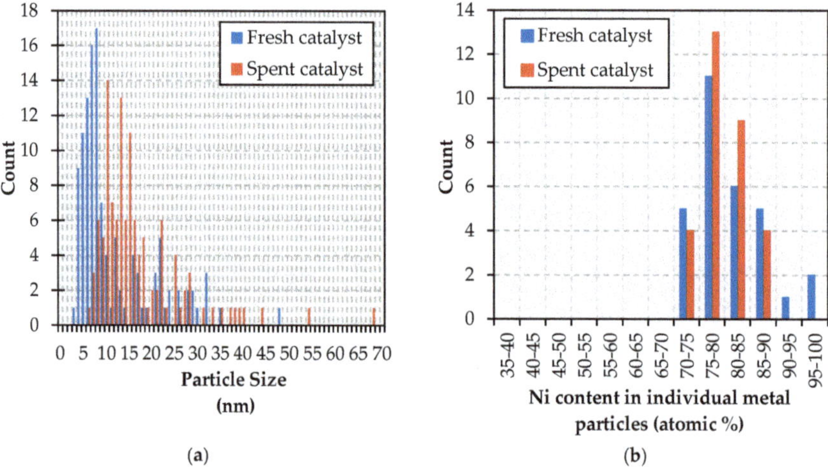

(a)

(b)

Figure 7. Particle size distribution (**a**) and metallic composition (**b**) histograms for the Ni-Fe catalyst.

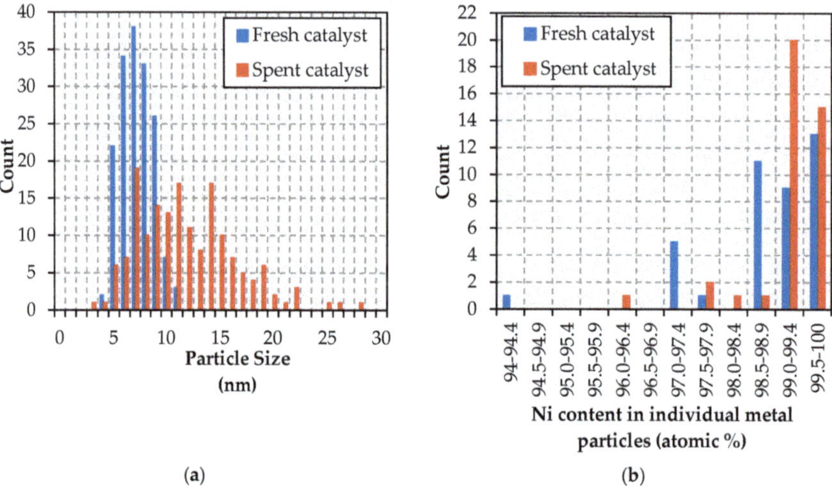

(a)

(b)

Figure 8. Particle size distribution (**a**) and metallic composition (**b**) histograms for the Ni-Pt catalyst.

2.3. Structural and Activity Changes Observed during Catalysts Aging and Regeneration

As mentioned in the preceding section, the similarity between the textural properties (surface area, pore volume, and pore diameter) of all fresh catalysts suggests that the effect of these properties on catalyst performance should be minimal, at least at the onset. However, any variations in these and other properties with TOS may influence catalyst performance. Indeed, based on TGA results, the loss of surface area and porosity attributable to coking and fouling should follow the trend Ni-Fe < Ni-Cu < Ni-Pt, the latter catalyst showing a higher amount of more recalcitrant (graphitic) carbonaceous deposits. Notably, complete deoxygenation of the feed was maintained throughout the experiment for each of the catalysts, which suggests that losses in surface area and porosity due to coking, fouling, and sintering are not sufficient to noticeably impact the deoxygenation activity of these formulations in the time period investigated. Thus, differences in the cracking activity of the catalysts offer better insights vis-à-vis structure-activity relationships.

Looking at both the composition of the liquid and gaseous products, the cracking activity of the Cu-promoted catalyst—although always relatively low—is highest between 8 and 30 h on stream and progressively drops between 30 and 72 h on stream, at which point it becomes stable. Somewhat similarly, the cracking activity of the Fe-promoted formulation is considerably higher in the first 8 h of the experiment, becomes moderate between 24 and 52 h on stream, and is both negligible and stable beyond t = 72 h. Contrastingly, the cracking activity of the Pt-promoted formulation is higher at the end of the run than at its onset. Therefore, at least in the case of the Ni-Pt catalyst, it appears that neither the loss of surface area due to coking, fouling, and sintering, nor changes in the bulk or surface composition of the metal particles, reduce the cracking activity of this formulation in the time period investigated. The opposite is the case for the Cu- and Fe-promoted catalysts, which display a lower cracking activity towards the end of the experiment, this effect being more pronounced for the Ni-Fe formulation.

TPR measurements performed on the spent catalysts after calcination (see Figure A12) provide valuable insights on the structural changes that occur during regeneration. The first thing worth noting is that all peaks (except for those assigned to $NiAl_2O_4$) are shifted to lower temperatures, which can be partially attributed to the formation of larger particles that are easier to reduce. However, the more substantial shifts (>100 °C) can only be fully explained by invoking a considerable increase in the association of Ni with the promoter metal. For the Cu-promoted catalyst, changes in the relative intensity of peaks attributed to the reduction of copper oxide, NiO-CuO, and NiO (with respect to that in the fresh formulation) point to a reduction in the amount of unalloyed Cu and the formation of Ni-Cu bimetallic particles with a Cu-rich surface. Indeed, the peaks with maxima at 245 and 375 °C likely correspond to the reduction of NiO-CuO at the surface and of NiO at the core of these particles, respectively. For the Fe-promoted formulation, the narrowness of the main peak relative to that in the fresh formulation indicates a closer association between the metals forming a Ni-Fe alloy, which is consistent with the disappearance of Ni-rich particles observed via TEM-EDS (see Figure 7b). In addition, the intensity of this peak relative to that of peaks shown by other catalysts is in line with the higher amount of surface Ni in the Fe-promoted formulation measured via XPS (*vide infra*). Lastly, for the Pt-promoted catalyst, the sizable shift (>200 °C) that the main peak displays between the fresh and the regenerated formulation, is indicative of a major increase in the association between Ni and Pt.

The results of the XPS measurements performed on calcined and re-reduced spent catalysts confirm these conclusions and offer additional insights. While most (if not all) of the coke is removed from Cu- and Fe-promoted spent catalysts, the regenerated Ni-Pt formulation displays both a significant amount of residual coke, as well as a possible metallic carbide phase. In addition, metal particle sintering takes place during the regeneration of spent Ni-Cu and Ni-Pt catalysts, likely explaining the lower amount of surface Ni and Ni^0 these formulations display relative to their Fe-promoted counterpart.

Finally, the regenerated (i.e., calcined in air at 450 °C for 5 h and re-reduced under H_2 at 400 °C for 3 h) spent catalysts were subjected to in-situ diffuse reflectance infrared Fourier transform spectroscopy after CO adsorption (CO-DRIFTS) to gain additional information on their structure post-regeneration (see Figure 9). Since the adsorption of CO on metals is highly temperature-dependent, adsorption was carried out at 25 °C in order to focus on the most active sites for CO adsorption and limit the confounding effects that the presence of multiple metals can have on M-CO spectra [40]. Irrespective of its state (fresh or spent and regenerated), the Cu-promoted catalyst displays a very intense band at 2119–2100 cm^{-1}, which is attributed to CO adsorbed on Cu sites [41,42]. However, this band increases in intensity and shifts to higher wavenumbers on the regenerated catalyst, which signals an increase in the total quantity of Cu sites, as well as a change in their electronic properties. While the increase in intensity is in line with the enrichment of the surface with Cu, the shift in wavenumber may result from a greater extent of Ni-Cu alloying and from the rise of Cu-hollow spaces [43]. Tellingly, while the fresh Ni-Fe catalyst shows a Ni-CO band at 2179 cm^{-1}, this band is absent from the corresponding spectrum post-regeneration. This suggests that the most coordinatively unsaturated Ni sites—which are the most active cracking sites [44]—are irreversibly deactivated during reaction/regeneration. Similarly, while the fresh Ni-Pt catalyst shows a well-defined peak at ~2180 cm^{-1} and a broad feature at ~2120 cm^{-1} associated with CO on metallic Ni sites, as well as a large and well-defined peak at ~2077 cm^{-1} assigned to CO on Pt sites [13], none of these signals are observed post-regeneration. Since XPS results demonstrate the presence of Ni^0 on the regenerated Ni-Pt formulation, the dearth of signals can be explained by residual coke blocking the sites responsible for low-temperature CO adsorption.

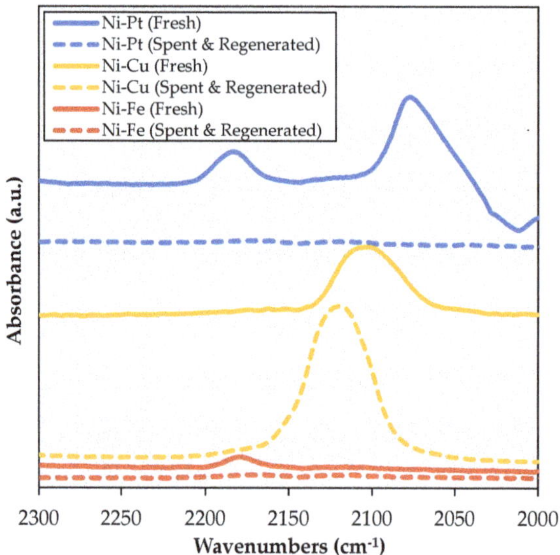

Figure 9. CO-DRIFTS of the catalysts studied.

These observations are reinforced by a recent report in which the performance of a regenerated Ni-Cu catalyst in fatty acid deoxygenation was found to be distinct from—and superior to—that of the fresh formulation [11]. Thus, the need for additional work in which the regenerated catalysts are tested to study their performance (and the evolution thereof) in a second cycle post-regeneration is clearly indicated, particularly since these tests stand to shed light on the recyclability of these formulations and unveil additional structure-activity relationships.

3. Materials and Methods

3.1. Catalyst Preparation

20% Ni–5% Cu/Al$_2$O$_3$, 20% Ni–5% Fe/Al$_2$O$_3$ and 20% Ni-0.5% Pt/Al$_2$O$_3$ catalysts were prepared by excess wetness impregnation using Ni(NO$_3$)$_2$·6H$_2$O (Alfa Aesar, Haverhill, MA, USA), Cu(NO$_3$)$_2$·3H$_2$O (Sigma-Aldrich, St. Louis, MO, USA), Fe(NO$_3$)$_3$·9H$_2$O (Alfa Aesar, Haverhill, MA, USA) and/or Pt(NH$_3$)$_4$(NO$_3$)$_2$ (Sigma Aldrich, St. Louis, MO, USA) as the metal precursors and γ-Al$_2$O$_3$ (Sasol, Johannesburg, South Africa; surface area of 216 m^2/g) as the support. Following impregnation, the materials were dried overnight at 60 °C under vacuum and then calcined at 500 °C for 3 h in static air. The catalysts and SiC diluent (Kramer Industries, Piscataway Township, NJ, USA) were sieved separately to the desired particle size (150–300 μm), and stored in a vacuum oven at 60 °C prior to their use.

3.2. Catalyst Characterization

A detailed description of the instrumentation and procedures employed for catalyst characterization—by means of N$_2$ physisorption, XRD, TGA, TEM-EDS, TPR, and DRIFTS—can be found in previous contributions [13,15,45]. Briefly, XRD measurements were performed on a Phillips X'Pert diffractometer using Cu Kα radiation (λ = 1.5406 Å) and a step size of 0.02°. TGA was performed on a TA instrument Q500 thermogravimetric analyzer under flowing air (50 mL min^{-1}) by ramping the temperature from room temperature to 1000 °C at a rate of 10 °C/min. TEM observations were conducted using a Thermo Scientific Talos F200X analytical electron microscope equipped with a SuperX EDS system consisting of 4 windowless silicon drift detectors (SDD) for quantitative chemical composition analysis and elemental distribution mapping. XPS analyses were performed using a PHI 5000 Versaprobe apparatus with monochromatic Al Kα1 X-Ray source (energy of 1486.6 eV, accelerating voltage of 15 kV, power of 50 W and spot size diameter of 200 µm). Pass energies of 187.5 eV and 58.7 eV were used for survey spectra and high-resolution windows, respectively. The signal for Al_2O_3 (Al2p at 74.4 eV) was employed for energy calibration purposes (measurements being performed with a neutralization system). Spectra were processed with the CasaXPS software package, ionization cross-sections from Landau being used to quantify the semi-empirical relative sensitivity factors. Prior to analysis, powders were deposited on a steatite sample holder made in house. This sample holder enables the transfer of samples between a pre-treatment chamber and the XPS analysis chamber without exposure to air. The pre-treatment chamber—which was also designed and manufactured in house and is equipped with a furnace that can heat samples up to 1050 °C—can be filled with pre-treatment gases (up to 1 bar) and be placed under vacuum, which is done prior to transferring pretreated samples to the XPS analysis chamber.

3.3. Continuous Fixed-Bed Deoxygenation Experiments

Used cooking oil upgrading experiments were performed in continuous mode using previously described equipment and procedures [11]. Briefly, a fixed-bed stainless-steel tubular reactor (1/2 in. o.d., Parr, Moline, IL, USA) with a stainless-steel porous frit to hold the bed—0.5 g of catalyst and 0.5 g of SiC as a diluent (or 1 g of SiC in the blank run)—in place was employed. Prior to each deoxygenation experiment, the catalyst to be tested was reduced in situ for 3 h at 400 °C under 40 bar of flowing H_2 (60 mL/min). The same pressure and H_2 flow were used during deoxygenation experiments, which were performed at 375 °C. The feed was introduced to the reactor—as a solution of 75 wt.% UCO in dodecane (>99% Alfa Aesar, Haverhill, MA, USA)—at a rate of 0.75 mL/h (equivalent to a WHSV of 1 h^{-1}) using a Harvard Apparatus (Holliston, MA, USA) syringe pump equipped with an 8 mL syringe. Liquid products were sampled from a liquid-gas separator (kept at 0 °C) placed downstream from the catalyst bed. Incondensable gases were directed to a dry test meter before being collected in Tedlar® gas sample bags. Gas sample bags were changed every time a liquid sample was taken to ensure that the gas samples analyzed and the liquid samples collected could be correlated. A blank (sans catalyst) experiment was conducted using 1 g of SiC. Representative experiments were performed in duplicate to ensure reproducibility. The highest average standard deviation values observed in the amount of diesel-like and heavier hydrocarbons formed were ±6.15% and ±5.66%, respectively.

3.4. Analysis of Reaction Products

Liquid products were analyzed using a combined simulated-distillation-GC and GC-MS approach. A detailed description of the development and application of this method is available elsewhere [46]. Briefly, the analyses were performed using an Agilent 7890B GC system equipped with an Agilent 5977A extractor MS detector and flame ionization detector (FID). The multimode inlet was run with an initial temperature of 100 °C. Upon injection, this temperature was immediately increased at a rate of 8 °C/min to 380 °C, which was maintained for the remainder of the analysis. The oven temperature was increased upon injection from 40 °C to 325 °C at a rate of 4 °C/min, followed by a ramp of 10 °C/min to 400 °C, which was maintained for 12.5 min. An Agilent J&W VF-5ht column (30 m × 250 µm × 0.1 µm) rated to 450 °C was used. Gaseous samples were analyzed using an Agilent (Santa Clara, CA, USA) 3000 Micro-GC equipped with 5 Å molecular sieve, PoraPLOT U, alumina, and OV-1 columns, as well as with a universal thermal conductivity detector (TCD). The GC was calibrated for all of the gaseous products obtained, including CO and CO_2, as well as straight-chain C1–C6 alkanes and alkenes.

4. Conclusions

In this contribution, an industrially relevant experimental approach was used to upgrade UCO to diesel-like hydrocarbons in order to compare the performance of Ni catalysts promoted with Cu, Fe or Pt. Results indicate that all catalysts tested display and retain the ability to fully deoxygenate the feed to hydrocarbons through the entirety of the time period investigated. However, the cracking activity of Ni-Cu is relatively low and stable throughout, that of Ni-Fe drops with TOS, and that of Ni-Pt is higher, variable, and is still high at the end of a 76 h run. Analysis of the fresh and spent catalysts helps explain these trends and their underlying structure-activity relationships. In the case of the Ni-Pt catalyst, neither coking, fouling, and metal particle sintering—nor changes in the bulk or surface composition of the metal particles—reduce the cracking activity of this formulation in the time period investigated. In contrast, the cracking activity of both the Ni-Cu and the Ni-Fe catalysts decrease with TOS, this decrease being more pronounced for the Fe-promoted formulation. Based on TEM-EDS data, this reduction in cracking activity can be attributed to an increased degree of alloying between Ni and Cu or Fe, the formation of Ni-Cu and Ni-Fe alloys disrupting the adjacency of Ni atoms required for C–C hydrogenolysis. In short, results suggest that Cu- and Fe-promoted catalysts are preferable to Ni-Pt formulations, which are rendered disadvantageous by the fact that their higher price and cracking activity would reduce the cost and carbon efficiency of a process designed to convert UCO to diesel-like hydrocarbons.

Author Contributions: Individual contributions are as follows: conceptualization—E.S.-J. and M.C.; methodology—G.C.R.S., E.S.-J., and M.C.; validation—G.C.R.S.; formal analysis—G.C.R.S., and O.H.; investigation—G.C.R.S., O.H., D.Q., and R.P.; resources—E.S.-J., M.C., and O.H.; data curation—O.H.; writing—original draft preparation—E.S.-J., and G.C.R.S.; writing—review and editing—E.S.-J., and M.C..; visualization—G.C.R.S., and E.S.-J.; supervision—M.C.; project administration—G.C.R.S., E.S.-J., and M.C.; funding acquisition—G.C.R.S., E.S.-J., M.C., and G.C. All authors have read and agreed to the published version of the manuscript.

Funding: This work was supported in part by the National Science Foundation (grant No. 1437604); by the J. William Fulbright Foreign Scholarship Board and the Bureau of Educational and Cultural Affairs of the United States Department of State through a Fulbright Scholarship awarded to G.S; and by the French National Center for Scientific Research (CNRS) Graduate School EIPHI (contract ANR-17-EURE-0002) and through a CNRS visiting researcher position granted to E.S.-J.

Acknowledgments: Sarah Cummins and the Redwood Cooperative School in Lexington, Kentucky, as well as Jennifer Wyatt and personnel from the Lexington-Fayette Urban Country Government, are thanked for collecting and delivering the used cooking oil used in this study. Tonya Morgan is thanked for her help with the GC-MS analysis of liquid reaction products.

Conflicts of Interest: The authors declare no conflict of interest. The funders had no role in the design of the study; in the collection, analyses, or interpretation of data; in the writing of the manuscript, or in the decision to publish the results.

Appendix A

Table A1. GC-MS analysis of the used cooking oil used in the catalytic upgrading experiments. Text in italics corresponds to compound classes comprising the individual compounds listed. Values shown are wt.%.

Compound	Used Cooking Oil
Fatty Acids	5.44
Oleic Acid (C18:1)	5.44
Triglycerides	94.56
Triolein	94.56

Table A2. GC-MS analysis of the liquid products from the catalytic upgrading of used cooking oil (75 wt.% in C12) over 20% Ni–5% Cu/Al$_2$O$_3$ at 375 °C and WHSV = 1.0 h^{-1}, in 100% H$_2$. Text in bold corresponds to compound classes comprising the individual compounds listed. Text in italics corresponds to compound sub-classes comprising the individual compounds detected. Values shown are wt.%.

Compound	4 h	8 h	24 h	28 h	48 h	52 h	56 h	72 h	76 h
Total Hydrocarbons	**89.92**	**91.42**	**87.48**	**91.94**	**107.34**	**83.93**	**94.30**	**88.50**	**86.58**
Normal Alkanes	*89.92*	*91.42*	*87.48*	*91.94*	*107.34*	*83.93*	*94.30*	*88.50*	*86.58*
Decane (C10)	2.64	2.09	-	2.15	-	-	2.68	2.36	-
Undecane (C11)	1.01	0.15	-	0.37	-	-	0.67	0.85	-
Tetradecane (C14)	1.14	1.38	-	0.95	-	-	1.02	1.30	-
Pentadecane (C15)	6.17	5.99	-	6.52	-	5.29	4.99	4.72	-
Hexadecane (C16)	5.42	6.01	5.01	5.81	-	4.36	5.69	5.11	4.95
Heptadecane (C17)	47.05	38.96	44.54	38.98	64.95	37.91	32.73	32.90	36.89
Octadecane (C18)	22.64	24.45	28.05	22.95	35.05	29.38	27.55	28.76	31.15
Nonadecane (C19)	-	2.17	-	2.38	-	-	2.72	2.34	-
Eicosane (C20)	2.06	1.26	2.38	1.28	-	3.20	1.74	1.55	2.21
Heneicosane (C21)	1.06	0.96	-	0.89	-	2.01	1.10	1.10	1.82
Docosane (C22)	-	0.96	0.62	-	1.17	-	1.57	0.86	1.14
Tricosane (C23)	-	0.52	1.40	-	1.70	-	1.02	2.06	2.24
Tetracosane (C24)	-	0.86	1.45	1.37	2.34	-	2.15	2.05	2.26
Pentacosane (C25)	-	0.62	1.78	2.40	2.13	-	1.95	2.04	2.01
Hexacosane (C26)	0.73	5.04	2.25	5.89	-	1.78	6.71	0.50	1.91
Olefins	***3.03***	***2.21***	***6.40***	***2.14***	**-**	***2.24***	***1.76***	***1.75***	***7.21***
Heptadecene (C17:1)	3.03	2.21	6.40	2.14	-	2.24	1.76	1.75	7.21
Unidentified	*1.92*	*1.64*	*2.47*	*1.29*	*-*	*1.98*	*1.84*	*1.61*	*-*

Table A3. GC-MS analysis of the liquid products from the catalytic upgrading of used cooking oil (75 wt.% in C12) over 20% Ni–5% Fe/Al$_2$O$_3$ at 375 °C and WHSV = 1.0 h^{-1}, in 100% H$_2$. Text in bold corresponds to compound classes comprising the individual compounds listed. Text in italics corresponds to compound sub-classes constituting the individual compounds detected. Values shown are wt.%.

Compound	4 h	8 h	24 h	28 h	48 h	52 h	56 h	72 h	76 h
Total Hydrocarbons	**93.39**	**91.16**	**90.59**	**95.10**	**97.61**	**94.20**	**99.68**	**97.72**	**98.84**
Normal Alkanes	*93.39*	*91.16*	*90.59*	*95.10*	*97.61*	*94.20*	*99.68*	*97.72*	*98.84*
Decane (C10)	12.23	11.06	4.19	3.59	3.42	-	-	3.99	-
Undecane (C11)	19.46	14.88	4.62	1.48	2.78	-	-	-	-
Tetradecane (C14)	6.85	5.86	2.69	1.15	1.55	-	-	-	-
Pentadecane (C15)	9.71	11.09	8.99	7.07	7.4	7.59	7.99	7.35	8.57
Hexadecane (C16)	10.06	11.58	8.81	6.71	6.61	5.81	8.29	7.68	7.20
Heptadecane (C17)	19.22	20.2	34.33	38.05	38.39	42.19	40.57	40.94	49.85
Octadecane (C18)	11.58	9.7	14.95	17.96	17.9	20.09	21.01	18.38	21.03
Nonadecane (C19)	-	1.69	2.29	2.52	2.6	2.68	2.32	3.06	-
Eicosane (C20)	-	0.9	1.45	1.98	1.45	1.70	2.63	1.39	-
Heneicosane (C21)	-	0.86	1.36	1.58	1.22	0.94	-	1.88	-
Docosane (C22)	0.47	-	0.99	1.48	1.67	1.51	0.94	-	1.42
Tricosane (C23)	0.49	-	1.39	2.6	2.59	2.47	3.14	2.65	2.26
Tetracosane (C24)	0.65	-	1.48	2.73	2.46	2.45	3.51	2.34	2.27
Pentacosane (C25)	0.66	1.03	1.17	2.9	3.14	2.75	3.41	3.53	2.49
Hexacosane (C26)	2.01	2.31	1.88	3.3	4.43	4.02	5.87	4.53	3.75
Nonacosane (C29)	-	-	-	-	-	-	-	-	-
Hentriacontane (C31)	-	-	-	-	-	-	-	-	-
Tetratriacontane (C34)	-	-	-	-	-	-	-	-	-
Olefins	**-**	***1.26***	***1.8***	***2.36***	***2.44***	***2.14***	***4.14***	***3.14***	***2.67***
Heptadecene (C17:1)	-	1.26	1.8	2.36	2.44	2.14	4.14	3.14	2.67

Table A4. GC-MS analysis of the liquid products from the catalytic upgrading of used cooking oil (75 wt.% in C12) over 20% Ni-0.5% Pt/Al$_2$O$_3$ at 375 °C and WHSV = 1.0 h^{-1}, in 100% H$_2$. Text in bold corresponds to compound classes comprising the individual compounds listed. Text in italics corresponds to compound sub-classes constituting the individual compounds detected. Values shown are wt.%.

Compound	4 h	8 h	24 h	28 h	48 h	52 h	56 h	72 h	76 h
Total Hydrocarbons	***84.68***	***13.29***	***100.00***	***99.99***	***94.58***	***85.31***	***98.44***	***96.29***	***88.63***
Normal Alkanes	*84.68*	*13.29*	*100.00*	*99.99*	*94.58*	*85.31*	*98.44*	*96.29*	*88.63*
Decane (C10)	4.35	-	-	-	9.02	14.92	27.33	31.76	31.38
Undecane (C11)	1.84	-	-	-	8.22	10.11	17.26	31.74	34.56
Tetradecane (C14)	1.62	-	-	-	2.61	2.68	3.75	5.24	7.80
Pentadecane (C15)	6.65	-	12.09	13.46	7.52	6.71	6.10	5.72	5.23
Hexadecane (C16)	6.89	-	13.54	15.30	8.69	6.41	5.39	4.16	2.72
Heptadecane (C17)	34.80	-	46.23	50.78	35.08	28.26	20.87	9.53	5.08
Octadecane (C18)	23.82	-	28.14	20.45	23.44	13.29	8.13	3.99	1.86
Nonadecane (C19)	2.48	-	-	-	-	1.39	-	-	-
Eicosane (C20)	1.32	-	-	-	-	0.96	-	-	-
Heneicosane (C21)	0.91	-	-	-	-	0.58	-	-	-
Docosane (C22)	-	0.82	-	-	-	-	0.84	-	-
Tricosane (C23)	-	1.94	-	-	-	-	1.50	-	-
Tetracosane (C24)	-	1.82	-	-	-	-	1.44	-	-
Pentacosane (C25)	-	2.74	-	-	-	-	1.56	1.18	-
Hexacosane (C26)	-	5.97	-	-	-	-	4.27	2.97	-
Olefins	*1.17*	-	-	-	*2.55*	*1.82*	*1.30*	*0.71*	*0.60*
Heptadecene (C17:1)	1.17	-	-	-	2.55	1.82	1.30	0.71	0.60

Table A5. GC-MS analysis of product mixtures in the thermal upgrading of yellow grease (75 wt.% in C12) over 1 g SiC at 375 °C and WHSV = 1.0 h^{-1}, in 100% H$_2$. Text in bold corresponds to compound classes comprising the individual compounds listed. Text in italics corresponds to compound sub-classes constituting the individual compounds detected. Values shown are wt.%.

Compound	4 h	8 h	24 h	28 h	48 h	52 h	56 h	72 h	76 h
Total Hydrocarbons	***15.21***	***20.50***	***14.99***	***11.95***	***15.74***	***15.21***	***15.74***	***19.54***	***17.50***
Normal Alkanes	*0*	*0*	*1.87*	*1.95*	*2.59*	*2.64*	*2.58*	*2.43*	*1.55*
Decane (C10)	0	0	1.87	1.95	2.59	2.64	2.58	2.43	1.55
Pentadecane (C15)	0	0	0	0	0	0	0	0	0
Olefins	*15.21*	*20.50*	*13.12*	*10.00*	*13.15*	*12.57*	*13.16*	*17.11*	*15.95*
Heptadecene (C17:1)	4.34	4.74	0.71	0.18	0.09	1.02	0.86	1.37	1.25
Hexacosene (C26:1)	0	0	1.14	0.91	1.01	0.62	1.00	1.36	0.50
Tricosene (C23:1)	10.87	15.76	11.27	8.91	12.05	10.93	11.30	14.38	14.20
Total Oxygenates	***84.79***	***79.50***	***85.02***	***88.05***	***84.26***	***84.79***	***84.27***	***80.46***	***82.5***
Fatty Acids	*84.79*	*79.50*	*83.24*	*86.74*	*81.93*	*82.9*	*82.14*	*78.52*	*80.46*
Palmitic Acid (C16)	8.51	5.27	6.68	6.97	7.47	7.50	7.24	6.71	6.54
Oleic Acid (C18:1)	76.28	74.23	66.32	67.65	63.36	64.83	62.84	59.65	62.81
Stearic Acid (C18)	0	0	10.24	12.12	11.1	10.57	12.06	12.16	11.11
Monolein	*0*	*0*	*1.78*	*1.31*	*2.33*	*1.89*	*2.13*	*1.94*	*2.04*

Appendix B

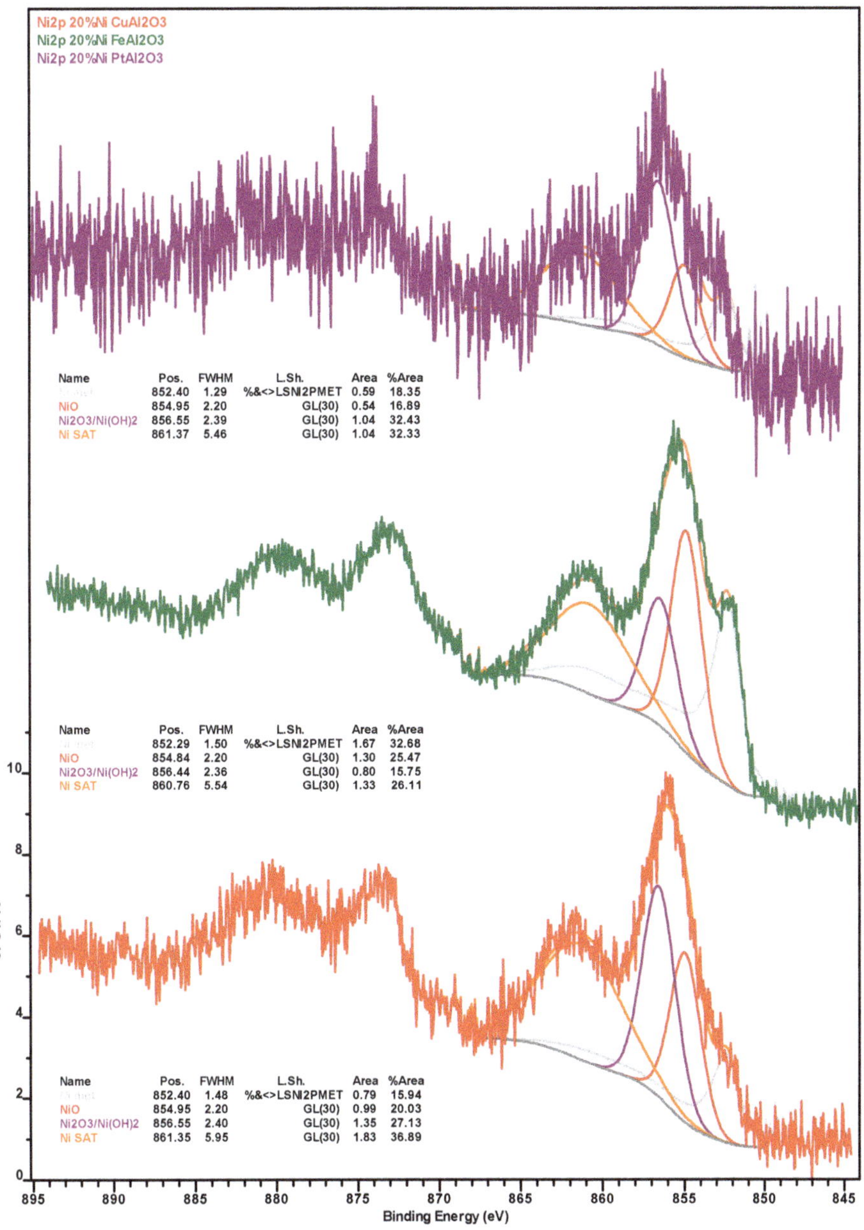

Figure A1. Nickel 2p X-ray photoelectron spectra of the spent Ni-Cu (red spectrum), Ni-Fe (green spectrum) and Ni-Pt (purple spectrum) catalysts recovered from deoxygenation experiments.

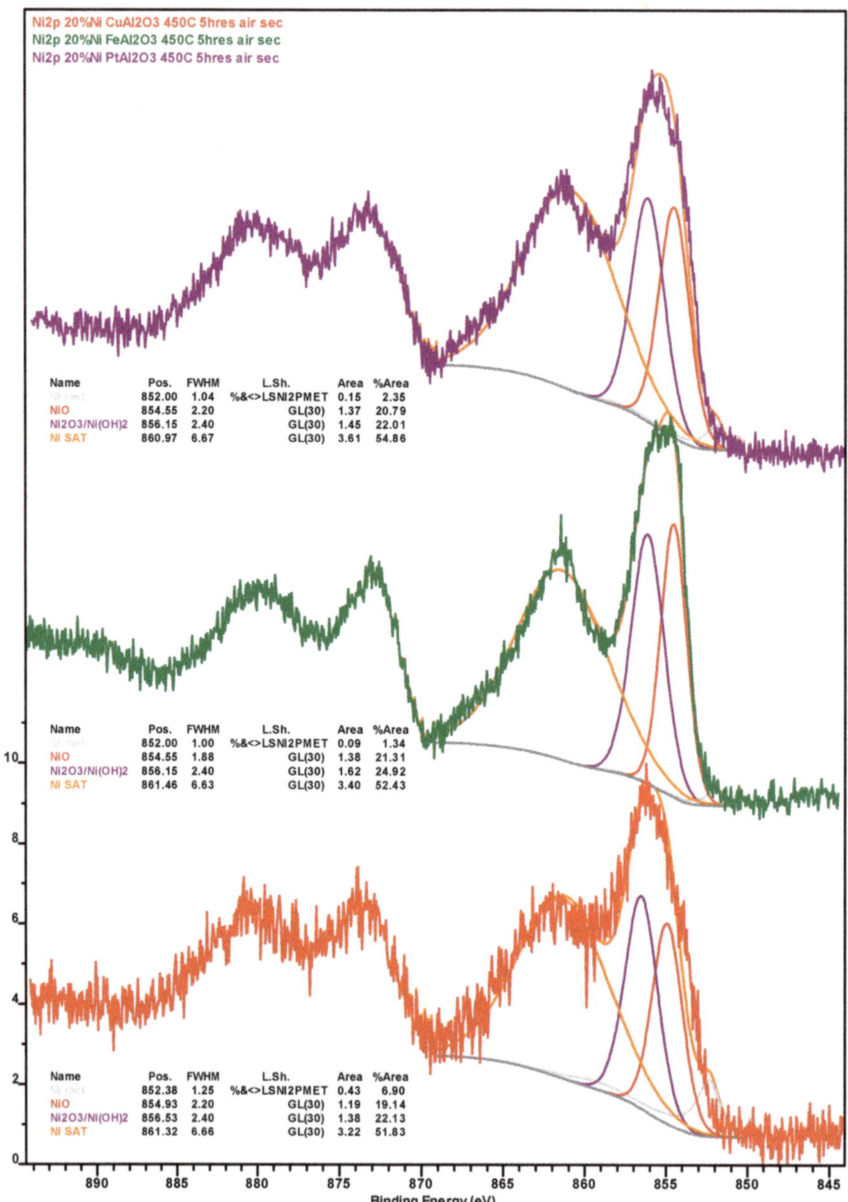

Figure A2. Nickel 2p X-ray photoelectron spectra of the spent Ni-Cu (red spectrum), Ni-Fe (green spectrum), and Ni-Pt (purple spectrum) catalysts recovered from deoxygenation experiments after calcination in air at 450 °C for 5 h.

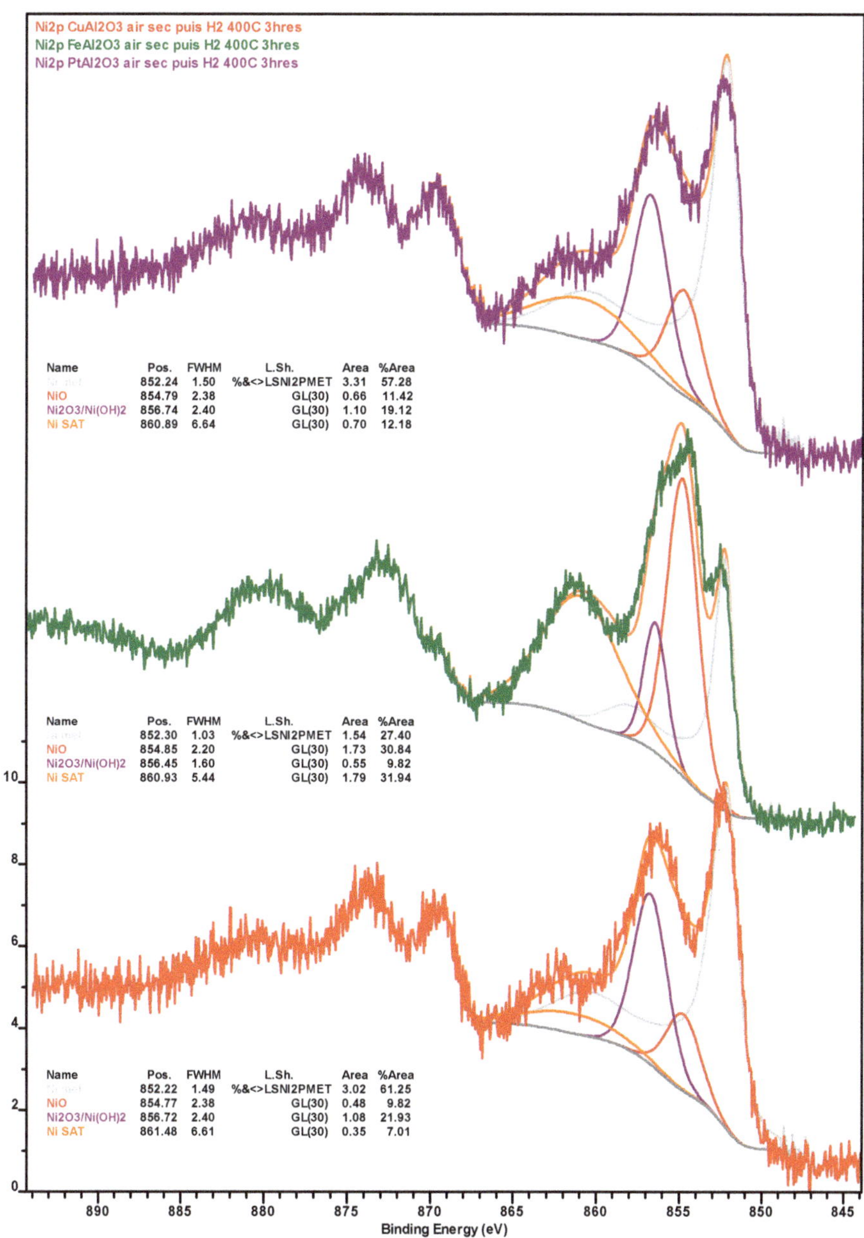

Figure A3. Nickel 2p X-ray photoelectron spectra of the spent Ni-Cu (red spectrum), Ni-Fe (green spectrum) and Ni-Pt (purple spectrum) catalysts recovered from deoxygenation experiments after calcination in air at 450 °C for 5 h and reduction under H_2 at 400 °C for 3 h.

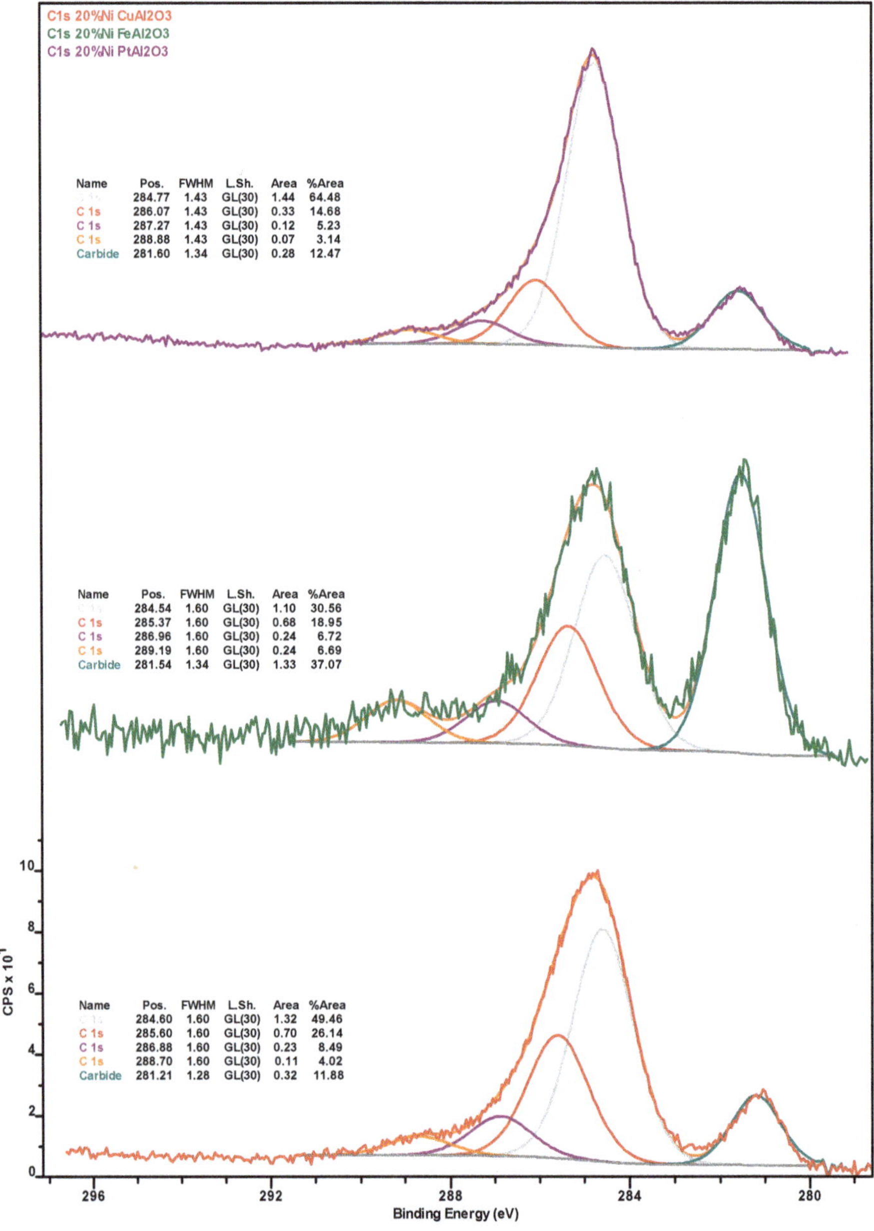

Figure A4. Carbon 1s X-ray photoelectron spectra of the spent Ni-Cu (red spectrum), Ni-Fe (green spectrum), and Ni-Pt (purple spectrum) catalysts recovered from the deoxygenation experiments.

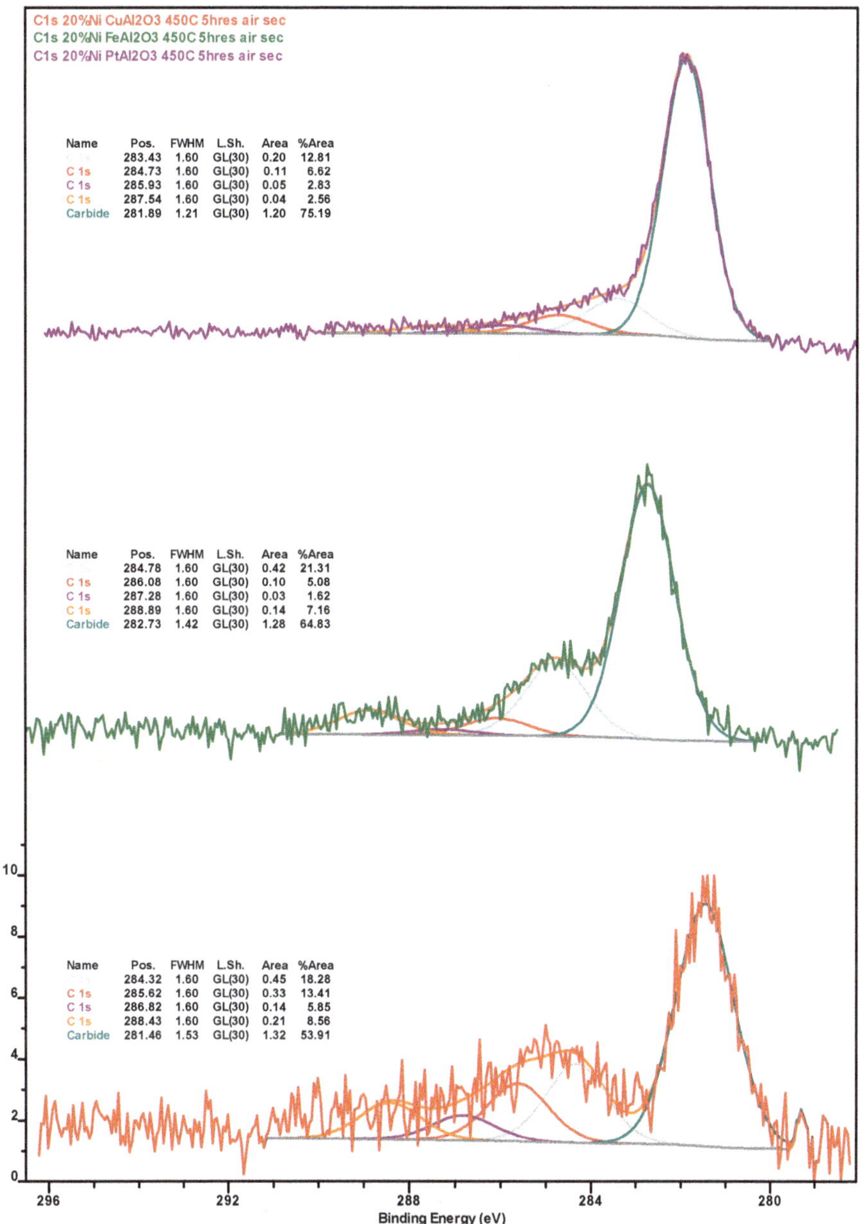

Figure A5. Carbon 1s X-ray photoelectron spectra of the spent Ni-Cu (red spectrum), Ni-Fe (green spectrum), and Ni-Pt (purple spectrum) catalysts recovered from the deoxygenation experiments after calcination in air at 450 °C for 5 h.

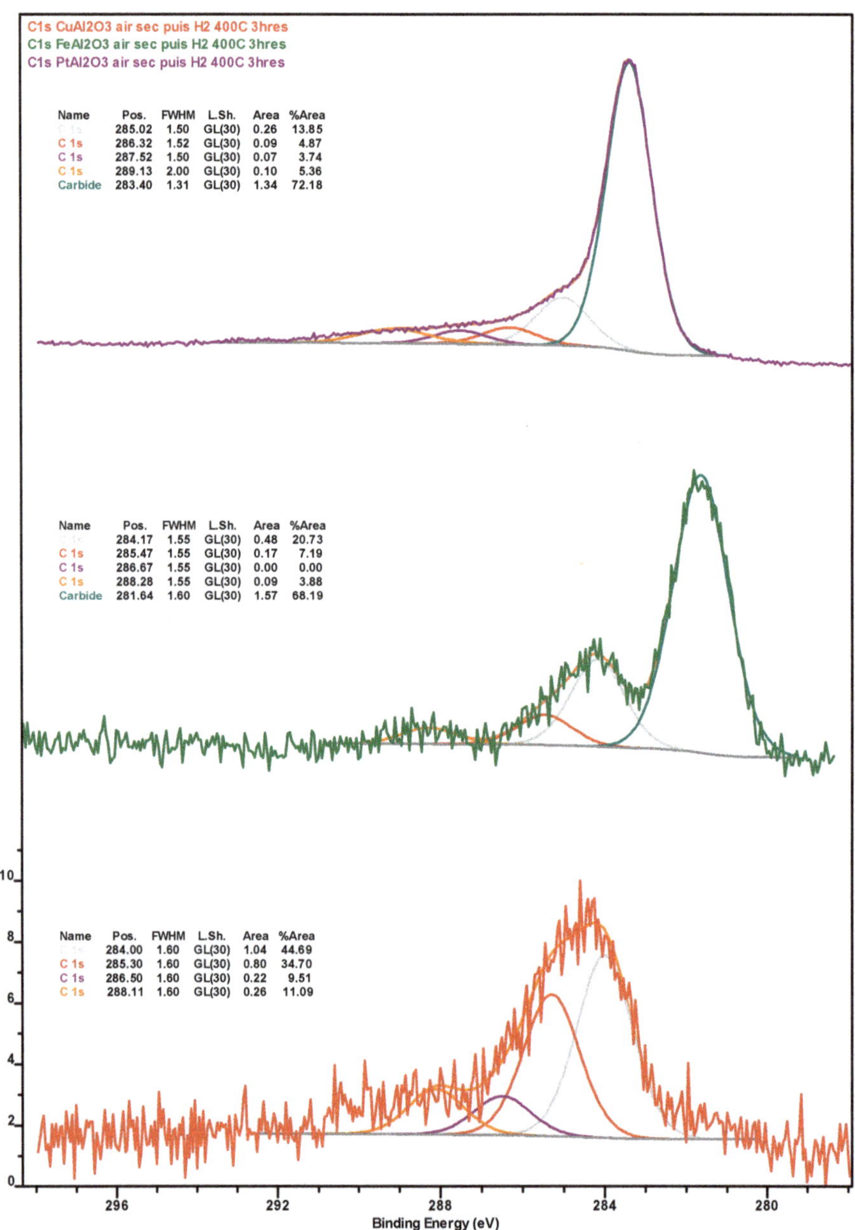

Figure A6. Carbon 1s X-ray photoelectron spectra of the spent Ni-Cu (red spectrum), Ni-Fe (green spectrum), and Ni-Pt (purple spectrum) catalysts recovered from the deoxygenation experiments after calcination in air at 450 °C for 5 h and reduction under H_2 at 400 °C for 3 h.

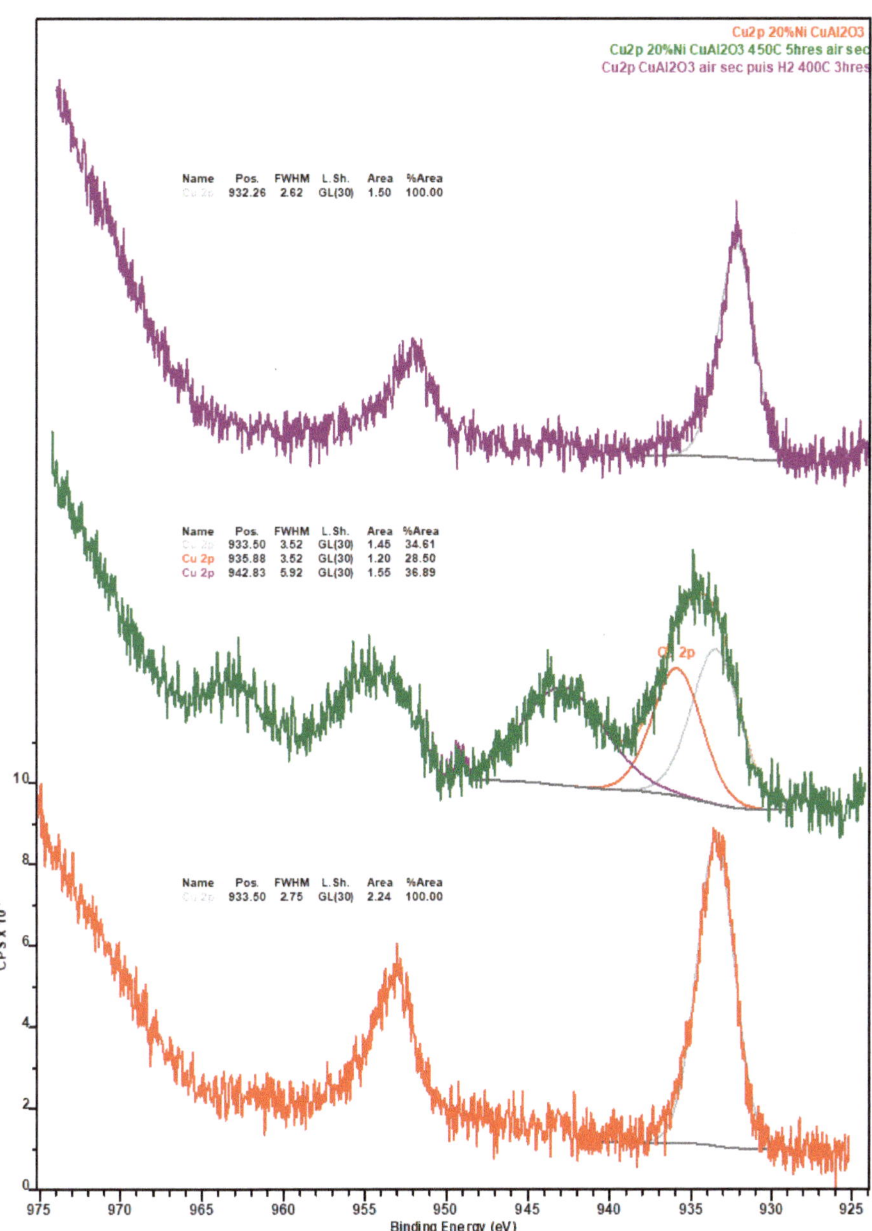

Figure A7. Copper 2p X-ray photoelectron spectra of the spent Ni-Cu catalyst after the deoxygenation experiment (red spectrum), after calcination in air at 450 °C for 5 h (green spectrum), and reduction under H_2 at 400 °C for 3 h (purple spectrum).

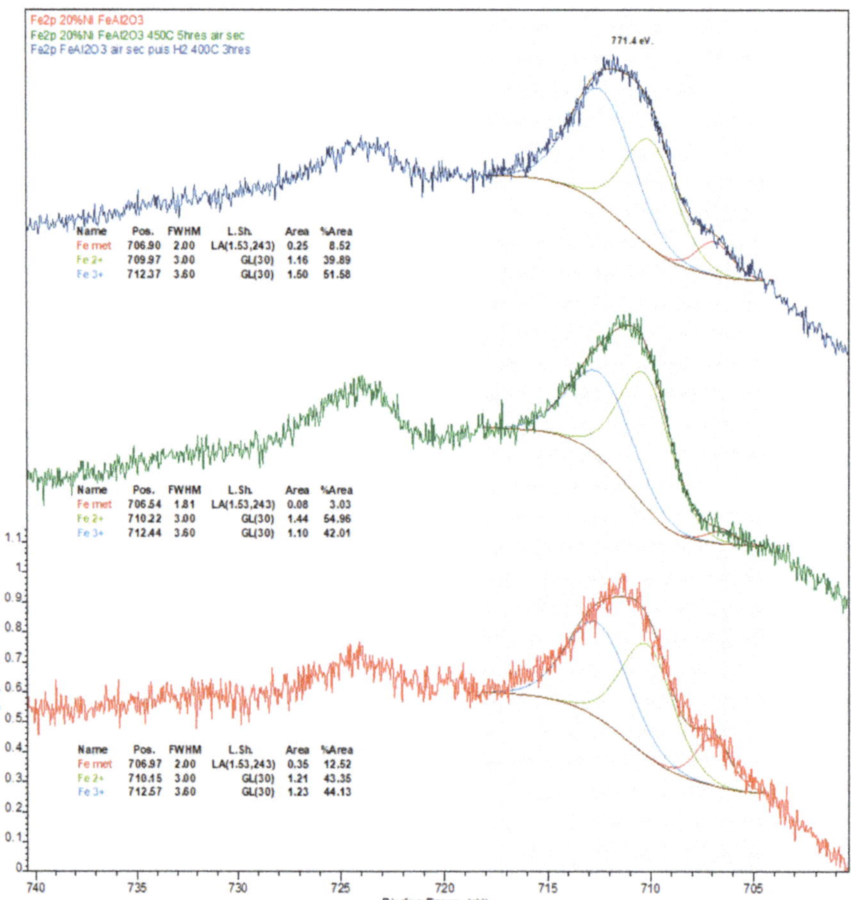

Figure A8. Iron 2p X-ray photoelectron spectra of the spent Ni-Fe catalyst after the deoxygenation experiment (red spectrum), after calcination in air at 450 °C for 5 h (green spectrum), and reduction under H_2 at 400 °C for 3 h (blue spectrum).

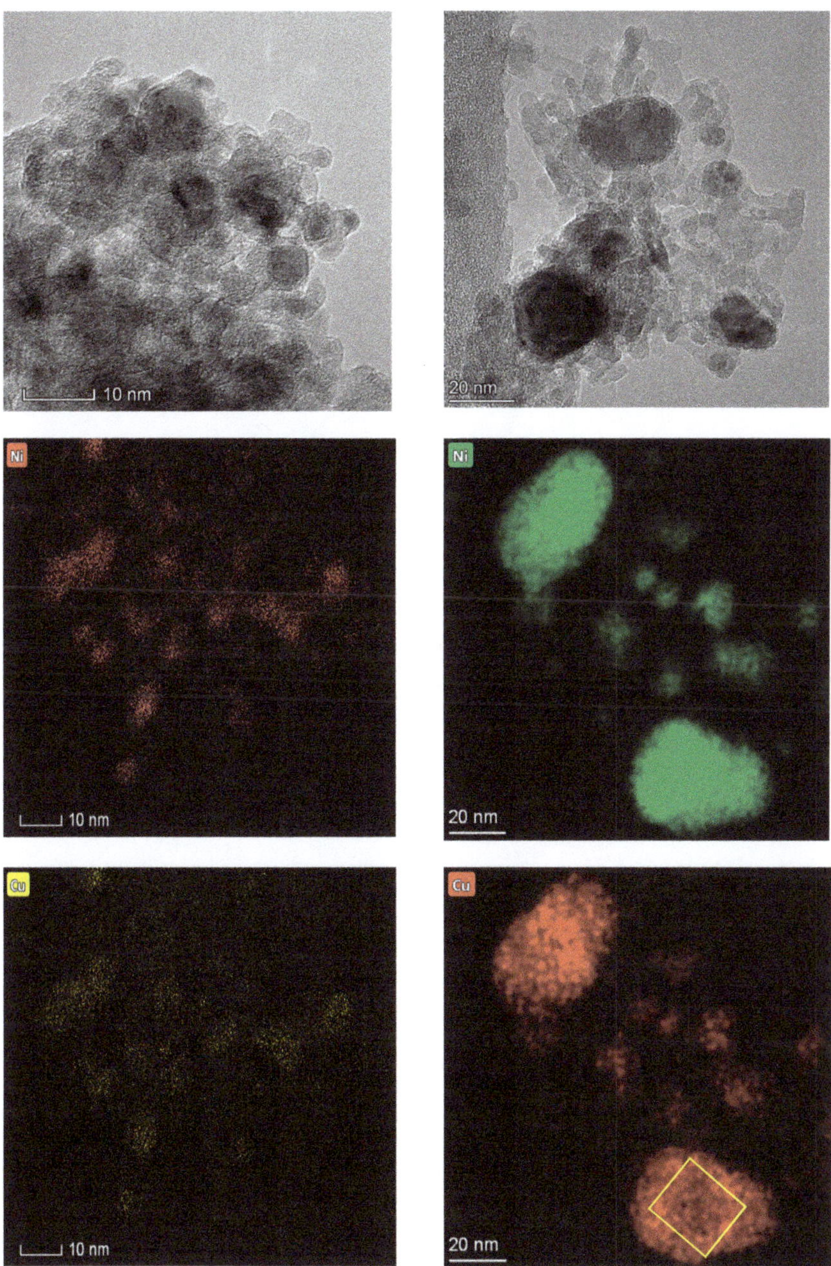

Figure A9. TEM micrographs and elemental maps of the fresh (**left**) and spent (**right**) Cu-promoted catalysts showing both the formation of large particles during catalyst aging and the close association of Ni and Cu irrespective of the catalyst state (fresh or spent). Note that micrographs and elemental maps correspond to different regions. The yellow box on the bottom right micrograph denotes the Cu-hollow space.

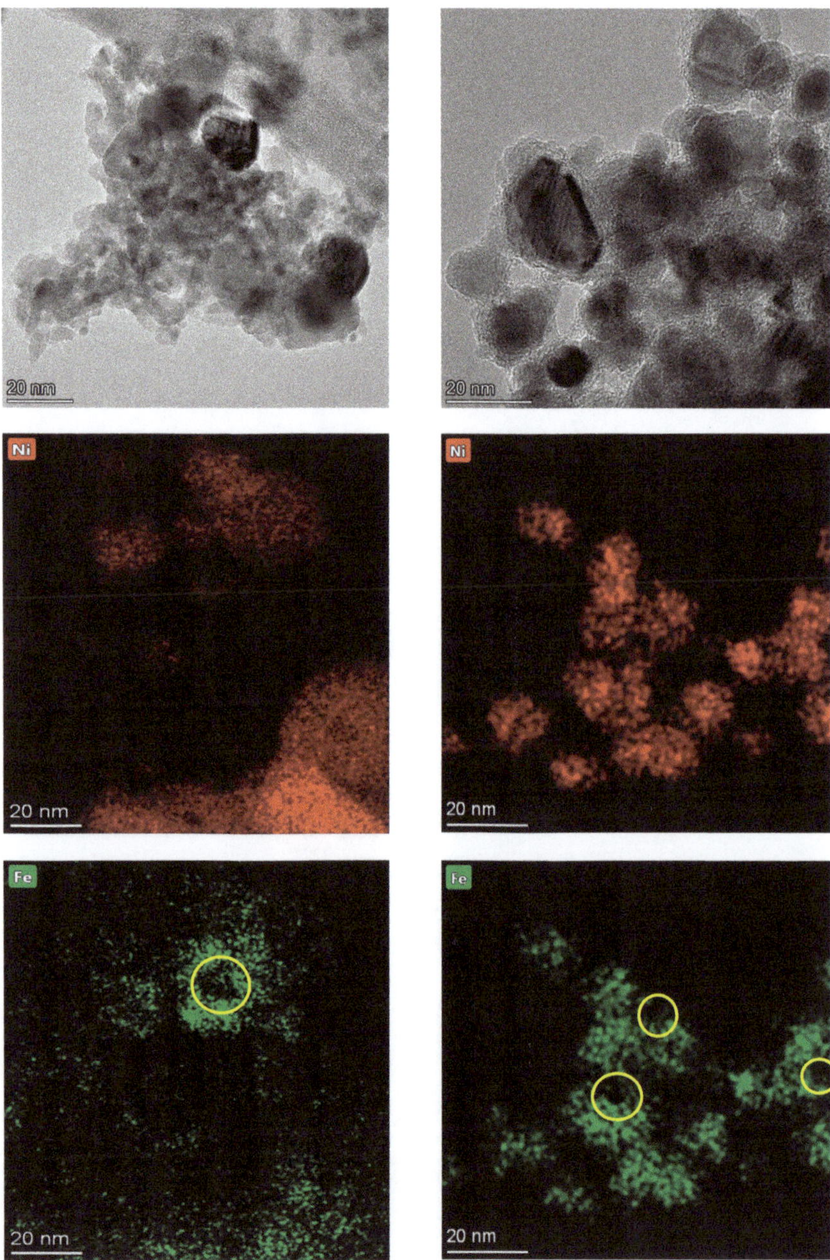

Figure A10. TEM micrographs and elemental maps of the fresh (**left**) and spent (**right**) Fe-promoted catalysts showing both the formation of large particles during catalyst aging and the degree of association between Ni and Fe in the fresh and spent formulations. Note that micrographs and elemental maps correspond to different regions. The yellow circles on the bottom micrographs denote the Fe-hollow spaces.

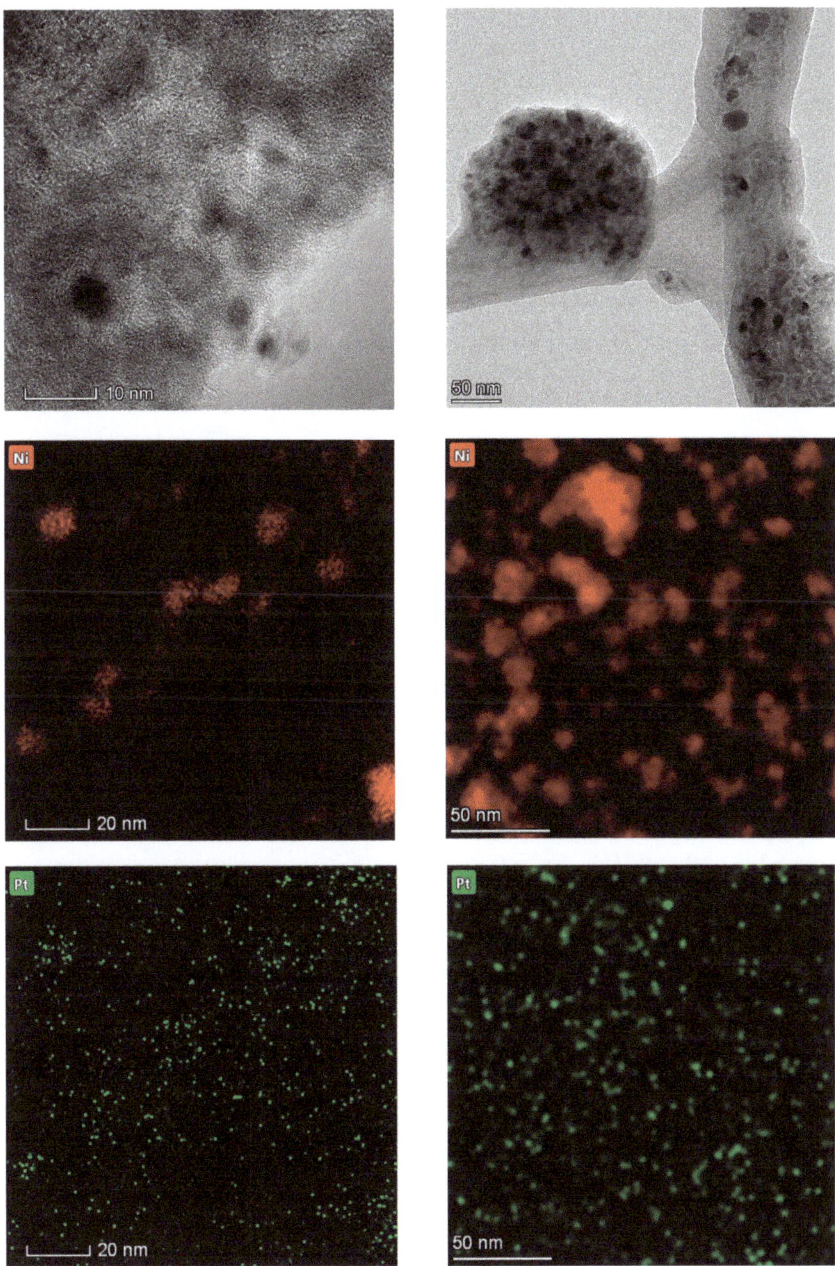

Figure A11. TEM micrographs and elemental maps of the fresh (**left**) and spent (**right**) Pt-promoted catalysts showing both the formation of large Ni particles during catalyst aging and the degree of association between Ni and Pt in the fresh and spent formulations. Note that micrographs and elemental maps correspond to different regions.

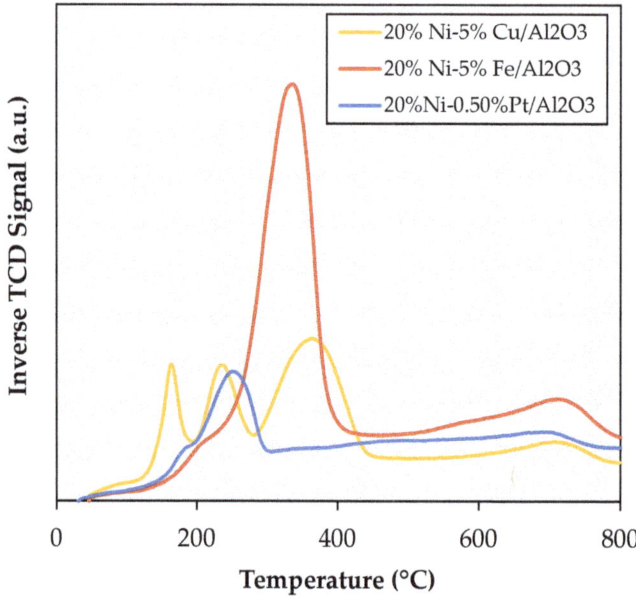

Figure A12. TPR profiles of the spent catalysts after calcination in air at 450 °C for 5 h. Variations in the baseline were magnified by the presence of the SiC diluent (which further reduced the limited amount of sample available for analysis); however, all reduction events are clearly resolved.

References

1. Kordulis, C.; Bourikas, K.; Gousi, M.; Kordouli, E.; Lycourghiotis, A. Development of nickel based catalysts for the transformation of natural triglycerides and related compounds into green diesel: A critical review. *Appl. Catal. B* **2016**, *181*, 156–196. [CrossRef]
2. Orozco, L.M.; Echeverri, D.A.; Sánchez, L.; Rios, L.A. Second-generation green diesel from castor oil: Development of a new and efficient continuous-production process. *Chem. Eng. J.* **2017**, *322*, 149–156. [CrossRef]
3. Pimerzin, A.A.; Tomina, N.N.; Nikul'shin, P.A.; Maksimov, N.M.; Mozhaev, A.V.; Ishutenko, D.I.; Vishnevskaya, E.E. Catalysts based on molybdenum and tungsten heteropoly compounds for the hydrotreatment of oil fractions. *Catal. Ind.* **2015**, *7*, 30–37. [CrossRef]
4. Douvartzides, S.L.; Charisiou, N.D.; Papageridis, K.N.; Goula, M.A. Green Diesel: Biomass Feedstocks, Production Technologies, Catalytic Research, Fuel Properties and Performance in Compression Ignition Internal Combustion Engines. *Energies* **2019**, *12*, 809. [CrossRef]
5. Hilbers, T.J.; Sprakel, L.M.; van den Enk, L.B.; Zaalberg, B.; van den Berg, H.; van der Ham, L.G. Green diesel from hydrotreated vegetable oil process design study. *Chem. Eng. Technol.* **2015**, *38*, 651–657. [CrossRef]
6. Chen, S.; Zhou, G.; Miao, C. Green and renewable bio-diesel produce from oil hydrodeoxygenation: Strategies for catalyst development and mechanism. *Renew. Sustain. Energy Rev.* **2019**, *101*, 568–589. [CrossRef]
7. *Biofuels and Bioproducts from Wet and Gaseous Waste Streams: Challenges and Opportunities*; EERE Publication and Product Library; U.S. Department of Energy: Washington, DC, USA, 2017.
8. Thinnakorn, K.; Tscheikuna, J. Biodiesel production via transesterification of palm olein using sodium phosphate as a heterogeneous catalyst. *Appl. Catal. A* **2014**, *476*, 26–33. [CrossRef]
9. Meher, L.C.; Sagar, D.V.; Naik, S. Technical aspects of biodiesel production by transesterification—A review. *Renew. Sustain. Energy Rev.* **2006**, *10*, 248–268. [CrossRef]
10. Santillan-Jimenez, E.; Crocker, M. Catalytic deoxygenation of fatty acids and their derivatives to hydrocarbon fuels via decarboxylation/decarbonylation. *J. Chem. Technol. Biotechnol.* **2012**, *87*, 1041–1050. [CrossRef]

11. Loe, R.; Lavoignat, Y.; Maier, M.; Abdallah, M.; Morgan, T.; Qian, D.; Pace, R.; Santillan-Jimenez, E.; Crocker, M. Continuous Catalytic Deoxygenation of Waste Free Fatty Acid-Based Feeds to Fuel-Like Hydrocarbons Over a Supported Ni-Cu Catalyst. *Catalysts* **2019**, *9*, 123. [CrossRef]

12. Kiméné, A.; Wojcieszak, R.; Paul, S.; Dumeignil, F. Catalytic decarboxylation of fatty acids to hydrocarbons over non-noble metal catalysts: The state of the art. *J. Chem. Technol. Biotechnol.* **2019**, *94*, 658–669. [CrossRef]

13. Loe, R.; Huff, K.; Walli, M.; Morgan, T.; Qian, D.; Pace, R.; Song, Y.; Isaacs, M.; Santillan-Jimenez, E.; Crocker, M. Effect of Pt Promotion on the Ni-Catalyzed Deoxygenation of Tristearin to Fuel-Like Hydrocarbons. *Catalysts* **2019**, *9*, 200. [CrossRef]

14. Pattanaik, B.P.; Misra, R.D. Effect of reaction pathway and operating parameters on the deoxygenation of vegetable oils to produce diesel range hydrocarbon fuels: A review. *Renew. Sustain. Energy Rev.* **2017**, *73*, 545–557. [CrossRef]

15. Morgan, T.; Grubb, D.; Santillan-Jimenez, E.; Crocker, M. Conversion of triglycerides to hydrocarbons over supported metal catalysts. *Top. Catal.* **2010**, *53*, 820–829. [CrossRef]

16. Santillan-Jimenez, E.; Morgan, T.; Lacny, J.; Mohapatra, S.; Crocker, M. Catalytic deoxygenation of triglycerides and fatty acids to hydrocarbons over carbon-supported nickel. *Fuel* **2013**, *103*, 1010–1017. [CrossRef]

17. Yu, X.; Chen, J.; Ren, T. Promotional effect of Fe on performance of Ni/SiO$_2$ for deoxygenation of methyl laurate as a model compound to hydrocarbons. *RSC Adv.* **2014**, *4*, 46427–46436. [CrossRef]

18. Loe, R.; Santillan-Jimenez, E.; Morgan, T.; Sewell, L.; Ji, Y.; Jones, S.; Isaacs, M.A.; Lee, A.F.; Crocker, M. Effect of Cu and Sn promotion on the catalytic deoxygenation of model and algal lipids to fuel-like hydrocarbons over supported Ni catalysts. *Appl. Catal. B* **2016**, *191*, 147–156. [CrossRef]

19. Han, Q.; Rehman, M.U.; Wang, J.; Rykov, A.; Gutiérrez, O.Y.; Zhao, Y.; Wang, S.; Ma, X.; Lercher, J.A. The synergistic effect between Ni sites and Ni-Fe alloy sites on hydrodeoxygenation of lignin-derived phenols. *Appl. Catal. B* **2019**, *253*, 348–358. [CrossRef]

20. Khromova, S.A.; Smirnov, A.A.; Bulavchenko, O.A.; Saraev, A.A.; Kaichev, V.V.; Reshetnikov, S.I.; Yakovlev, V.A. Anisole hydrodeoxygenation over Ni–Cu bimetallic catalysts: The effect of Ni/Cu ratio on selectivity. *Appl. Catal. A* **2014**, *470*, 261–270. [CrossRef]

21. Peng, B.; Zhao, C.; Kasakov, S.; Foraita, S.; Lercher, J.A. Manipulating catalytic pathways: Deoxygenation of palmitic acid on multifunctional catalysts. *Chem. Eur. J.* **2013**, *19*, 4732–4741. [CrossRef] [PubMed]

22. Santillan-Jimenez, E.; Loe, R.; Garrett, M.; Morgan, T.; Crocker, M. Effect of Cu promotion on cracking and methanation during the Ni-catalyzed deoxygenation of waste lipids and hemp seed oil to fuel-like hydrocarbons. *Catal. Today* **2018**, *302*, 261–271. [CrossRef]

23. Santillan-Jimenez, E.; Morgan, T.; Loe, R.; Crocker, M. Continuous catalytic deoxygenation of model and algal lipids to fuel-like hydrocarbons over Ni–Al layered double hydroxide. *Catal. Today* **2015**, *258*, 284–293. [CrossRef]

24. Vizcaíno, A.J.; Carrero, A.; Calles, J.A. Hydrogen production by ethanol steam reforming over Cu–Ni supported catalysts. *Int. J. Hydrogen Energy* **2007**, *32*, 1450–1461. [CrossRef]

25. Wang, L.; Li, D.; Koike, M.; Koso, S.; Nakagawa, Y.; Xu, Y.; Tomishige, K. Catalytic performance and characterization of Ni-Fe catalysts for the steam reforming of tar from biomass pyrolysis to synthesis gas. *Appl. Catal. A* **2011**, *392*, 248–255. [CrossRef]

26. Gheisari, K.; Javadpour, S.; Oh, J.T.; Ghaffari, M. The effect of milling speed on the structural properties of mechanically alloyed Fe–45%Ni powders. *J. Alloys Compd.* **2009**, *472*, 416–420. [CrossRef]

27. Carrero, A.; Calles, J.; Vizcaíno, A. Hydrogen production by ethanol steam reforming over Cu-Ni/SBA-15 supported catalysts prepared by direct synthesis and impregnation. *Appl. Catal. A* **2007**, *327*, 82–94. [CrossRef]

28. Lee, J.-H.; Lee, E.-G.; Joo, O.-S.; Jung, K.-D. Stabilization of Ni/Al$_2$O$_3$ catalyst by Cu addition for CO$_2$ reforming of methane. *Appl. Catal. A* **2004**, *269*, 1–6. [CrossRef]

29. Miranda, B.C.; Chimentão, R.J.; Szanyi, J.; Braga, A.H.; Santos, J.B.O.; Gispert-Guirado, F.; Llorca, J.; Medina, F. Influence of copper on nickel-based catalysts in the conversion of glycerol. *Appl. Catal. B* **2015**, *166–167*, 166–180. [CrossRef]

30. Li, Y.; Chen, J.; Chang, L.; Qin, Y. The Doping Effect of Copper on the Catalytic Growth of Carbon Fibers from Methane over a Ni/Al$_2$O$_3$ Catalyst Prepared from Feitknecht Compound Precursor. *J. Catal.* **1998**, *178*, 76–83. [CrossRef]

31. Rynkowski, J.M.; Paryjczak, T.; Lenik, M. On the nature of oxidic nickel phases in NiO/γ-Al_2O_3 catalysts. *Appl. Catal. A* **1993**, *106*, 73–82. [CrossRef]

32. Zieliński, J. Morphology of nickel/alumina catalysts. *J. Catal.* **1982**, *76*, 157–163. [CrossRef]

33. Tanksale, A.; Beltramini, J.; Dumesic, J.; Lu, G. Effect of Pt and Pd promoter on Ni supported catalysts—A TPR/TPO/TPD and microcalorimetry study. *J. Catal.* **2008**, *258*, 366–377. [CrossRef]

34. Shi, D.; Wojcieszak, R.; Paul, S.; Marceau, E. Ni Promotion by Fe: What Benefits for Catalytic Hydrogenation? *Catalysts* **2019**, *9*, 451. [CrossRef]

35. El Mel, A.-A.; Nakamura, R.; Bittencourt, C. The Kirkendall effect and nanoscience: Hollow nanospheres and nanotubes. *Beilstein J. Nanotechnol.* **2015**, *6*, 1348–1361. [CrossRef]

36. Van Stiphout, P.; Stobbe, D.; Scheur, F.T.V.; Geus, J. Activity and stability of nickel-copper/silica catalysts prepared by deposition-precipitation. *Appl. Catal.* **1988**, *40*, 219–246. [CrossRef]

37. Webber, P.; Rojas, C.; Dobson, P.; Chadwick, D. A combined XPS/AES study of Cu segregation to the high and low index surfaces of a Cu-Ni alloy. *Surf. Sci.* **1981**, *105*, 20–40. [CrossRef]

38. Lv, L.; Li, Z.; Ruan, Y.; Chang, Y.; Ao, X.; Li, J.-G.; Yang, Z.; Wang, C. Nickel-iron diselenide hollow nanoparticles with strongly hydrophilic surface for enhanced oxygen evolution reaction activity. *Electrochim. Acta* **2018**, *286*, 172–178. [CrossRef]

39. Lv, L.; Li, Z.; Xue, K.-H.; Ruan, Y.; Ao, X.; Wan, H.; Miao, X.; Zhang, B.; Jiang, J.; Wang, C.; et al. Tailoring the electrocatalytic activity of bimetallic nickel-iron diselenide hollow nanochains for water oxidation. *Nano Energy* **2018**, *47*, 275–284. [CrossRef]

40. Xu, W.; Si, R.; Senanayake, S.D.; Llorca, J.; Idriss, H.; Stacchiola, D.; Hanson, J.C.; Rodriguez, J.A. In situ studies of CeO_2-supported Pt, Ru, and Pt–Ru alloy catalysts for the water–gas shift reaction: Active phases and reaction intermediates. *J. Catal.* **2012**, *291*, 117–126. [CrossRef]

41. Gamarra, D.; Martínez-Arias, A. Preferential oxidation of CO in rich H_2 over CuO/CeO_2: Operando-DRIFTS analysis of deactivating effect of CO_2 and H_2O. *J. Catal.* **2009**, *263*, 189–195. [CrossRef]

42. Boccuzzi, F.; Chiorino, A.; Martra, G.; Gargano, M.; Ravasio, N.; Carrozzini, B. Preparation, Characterization, and Activity of Cu/TiO_2 Catalysts. I. Influence of the Preparation Method on the Dispersion of Copper in Cu/TiO_2. *J. Catal.* **1997**, *165*, 129–139. [CrossRef]

43. Chandler, B.D.; Pignolet, L.H. DRIFTS studies of carbon monoxide coverage on highly dispersed bimetallic Pt-Cu and Pt-Au catalysts. *Catal. Today* **2001**, *65*, 39–50. [CrossRef]

44. Parizotto, N.; Zanchet, D.; Rocha, K.; Marques, C.; Bueno, J. The effects of Pt promotion on the oxi-reduction properties of alumina supported nickel catalysts for oxidative steam-reforming of methane: Temperature-resolved XAFS analysis. *Appl. Catal. A* **2009**, *366*, 122–129. [CrossRef]

45. Morgan, T.; Santillan-Jimenez, E.; Harman-Ware, A.E.; Ji, Y.; Grubb, D.; Crocker, M. Catalytic deoxygenation of triglycerides to hydrocarbons over supported nickel catalysts. *Chem. Eng. J.* **2012**, *189–190*, 346–355. [CrossRef]

46. Morgan, T.; Santillan-Jimenez, E.; Huff, K.; Javed, K.R.; Crocker, M. Use of dual detection in the gas chromatographic analysis of oleaginous biomass feeds and biofuel products to enable accurate simulated distillation and lipid profiling. *Energy Fuels* **2017**, *31*, 9498–9506. [CrossRef]

Article

Selective Conversion of Glucose to 5-Hydroxymethylfurfural by Using L-Type Zeolites with Different Morphologies

María José Ginés-Molina [1,†], Nur Hidayahni Ahmad [2,†], Sandra Mérida-Morales [1,†], Cristina García-Sancho [1], Svetlana Mintova [3], Ng Eng-Poh [2,*] and Pedro Maireles-Torres [1,*]

[1] Departamento de Química Inorgánica, Cristalografía y Mineralogía (Unidad Asociada al ICP-CSIC) Facultad de Ciencias, Universidad de Málaga, Campus de Teatinos, 29071 Málaga, Spain; mariaj.gimo@uma.es (M.J.G.-M.); sandra_merida@uma.es (S.M.-M.); cristinags@uma.es (C.G.-S.)
[2] School of Chemical Sciences, Universiti Sains Malaysia, USM, Penang 11800, Malaysia; nurhidayahni@gmail.com
[3] Normandie Université, ENSICAEN, UNICAEN, CNRS, LCS, 14000 Caen, France; mintova@ensicaen.fr
*   Correspondence: epng@usm.my (E.-P.N.); maireles@uma.es (P.M.T.)
†   These authors contributed equally to this work as first authors.

Received: 25 November 2019; Accepted: 13 December 2019; Published: 16 December 2019

Abstract: In the present work, the morphology of L-type zeolite (LTL topology) has been modified in order to evaluate the influence of several protonated-form LTL-zeolites with different morphologies on their stability and catalytic performance in the conversion of glucose into 5-hydroxymethylfurfural (5-HMF). Physico-chemical characterization of the LTL-based catalysts has revealed that the three types of morphologies (needle, short rod and cylinder) are active, providing complete glucose conversion and high 5-HMF yield values. The addition of $CaCl_2$ had a positive influence on the catalytic performance. It was found that morphology influences the textural and acid properties of LTL-zeolites, and hence their catalytic performance. The best catalytic results have been obtained with the NEEDLE-LTL, showing nanoparticles with a length of 4.46 μm and a width of 0.63 μm, which attains a 5-HMF yield of 63%, at 175 °C after 90 min of reaction, and a glucose conversion of 88%. The reusability study has revealed a progressive decrease in 5-HMF yield after each catalytic cycle. Different regeneration methods have been essayed without recovering the initial catalytic activity. The presence of organic molecules in micropores has been demonstrated by TG analysis, which are difficult to remove even after a regeneration process at 550 °C.

Keywords: biomass; glucose; 5-hydroxymethylfurfural; LTL-zeolites; heterogeneous catalysis

---

1. Introduction

In recent years, much attention is being paid to the development of environmentally and economically viable synthetic routes and technologies for producing chemicals and fuels from non-fossil carbon sources as alternative to fossil raw materials [1]. In this context, biomass is emerging as a very promising sustainable feedstock, being the only widely available and renewable carbon source [2,3]. Lignocellulosic biomass, mainly composed by lignin, cellulose and hemicellulose, with an estimated annual production about 2×10^{11} metric tons, is the most abundant source of carbohydrates, but physico-chemical treatments are required for its use as a raw material [4]. Although lignocellulose is a sustainable resource for production of biofuels and chemicals, it is necessary that this does not interfere with the food chain. The hydrolysis of cellulose and hemicellulose leads to monomeric C_5 and C_6 sugars, which can be converted into important platform molecules, such as furfural and 5-hydroxymethylfurfural (5-HMF), respectively, which are the starting point for the synthesis of a large variety of biofuels and chemicals [5,6]. For instance, 5-HMF can be transformed into 2,5-dimethylfuran [7] or levulinic acid [8,9], among others, which are key intermediates for the synthesis of pharmaceuticals, polymers or biofuels.

Although the dehydration of fructose to 5-HMF has largely been reported in the literature, glucose is preferred due to its abundance and low price [10]. There is not a general consensus about the mechanism of glucose dehydration to 5-HMF, even though a generally accepted route based on: (i) isomerization of glucose to fructose, and (ii) dehydration of fructose to 5-HMF [4]. The first step is considerably difficult and requires Lewis acid or basic sites, being the limiting factor for 5-HMF production. The reaction may be performed either in water, organic solvents or ionic liquids, in particular polar aprotic solvents. Homogeneous catalysts such as sulphuric or hydrochloric acids can be effective for the hydrolysis of cellulose to glucose, and even for the dehydration of fructose to 5-HMF. However, due to their corrosive properties which are hazardous for equipment, they are gradually replaced by heterogeneous catalysts. Besides, heterogeneous catalysts allow their easy separation from solution, recovery and reuse [11–13]. Different solid acid catalysts have been tested for dehydration of glucose to 5-HMF, such as γ-Al_2O_3 [14], zeolites [15], metal oxides like TiO_2 or ZrO_2 [12–15], mesoporous solids [16,17], inter alia.

Nevertheless, both solvent and catalyst must be considered as two key factors to attain high 5-HMF yields from C_6 carbohydrates. A common strategy for 5-HMF production is the use of biphasic systems because this approach gives higher 5-HMF yields than systems employing only water. Usually, the biphasic medium is formed by the addition of organic solvents (toluene, methyl isobutyl ketone, among other) to an aqueous solution, or the addition of miscible organic solvents like tetrahydrofuran (THF) [18], to a saturated salt solution, which allows to extract the 5-HMF formed from the aqueous phase, preventing its further degradation and condensation [4].

In order to prevent these side reactions, Román-Leshkov et al. [19] employed inorganic salts in a biphasic system for dehydration of fructose to 5-HMF and concluded that the salting-out effect leading to a higher partition coefficient, limiting the individual cationic or anionic contributions, so then it is feasible to correlate to the interaction of all ionic species. In this context, it has also been reported that divalent cations interact more strongly with saccharides than the monovalent ones [20]. Thus, Combs et al. [21] observed that alkaline earth metal cations can form bidentate complexes with glucose, which accelerated its transformation. Later, our research group studied the beneficial effects of $CaCl_2$ on glucose dehydration to 5-HMF in the presence of Al_2O_3 as catalyst, in such a way that the addition of $CaCl_2$ to the reaction medium notably improved the catalytic performance, even at very short reaction times, due to the interaction between Ca^{2+} ions and glucose molecules, which favored the α-D-glucopyranose formation [14].

Concerning the use of zeolites for glucose dehydration, different acidic ZSM-5-zeolites (H-, Fe- and Cu-ZSM-5) were prepared and studied by modifying several experimental variables [15]. It was demonstrated the positive effect of the addition of inorganic salt (NaCl) to a biphasic water/methyl isobutyl ketone (MIBK) system, since a glucose conversion of 80%, with a HMF yield of 42% was attained at 195 °C, after 30 min, by using a H-ZSM-5-zeolite, which had the lowest Lewis/Brönsted ratio among the studied zeolites. Later, the H-ZSM-5-zeolite was compared with H-Y and H-β-zeolites, in order to assess the influence of the textural properties on the catalytic performance in glucose dehydration [22]. Under similar experimental conditions, by using a H-β-zeolite, the highest 5-HMF yield (56%) was reached, thus demonstrating the benefit of mesoporosity in this catalytic process.

Recently, by using a bifunctional Cr/β zeolite, a high selectivity to 5-HMF with a yield of 72%, was found at 150 °C, after 90 min, by adding NaCl to a biphasic H_2O/THF system [23]. After three consecutive catalytic cycles, the catalytic activity slightly decreased, but after a thermal treatment was almost recovered.

Morphological or textural characteristics play a highly important role when discussing catalytic activity in a chemical reaction [24]. Nevertheless, the morphological roles of zeolite in glucose dehydration are still not completely understood, hence further investigation to reveal this effect in the dehydration reaction is of the utmost importance. The aim of this work is a thorough study of glucose dehydration for 5-HMF production (Scheme 1) using protonated L-type (H-LTL) zeolites with different morphologies, which have been characterized and their catalytic performance has been correlated with their textural and acid-base properties.

Scheme 1. Reaction pathway of glucose to 5-hydroxymethylfurfural.

2. Results and Discussion

2.1. Catalyst Characterization

The choice of a LTL zeolitic framework to evaluate the influence of the morphology (short rod, cylinder and needle) on the catalytic performance in glucose dehydration to HMF arises from the dimensions of pore mouth and channels of this crystallographic structure (Figure 1). These allow the entrance of glucose molecules to reach active acid sites, and 5-hydroxymethylfurfural can easily go out, leaving active sites available for new sugar molecules.

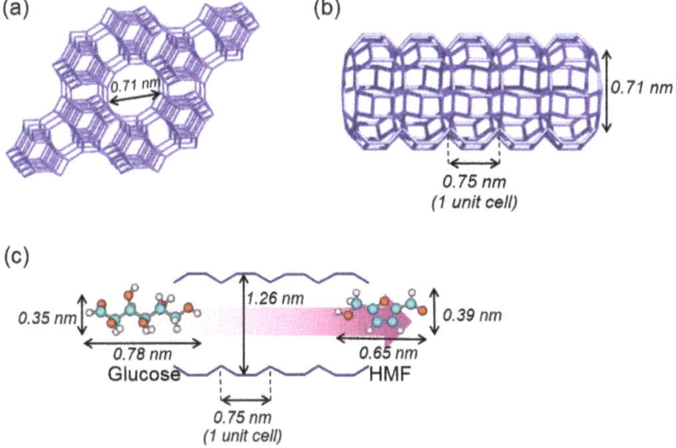

Figure 1. (a) Structure of LTL-zeolite view along [001] plane illustrating its hexagonal framework. (b) An LTL channel view normal to [001] plane that consists of 0.75 nm unit cells with a pore opening of 0.71 nm. (c) Schematic illustration of diffusion of glucose as reactant and HMF as product in the 12-membered ring channel of LTL zeolite.

The X-ray diffraction patterns of H-LTL-zeolites exhibit many narrow diffraction peaks at $2\theta = 5.55°$ (100), 11.10° (200), 11.77° (001), 18.89° (210), 15.23° (111), 19.31° (220), 20.09° (310), 20.47° (301), etc. which can be assigned to the hexagonal LTL-type framework (PDF 98-007-4170) (Figure 2). In all cases, no additional crystalline phase is detected and the crystallinity of the solids is preserved after ion-exchange and calcination treatments.

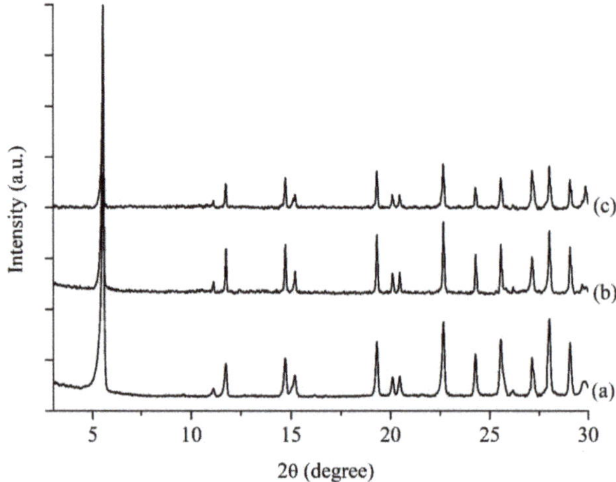

Figure 2. XRD patterns of (**a**) ROD-LTL, (**b**) NEEDLE-LTL and (**c**) CYL-LTL zeolites.

Concerning the chemical composition of LTL-zeolites, the Si/Al molar ratio values obtained from ICP-OES are close to 3, which are lower than that used in the synthesis gel (10) (Table 1). The comparison between the bulk and surface chemical composition data, deduced from ICP-OES and XPS, respectively, points an enrichment of Si on the surface of LTL-zeolites. This could be explained by the higher Si/Al molar ratio (10) used in the synthesis of these zeolites. However, the inherent error associated with the semi-quantitative analysis by XPS must be taken into account. After the ion exchange process between K^+ and NH_4^+ ions, not all the alkaline cations are removed, which can be due to the existence of strong acid sites remaining dissociated and neutralized by K^+ cations, even in the presence of an excessive ammonium solution. Therefore, the K/Al molar ratio is lower than 1 for NEEDLE-LTL and ROD-LTL, indicating that, although acid sites are protonated, a fraction of sites is still occupied by K^+ ions. This fraction is even higher for the CYL-LTL, where its value closed to 1 would point out that this morphology is the less favorable for the ion-exchange process.

Table 1. XPS and ICP-OES (* in parentheses) data of LTL-zeolites.

	Binding Energy (eV)				Si/Al Molar Ratio *	K/Al Molar Ratio *
	Si 2p	Al 2p	O 1s	K 2p$_{3/2}$		
NEEDLE-LTL	102.8	74.3	532.1	293.3	4.23 (3.00)	0.45 (0.63)
ROD-LTL	102.9	74.5	532.2	293.4	4.54 (2.90)	0.45 (0.55)
CYL-LTL	103.2	74.6	532.4	293.7	4.43 (3.45)	1.28 (1.07)

X-ray photoelectron spectroscopy (XPS) has been used to get insights into the surface nature of LTL- zeolites. In all cases, the binding energies of Si 2p (102.8–103.2 eV), Al 2p (74.3–74.6 eV) and O 1s (532.1–532.4 eV) are typical of these elements forming part of microporous aluminosilicates [25]. As regards the K 2p spectra, they exhibit the characteristic doublet with the K 2p$_{3/2}$ at 293.3-293.7 eV, and a spectral separation of 2.8 eV, typical of K^+ ions [26].

On the other hand, the textural properties reveal that all zeolites maintain high surface area values, being the largest one for short rod morphology (Table 2). As expected, these zeolites are mainly microporous solids, with a percentage of microporous surface area higher than 95%.

Table 2. Textural and acid properties of LTL-zeolites.

	$S_{Langmuir}$ $(m^2\,g^{-1})$	S_{micro} $(m^2\,g^{-1})$	V_p $(cm^3\,g^{-1})$	Length (μm)	Width (μm)	Total Acidity $(\mu mol\,m^{-2})$ *
NEEDLE-LTL	248	235	0.100	4.46	0.63	6.73
ROD-LTL	303	293	0.120	1.43	1.38	6.17
CYL-LTL	219	199	0.090	3.70	0.97	4.88

* Determined from NH_3-TPD.

The microporous nature of these LTL-based zeolites can be easily confirmed by the shape of their adsorption-desorption isotherms of N_2 at $-196\ ^\circ$C, which are Type I in the IUPAC classification, typical of microporous solids (Figure 3). The slight hysteresis loop could be associated to some mesoporosity generated during treatments used for synthesizing their protonated forms, probably associated to interparticular voids [27,28], which is corroborated by the pore size distribution curves (Figure 3, right).

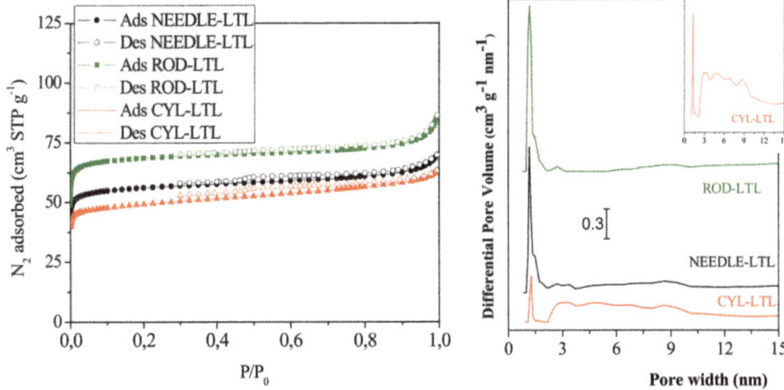

Figure 3. N_2 adsorption-desorption isotherms (**left**) and pore size distributions (**right**) of LTL-zeolites at $-196\ ^\circ$C.

The different morphologies are more clearly appreciated in the scanning electron micrographs (Figure 4), where micrometric particles are observed, whose dimensions and shapes fit in really well with short rods, needles and cylinders (Table 2).

On the other hand, the chemical environment of aluminium has been analyzed by ^{27}Al MAS-NMR spectroscopy (Figure 5). Extra-framework octahedral Al species are associated to a resonance signal at a chemical shift near 0 ppm, whereas tetrahedral Al in crystallographic sites in the zeolite framework appears at about 60 ppm. This latter signal is observed in all cases, and the absence of broadening effect at lower chemical shift values could discard the existence of pentacoordinated or distorted tetrahedral aluminium, reported in other microporous aluminosilicates, which give rise to signals at 30 and 47 ppm, respectively [29–31]. The small contribution at 0 ppm is associated to extra-framework Al species, but this is absent in the MAS-NMR spectrum of CYL-LTL, where all Al seems to be in tetrahedral coordination.

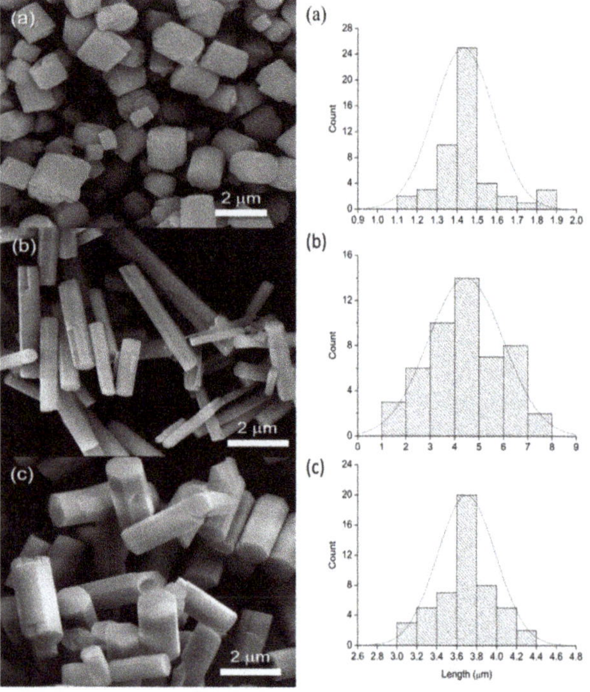

Figure 4. SEM images (left) and particle size distribution histograms (right) of LTL-zeolite crystals synthesized of (**a**) ROD-LTL, (**b**) NEEDLE-LTL and (**c**) CYL-LTL. Scale bar = 2 μm.

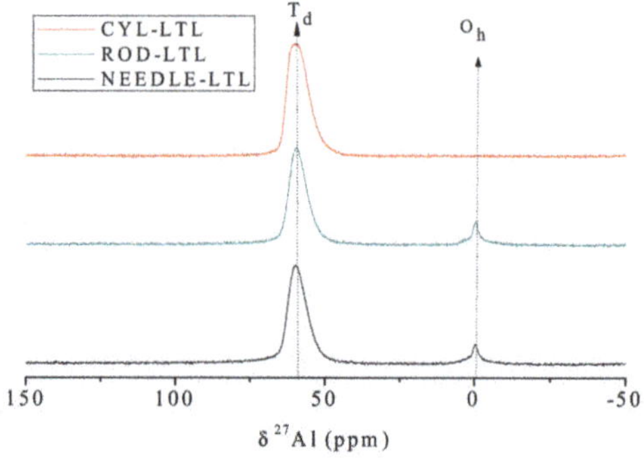

Figure 5. Solid state ^{27}Al-NMR spectra of LTL-zeolites.

The total acidity of catalysts was determined from ammonia temperature-programmed desorption (NH$_3$-TPD), whereas the nature (Brönsted or Lewis) of acid sites was studied from pyridine adsorption coupled to FTIR spectroscopy. Figure 6 displays the amount of ammonia desorbed in different temperature ranges, which have been assigned to weak (100-200 °C), medium (200-300 °C) and strong (300–550 °C) acid sites. The total acidity follows the order: ROD-LTL (1874 μmol g^{-1}) > NEEDLE-LTL (1672 μmol g^{-1}) > CYL-LTL (1065 μmol g^{-1}), being strong acid sites predominant in all cases. This acidity order is the same found for Langmuir surface area values (Table 2).

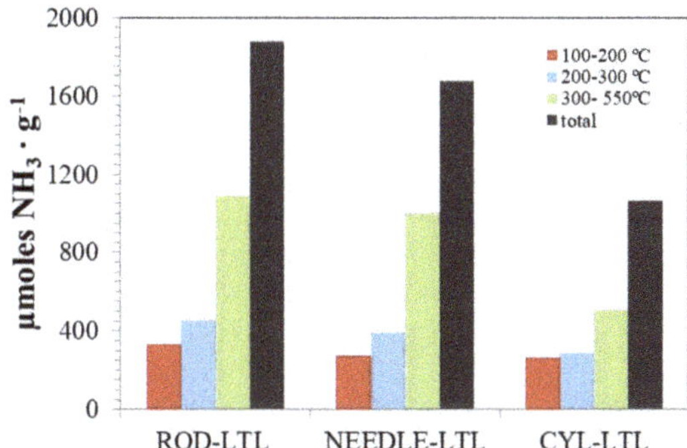

Figure 6. NH$_3$ desorption of LTL-based zeolites, as a function of the strength: weak (100–200 °C), medium (200–300 °C) and strong (300–550 °C).

The nature of acid sites (Brönsted and/or Lewis) has been studied by pyridine adsorption coupled to FTIR spectroscopy. The concentration of Lewis and Brönsted acid sites after adsorption and subsequent desorption of pyridine at 150 and 300 °C is given in Table 3. The lowest concentration of both Lewis and Brönsted acid sites is found for the ROD-LTL-zeolite, which after evacuation at 300 °C does not exhibit Lewis acidity. The other two zeolites (CYL- and NEEDLE-LTL) exhibit similar concentrations of strong Brönsted and Lewis acid sites, calculated from pyridine amount remaining on catalysts after evacuation at 300 °C. Moreover, these two zeolites possess a similar B/L molar ratio (4.76 and 4.86, respectively).

It has been previously reported that Brönsted acid sites could be associated to distorted tetrahedral Al species located in the zeolite framework, whereas distorted pentacoordinated Al species have been associated to Lewis acid sites in zeolites [32].

Table 3. Surface acidity of LTL-zeolites with different morphology at different temperature.

	Py-FTIR Acidity (µmol/g)						B/L	
	Lewis (L)		Brönsted (B)		Total (L + B)			
	150 °C	300 °C	150 °C	300 °C	150 °C	300 °C	150 °C	300 °C
ROD-LTL	89.7	24.40	101.1	84.4	152.23	84.51	3.52	-
CYL-LTL	126.08	32.08	128.07	152.85	254.15	184.93	1.02	4.76
NEEDLE-LTL	120.55	31.76	161.39	154.50	281.95	186.26	1.34	4.86

2.2. Catalytic Study

These zeolites have been evaluated as acid catalysts in the dehydration of glucose to 5-hydroxymethylfurfural, an important platform molecule for the synthesis of biofuels and high value-added chemicals [33]. Although the mechanism involved in glucose dehydration to 5-HMF seems to depend on type of catalyst, nature of solvent, among other parameters, it is broadly accepted that the rate-determining step is the isomerization of glucose to fructose [4]. For this reason, most of studies have used fructose as starting sugar. The isomerization process is catalyzed by Lewis acid or basic sites, whereas dehydration of fructose to 5-HMF requires the participation of Brönsted acid sites. The production of 5-HMF from glucose has two main drawbacks associated to the catalytic process, that is, the formation of by-products, such as soluble/insoluble polymers and humins, and the rehydration of 5-HMF for giving rise to levulinic and formic acids. These processes decrease the 5-HMF yield and, in some cases, provoke the catalyst deactivation. In this sense, the use of organic co-solvents together with the aqueous phase containing carbohydrates has been usually reported as a suitable strategy for improving 5-HMF yield. Among these organic solvents, MIBK is one of the most solvents often used due to their benign physico-chemical properties [34]. In this work, a biphasic water-MIBK system was employed to study the dehydration of glucose to 5-HMF.

Moreover, previous studies have demonstrated that the addition of inorganic salts to biphasic phases ameliorates the extraction of 5-HMF from biphasic systems [14,16,19]. In particular, $CaCl_2$ exerts a positive effect on the catalytic performance, even at very short reaction times [14,35]. This was explained by the interaction between Ca^{2+} ions and carbohydrates (glucose and xylose), modifying the corresponding anomeric equilibrium towards the anomer α more prone for the dehydration process, as was concluded from ^{1}H-NMR spectroscopy data.

In this sense, the catalytic activity of the LTL-zeolites was studied with and without $CaCl_2$ addition (Figure 7). The positive effect is clearly observed, since at 150 °C, after 60 min of reaction and in the absence of inorganic salt, the conversion of glucose is lower than 20% and in this case the formation of 5-HMF is barely detected (5-HMF yield < 1%). However, the addition of $CaCl_2$ outstandingly improves the glucose conversion until values higher than 95% for ROD-LTL, with a 5-HMF yield of 58.5%, whereas the rest of catalysts attain conversion and 5-HMF yield values higher than 75% and 25%, respectively. Therefore, these data confirm previous results attained with other families of catalysts in the presence of inorganic salts [15,19].

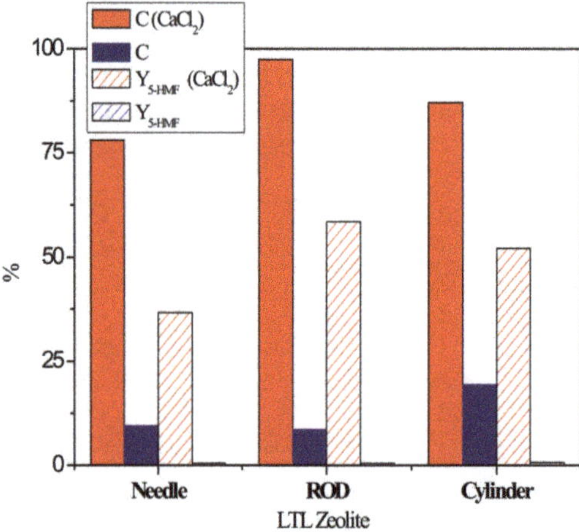

Figure 7. Effect of the addition of $CaCl_2$ on the catalytic performance (C: glucose conversion; Y: yield) (Experimental conditions: 0.15 g glucose, 0.05 g catalyst, 1.5 mL water, 3.5 mL MIBK, Temperature = 150 °C; time = 60 min, $CaCl_2$ = 0.65 g g_{H2O}^{-1}).

The improving effect of $CaCl_2$ addition on the catalytic performance of this series of LTL-based catalysts is evident. However, an additional catalytic study was performed in order to evaluate the contribution of $CaCl_2$ to the overall catalytic activity. For this, $CaCl_2$ (0.65 g g_{H2O}^{-1}) was put in contact with glucose in the biphasic system, and compared with the same system to which the NEEDLE-LTL catalyst was added. The data reveal that, in the absence of catalyst, after 3 h at 150 °C, the glucose conversion was 47.1%, with a 5-HMF yield of 16.8% (Figure 8).

However, under similar experimental conditions, this catalyst exhibits better catalytic performance, attaining a conversion of 86.9% and a 5-HMF yield of 44.0%. The catalytic activity in the presence of $CaCl_2$ might be explained by the formation of α-anomer in the presence of this salt, as it has been demonstrated in previous studies [14], which could easily be dehydrated due to the thermal contribution (non-catalyzed process) to the overall catalytic performance.

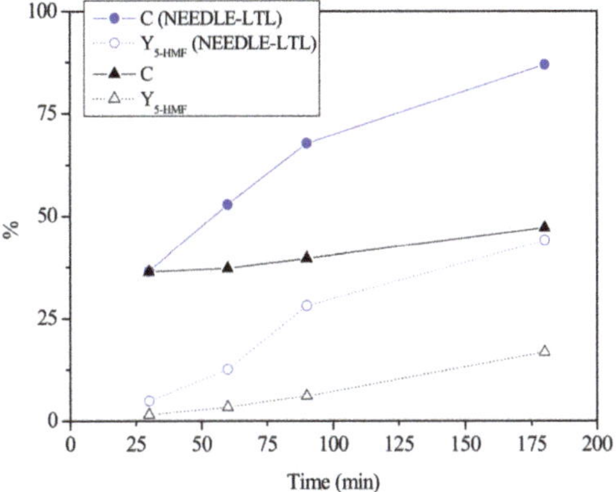

Figure 8. Catalytic performance of CaCl$_2$ and CaCl$_2$/NEEDLE-LTL (C: glucose conversion; Y: yield) (Experimental conditions: 0.15 g glucose, straight line: 0.65 g$_{H2O}$$^{-1}$ and 0.05 g NEEDLE-LTL and dash line: 0.65 g g$_{H2O}$$^{-1}$, 1.5 mL water, 3.5 mL MIBK, Temperature = 150 °C).

The kinetics of the dehydration process with LTL-based catalysts were studied at 150 °C (Figure 9), and similar conversion curves were obtained for the three catalysts, in spite of their different acid properties (Table 2) and morphologies. However, at intermediate reaction times, the 5-HMF yield values are lower for CYL-LTL, which could be explained by considering its lower acidity, although its mesoporous character could compensate the lower concentration of acid sites in this catalyst. In this study, the highest 5-HMF yield (50.3%) was attained after 300 min with the ROD-LTL catalyst.

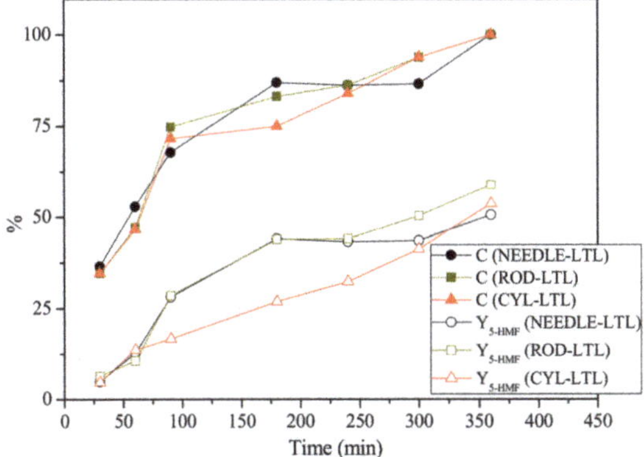

Figure 9. Kinetics of glucose dehydration (C: glucose conversion; Y: yield) (Experimental conditions: 0.15 g glucose, 0.05 g catalyst, 1.5 mL water, 3.5 mL MIBK, Temperature = 150 °C; CaCl$_2$ = 0.65 g g$_{H2O}$$^{-1}$).

By increasing the temperature to 175 °C, the reaction time for attaining the highest 5-HMF yield is shortened, making it possible to obtain a 5-HMF yield of 63.1% for a glucose conversion of 87.9% with the NEEDLE-LTL catalyst, after only 90 min (Figure 10). The figure shows that glucose conversion increases progressively with the reaction time, and values of 100% are already reached after 60 min of reaction time. However, 5-HMF yield values attain a maximum, and then, despite the raise of conversion, degradation processes lead to a decrease in 5-HMF yield.

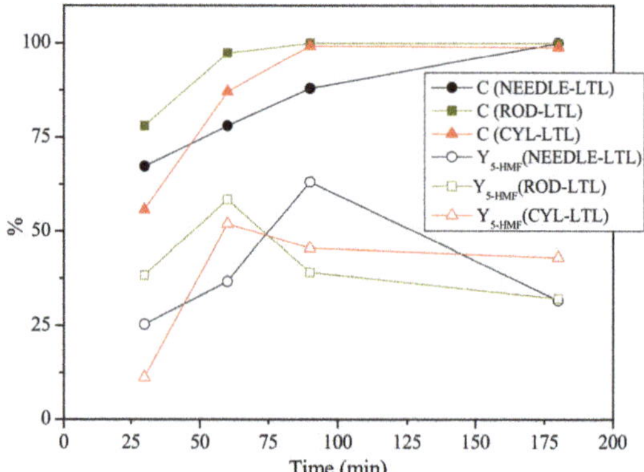

Figure 10. Kinetics of glucose dehydration (C: glucose conversion; Y: yield) (Experimental conditions: 0.15 g glucose, 0.05 g catalyst, 1.5 mL water, 3.5 mL MIBK, Temperature = 175 °C; $CaCl_2$ = 0.65 g g_{H2O}^{-1}).

Another experimental parameter that has been evaluated is the glucose:catalyst weight ratio. The ratio has been varied between 1:1 and 10:1, by maintaining the amount of glucose and adding different amounts of NEEDLE-LTL. It can be observed that the conversion of glucose rises with the catalyst loading until a weight ratio of 3:1, which can be explained by the increment of available acid sites for glucose dehydration (Figure 11). Nevertheless, a higher amount of catalyst (ratio of 1:1) does not improve the conversion, although a slightly higher 5-HMF yield is attained. This could be explained by considering diffusional problems associated to the location of active sites in micropores.

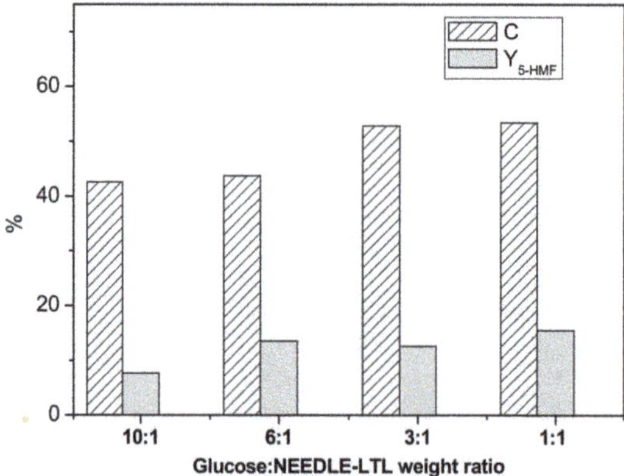

Figure 11. Influence of the glucose:catalyst (NEEDLE-LTL) weight ratio (C: glucose conversion; Y: yield) (Experimental conditions: 0.15 g glucose, 1.5 water, 3.5 mL MIBK, Temperature: 150 °C; time: 60 min, $CaCl_2$ = 0.65 g g_{H2O}^{-1}).

A key aspect in heterogeneous catalysis is the recovery of the solid catalyst to be used in successive catalytic runs. In order to carry out this study, a glucose:catalyst weight ratio of 1:1 was used, since a higher catalyst loading would minimize the loss of catalyst between cycles, produced by catalyst handling. The reaction was studied at 150 °C, for 60 min (Figure 12).

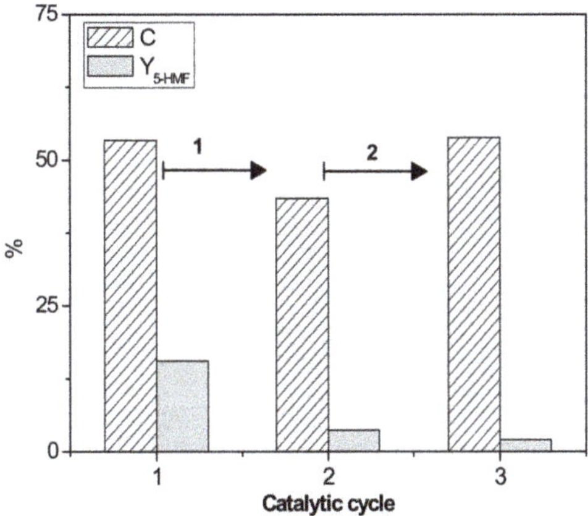

Figure 12. Reusing study of NEEDLE-LTL after (1) drying at 65 °C and (2) washing with water/acetone (C: glucose conversion; Y: yield) (Experimental conditions: 0.15 g glucose, 0.15 g catalyst, 1.5 water, 3.5 mL MIBK, Temperature: 150 °C; time: 60 min, $CaCl_2$ = 0.65 g_{H2O}^{-1}).

After the first catalytic cycle, the solid catalyst was filtered and dried at 65 °C. The catalytic data reveal a decrease in glucose conversion from 53.4 to 43.4%, although active sites involved in glucose dehydration to 5-HMF seem to be almost deactivated, since 5-HMF yield drastically diminishes from 15.5 to 3.7%. After the second run, a water/acetone washing and filtration of the used solid was also carried out, recovering the glucose conversion, but the 5-HMF yield is still lower. Finally, the used catalyst was calcined until 550 °C for 2 h after the first catalytic reaction cycle in order to remove the organic fraction. However, the initial catalytic performance continued without recovery, with values of glucose conversion and 5-HMF yield of 100% and 4.52%, respectively. In order to explain the reason of this behavior, the thermogravimetric analyses of used and fresh NEEDLE-LTL zeolite catalysts were performed (Figure 13). The fresh zeolite shows a weight loss of 10% between room temperature and 200 °C, associated to the removing of adsorbed water, without any remarkable weight variation onwards until 900. However, the used catalyst exhibits a high weight loss in the same temperature range, but with two distinguishable steps, which could be associated to the elimination of adsorbed water and MIBK, account for 20%. Then, a progressive weight loss is observed, but with an important sloop change between 700 and 900 °C. Thus, a weight loss of 22% is observed between 200 and 900 °C, and the high temperature required to remove organic molecules could be explained by the strong interaction between the organic guest molecules with the catalyst surface inside the micropores. In this sense, a regeneration temperature of 550 °C was not enough to remove organic species, which cover the active sites responsible of glucose dehydration to 5-HMF. The covering of active sites by carbonaceous deposits, as main cause of the decrease in catalytic activity, has been already reported for zeolites [23].

Therefore, it can be concluded that the LTL-zeolites can be used as heterogeneous acid catalysts for the dehydration of glucose to 5-hydroxymethylfurfural, attaining a maximum 5-HMF yield of 63.1% for a glucose conversion of 87.9%, at 175 °C after 90 min, with the NEEDLE-LTL catalyst, formed by nanoparticles with a length of 4.46 μm and a width of 0.63 μm. The diffusion of glucose molecules to the active sites present in LTL zeolites is favored by using needle and rod morphologies, but it is necessary to prepare hierarchical zeolites, where mesoporosity is a key feature for facilitating the access of reactants and the regeneration of active sites, thus delaying the catalyst deactivation.

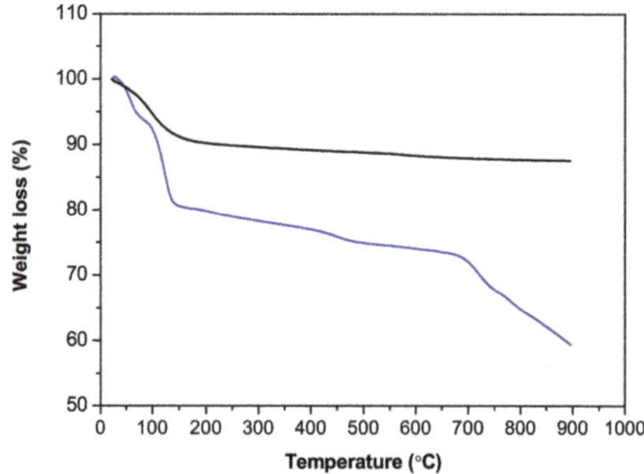

Figure 13. TG curves of fresh (black line) and used (blue line) (glucose:catalyst weight ratio = 1:1, T = 150 °C, time = 60 min) NEEDLE-LTL-zeolite.

3. Materials and Methods

3.1. Chemicals and Mterials

Potassium hydroxide pellets (85%) and ammonium nitrate were purchased from Merck (Darmstadt, Germany). Colloidal silica HS-40 was supplied by Sigma-Aldrich (Darmstadt, Germany). Aluminium sulfate hexadecahydrate ($Al_2(SO_4)_3$. $16H_2O$, 97%) was obtained from BDH Chemical Ltd, (Poole, UK). All chemicals were used without further purification. For the catalytic tests, the following chemicals have been utilized: glucose (Sigma-Aldrich, >99%), fructose (Sigma-Aldrich, >99%) and calcium chloride (VWR, Radnor, PA, USA, 97%). Deionized water and methyl isobutyl ketone (MIBK, VWR, 98%) have been used as solvents.

3.2. Synthesis of L-Type Zeolites

Short-rod shape L-type zeolite was synthesized by dissolving KOH (3.018 g) and $Al_2(SO_4)_3.16H_2O$ (1.442 g) with distilled water (18.208 g) in a polypropylene (PP) bottle. The mixture was magnetically stirred at room temperature to form a slightly cloudy solution. A clear silicate solution was prepared by mixing HS-40 (6.875 g) with distilled water (9.958 g) under stirring for 5 min. To avoid gelation, the silicate solution was added dropwise into the aluminate solution under vigorous stirring for 5 min. The solution with a final molar ratio of 10.0 K_2O:1 Al_2O_3:20 SiO_2:800 H_2O was further aged at room temperature for 18 h under stirring prior to crystallization at 180 °C for 3 days. The solid product was then filtered and purified with distilled water until pH 7 prior to freeze-drying. The zeolite powder in K-form (3.000 g) was then ion exchanged with ammonium nitrate (1.5 M, 100 mL) at 25 °C for 18 h before subjecting calcination at 480 °C for 4 h to produce protonated L-type zeolite (ROD-LTL). Similar procedures were used for preparing L-type zeolites with other morphologies (needle and cylinder) but using hydrogels of different molar compositions as stated in Table 4. The protonated L-type zeolite crystalline solids with short-rod, cylindrical and needle shapes were denoted as ROD-LTL, CYL-LTL and NEEDLE-LTL, respectively.

Table 4. Experimental conditions for the synthesis of L-type zeolites with different morphologies.

	Molar Composition				Aging (h)	Synthesis Temperature (°C)	Synthesis Time (h)
	K_2O	Al_2O_3	SiO_2	H_2O			
ROD-LTL	10.0	1	20	800	18	180	72
NEEDLE-LTL	10.2	1	20	1100	18	180	72
CYL-LTL	10.2	1	20	1030	18	180	72

3.3. Characterization of LTL Zeolites

XRD patterns were recorded using an AXS D8 diffractometer (Bruker, Rheinstetten, Germany) operating at 40 kV and 10 mA (Cu K$_\alpha$ radiation, λ = 0.15418 nm). The surface morphology of samples was observed using a JSM-6701F FESEM microscope (JEOL, Tokio, Japan). The infrared (IR) spectra were acquired using a System 2000 instrument (Perkin-Elmer, Waltham, MA, USA) using the KBr method (sample:KBr ratio = 1:50). The chemical composition of zeolites was also determined with an Optima 8300 inductively coupled plasma-optical emission spectrometer (ICP-OES). Textural properties were determined by an ASAP 2010 nitrogen adsorption analyzer (Micrometrics, Norcross, GA, USA) at −196 °C. First, the sample (ca. 0.08 g) was degassed under vacuum at 300 °C overnight. The surface area and pore size distribution of samples were estimated using the Langmuir and DFT models, respectively. The total pore volume of the solids was determined from the nitrogen adsorbed volume at P/P$_o$ = 0.990.

The FTIR spectra after pyridine adsorption were acquired using a 2000 FTIR spectrometer (Nicolet, Thermo Fisher Scientific, Waltham, MA, USA). Initially, the zeolite powder (ca. 0.01 g) was ground and pressed into a self-supporting wafer (area 2 cm^2) at 6.0 ton. The wafer was introduced into an IR vacuum cell and activated at 400 °C for 5 h under vacuum (10^{-3} mbar). The sample was cooled to 25 °C before the background spectrum of zeolite was recorded. Pyridine vapor was adsorbed onto the sample for 3 min before the excess pyridine vapor was evacuated. The FTIR spectra were recorded at 25 °C using 200 scans with a resolution of 6 cm^{-1}. The wafer was then heated at 150 °C for 1 h to remove weakly bound pyridine before the second IR spectrum was recorded. The wafer was heated again to 300 °C for 1 h before it was cooled to 25 °C and scanned again with an IR spectrometer. The concentration of Lewis and Brönsted acid sites were calculated by using the molar integral extinction coefficients of $\varepsilon_{Br\ddot{o}nsted}$ = 3.03 cm µmol^{-1} and ε_{Lewis} = 3.80 cm µmol^{-1} [36].

X-ray photoelectron spectra were obtained with a PHI 5700 spectrometer (Physical Electronics, Eden Prairie, Minnesota, USA) with non-monochromatic Mg Kα radiation (300 W, 15 kV, and 1253.6 eV) with a multi-channel detector. Spectra were recorded in the constant pass energy mode at 29.35 eV, using a 720 µm diameter analysis area. Charge referencing was measured against adventitious carbon (C 1s at 284.8 eV). The PHI ACCESS ESCA-V6.0 F software package was used for acquisition and data analysis. A Shirley-type background was subtracted from the signals. Recorded spectra were always fitted using Gaussian–Lorentzian curves in order to determine accurately the binding energies of the different element core levels.

^{27}Al MAS-NMR experiments were performed on an AV-400 (9.4 T) spectrometer (Bruker, Rheinstetten, Germany), using a BL-4 probe with zirconia rotors. The spectra were obtained using a spinning speed of vR = 10 kHz, a pulse width of 1 µs corresponding to a $\pi/12$ rad. Pulse length, a relaxation delay of 1 s, and typically 1200 scans. The temperature-programmed desorption of ammonia (NH$_3$-TPD) was carried out to evaluate the total surface acidity of catalysts. After cleaning of catalysts (0.08 g) with helium up to 550 °C and subsequent adsorption of ammonia at 100 °C, the NH$_3$-TPD was performed by raising the temperature from 100 to 550 °C, under a helium flow of 40 mL min^{-1}, with a heating rate of 10 °C min^{-1} and maintained at 550 °C for 15 min. The evolved ammonia was analyzed by using a TCD detector of a gas chromatograph (Shimadzu GC-14A).

Thermogravimetric analyses (TGA) were performed with a TGA/DSC model 1 instrument (Mettler-Toledo, Columbus, OH, USA) heating from room temperature until 900 °C with a heating ramp of 10 °C min^{-1} under air flow of 50 mL min^{-1}. The carbon content of spent catalysts was measured with a CHNS 932 analyzer (LECO, Madrid, Spain).

3.4. Catalytic Reaction

The catalytic performance of LTL-type zeolites in glucose dehydration was evaluated under batch conditions, by using a glass pressure tube equipped with threaded bushing (15 mL, pressure rated to 10 bars, Ace Glass, Vineland, NJ, USA) under magnetic stirring. In a typical experiment, 0.15 g of glucose, 0.05 g of catalyst, 1.5 mL of deionized water and 3.5 mL of methyl isobutyl ketone (MIBK) were introduced into the reactor. Reactors were always purged with nitrogen, prior to the catalytic study in order to minimize side reactions of HMF which decrease its yield. The mixture was heated with a thermostatically controlled oil bath. The reaction was quenched by submerging the reactor in a cool water bath, and the liquid phases were separated, filtered and the analysis of products was performed in both phases by high performance liquid chromatography (HPLC). A JASCO (Easton, MD, USA) instrument equipped with quaternary gradient pump (PU-2089), multiwavelength detector (MD-2015), autosampler (AS-2055), column oven (co-2065) using a Luna C18 reversed-phase column (250 mm × 4.6 mm, 5 μm) and a Rezex ROA-Organic Acid H^+ column (8%, 300 mm × 7.8 mm, 5 μm) (both supplied by Phenomenex (Torrance, CA, USA). Both glucose and fructose were monitored using a refractive index detector for aqueous phase, while 5-HMF production was monitored using a UV detector in both phases. The mobile phases consisted of pure methanol (flow rate 0.5 mL min^{-1}) for the Luna C18, and deionized water (flow rate 0.35 mL min^{-1}) for the Rezex ROA column, being the columns at room temperature and 40 °C, respectively.

Author Contributions: Experiment: M.J.G.M., N.H.A., S.M.M.; Data Curation: C.G.S., S.M.; Writing-Original Draft Preparation: C.G.S., E.-P.N., P.M.T.; Writing-Review & Editing: C.G.S., S.M., E.-P.N., P.M.T.; Supervision: E.-P.N., P.M.T.; Funding Acquisition: E.-P.N., P.M.T.

Funding: This research was funded by the Spanish Ministry of Economy and Competitiveness (RTI2018-94918-B-C44), FEDER (European Union) funds and FRGS (203/PKIMIA/6711642).

Acknowledgments: The authors thank the Servicios Centrales de Apoyo a la Investigación at University of Málaga for technical support. C.G.S. acknowledges the University of Málaga for a postdoctoral contract.

Conflicts of Interest: The authors declare no conflict of interest.

References

1. Werpy, T.; Petersen, G. *Top Value Added Chemicals from Biomass Volume I—Results of Screening for Potential Candidates from Sugars and Synthesis Gas*; National Renewable Energy Lab.: Golden, CO, USA, 2004; Volume 1.
2. Van Putten, R.J.; Van Der Waal, J.C.; De Jong, E.D.; Rasrendra, C.B.; Heeres, H.J.; De Vries, J.G. Hydroxymethylfurfural, A Versatile Platform Chemical Made from Renewable Resources. *Chem. Rev.* **2013**, *113*, 1499–1597. [CrossRef]
3. Alonso, D.M.; Bond, J.Q.; Dumesic, J.A. Catalytic conversion of biomass to biofuels. *Green Chem.* **2010**, *12*, 1493–1513. [CrossRef]
4. Wang, T.; Nolte, M.W.; Shanks, B.H. Catalytic dehydration of C 6 carbohydrates for the production of hydroxymethylfurfural (HMF) as a versatile platform chemical. *Green Chem.* **2014**, *16*, 548–572. [CrossRef]
5. Delidovich, I.; Leonhard, K.; Palkovits, R. Cellulose and hemicellulose valorisation: An integrated challenge of catalysis and reaction engineering. *Energy Environ. Sci.* **2014**, *7*, 2803. [CrossRef]
6. Corma, A.; Iborra, S.; Velty, A.; Corma Canos, A.; Iborra, S.; Velty, A. Chemical routes for the transformation of biomass into chemicals. *Chem. Rev.* **2007**, *107*, 2411–2502. [CrossRef] [PubMed]
7. Hu, L.; Tang, X.; Xu, J.; Wu, Z.; Lin, L.; Liu, S. Selective transformation of 5-hydroxymethylfurfural into the liquid fuel 2,5-dimethylfuran over carbon-supported ruthenium. *Ind. Eng. Chem. Res.* **2014**, *53*, 3056–3064. [CrossRef]
8. Gallezot, P. Conversion of biomass to selected chemical products. *Chem. Soc. Rev.* **2012**, *41*, 1538–1558. [CrossRef]
9. Rackemann, D.W.; Doherty, W.O. The conversion of lignocellulosics to levulinic acid. *Biofuels Bioprod. Bioref.* **2011**, *5*, 198–214. [CrossRef]
10. Nikolla, E.; Roman-Leshkov, Y.; Moliner, M.; Davis, M.E. "One-Pot" Synthesis of 5-(Hydroxymethyl)furfural from Carbohydrates using Tin-Beta Zeolite. *ACS Catal.* **2011**, *1*, 408–410. [CrossRef]
11. Yabushita, M.; Kobayashi, H.; Fukuoka, A. Catalytic transformation of cellulose into platform chemicals. *Appl. Catal. B Environ.* **2014**, *145*, 1–9. [CrossRef]
12. Climent, M.J.; Corma, A.; Iborra, S. Converting carbohydrates to bulk chemicals and fine chemicals over heterogeneous catalysts. *Green Chem.* **2011**, *13*, 520–540. [CrossRef]
13. Gandarias, I.; Arias, P.L. Heterogeneous acid-catalysts for the production of furan-derived compounds (furfural and hydroxymethylfurfural) from renewable carbohydrates: A review. *Catal. Today* **2014**, *234*, 42–58.

14. García-Sancho, C.; Fúnez-Núñez, I.; Moreno-Tost, R.; Santamaría-González, J.; Pérez-Inestrosa, E.; Fierro, J.L.G.; Maireles-Torres, P. Beneficial effects of calcium chloride on glucose dehydration to 5-hydroxymethylfurfural in the presence of alumina as catalyst. *Appl. Catal. B Environ.* **2017**, *206*, 617–625. [CrossRef]

15. Moreno-Recio, M.; Santamaría-González, J.; Maireles-Torres, P. Brönsted and Lewis acid ZSM-5 zeolites for the catalytic dehydration of glucose into 5-hydroxymethylfurfural. *Chem. Eng. J.* **2016**, *303*, 22–30. [CrossRef]

16. Jiménez-Morales, I.; Moreno-Recio, M.; Santamaría-González, J.; Maireles-Torres, P.; Jiménez-López, A. Production of 5-hydroxymethylfurfural from glucose using aluminium doped MCM-41 silica as acid catalyst. *Appl. Catal. B Environ.* **2015**, *164*, 70–76. [CrossRef]

17. Jiménez-Morales, I.; Moreno-Recio, M.; Santamaría-González, J.; Maireles-Torres, P.; Jiménez-López, A. Mesoporous tantalum oxide as catalyst for dehydration of glucose to 5-hydroxymethylfurfural. *Appl. Catal. B Environ.* **2014**, *154*, 190–196. [CrossRef]

18. Yang, Y.; Hu, C.W.; Abu-Omar, M.M. Conversion of carbohydrates and lignocellulosic biomass into 5-hydroxymethylfurfural using AlCl$_3$ center dot 6H(2)O catalyst in a biphasic solvent system. *Green Chem.* **2012**, *14*, 509–513. [CrossRef]

19. Román-Leshkov, Y.; Dumesic, J.A. Solvent effects on fructose dehydration to 5-hydroxymethylfurfural in biphasic systems saturated with inorganic salts. *Top. Catal.* **2009**, *52*, 297–303. [CrossRef]

20. Teychené, J.; Roux-De Balman, H.; Galier, S. Role of the triple solute/ion/water interactions on the saccharide hydration: A volumetric approach. *Carbohydr. Res.* **2017**, *448*, 118–127. [CrossRef]

21. Combs, E.; Cinlar, B.; Pagan-Torres, Y.; Dumesic, J.A.; Shanks, B.H. Influence of alkali and alkaline earth metal salts on glucose conversion to 5-hydroxymethylfurfural in an aqueous system. *Catal. Commun.* **2013**, *30*, 1–4. [CrossRef]

22. Moreno-Recio, M.; Jiménez-Morales, I.; Arias, P.L.; Santamaría-González, J.; Maireles-Torres, P. The Key Role of Textural Properties of Aluminosilicates in the Acid-Catalysed Dehydration of Glucose into 5-Hydroxymethylfurfural. *ChemistrySelect* **2017**, *2*, 2444–2451. [CrossRef]

23. Xu, S.; Pan, D.; Hu, F.; Wu, Y.; Wang, H.; Chen, Y.; Yuan, H.; Gao, L.; Xiao, G. Highly efficient Cr/β-zeolite catalyst for conversion of carbohydrates into 5-hydroxymethylfurfural: Characterization and performance. *Fuel Process. Technol.* **2019**, *190*, 38–46. [CrossRef]

24. Huang, Y.J.; Qi, G.R.; Chen, L.S. Effects of morphology and composition on catalytic performance of double metal cyanide complex catalyst. *Appl. Catal. A Gen.* **2003**, *240*, 263–271. [CrossRef]

25. Remy, M.J.; Stanica, D.; Poncelet, G.; Feijen, E.J.P.; Grobet, P.J.; Martens, J.A.; Jacobs, P.A. Dealuminated H-Y zeolites: Relation between physicochemical properties and catalytic activity in heptane and decane isomerization. *J. Phys. Chem.* **1996**, *100*, 12440–12447. [CrossRef]

26. Moulder, J.F.; Stickle, W.F.; Sobol, P.E.; Bombem, K.D. *Handbook of X-ray Photoelectron Spectroscopy*; Chastain, J., Ed.; Perkin-Elmer Corporation: Eden Prairie, MN, USA, 1992.

27. Ng, E.P.; Awala, H.; Tan, K.H.; Adam, F.; Retoux, R.; Mintova, S. EMT-type zeolite nanocrystals synthesized from rice husk. *Microporous Mesoporous Mater.* **2015**, *204*, 204–209. [CrossRef]

28. Valtchev, V.P.; Bozhilov, K.N. Transmission electron microscopy study of the formation of FAU-type zeolite at room temperature. *J. Phys. Chem. B* **2004**, *108*, 15587–15598. [CrossRef]

29. Xu, B.; Bordiga, S.; Prins, R.; van Bokhoven, J.A. Effect of framework Si/Al ratio and extra-framework aluminum on the catalytic activity of Y zeolite. *Appl. Catal. A Gen.* **2007**, *333*, 245–253. [CrossRef]

30. Mu, M.; Harvey, G.; Prins, R. Quantitative multinuclear MAS NMR studies of zeolites. *Microporous Mesoporous Mater.* **2000**, *34*, 281–290.

31. Chen, F.R.; Davis, J.G.; Fripiat, J.J. Aluminum coordination and Lewis acidity in transition aluminas. *J. Catal.* **1992**, *133*, 263–278. [CrossRef]

32. Galadima, A.; Muraza, O. Zeolite catalyst design for the conversion of glucose to furans and other renewable fuels. *Fuel* **2019**, *258*, 115851. [CrossRef]

33. Tang, Z.; Su, J. Direct conversion of cellulose to 5-hydroxymethylfurfural (HMF) using an efficient and inexpensive boehmite catalyst. *Carbohydr. Res.* **2019**, *481*, 52–59. [CrossRef] [PubMed]

34. Prat, D.; Wells, A.; Hayler, J.; Sneddon, H.; McElroy, C.R.; Abou-Shehada, S.; Dunn, P.J. CHEM21 selection guide of classical- and less classical-solvents. *Green Chem.* **2015**, *18*, 288–296. [CrossRef]

35. Fúnez-Núñez, I.; García-Sancho, C.; Cecilia, J.A.; Moreno-Tost, R.; Pérez-Inestrosa, E.; Serrano-Cantador, L.; Maireles-Torres, P. Synergistic effect between $CaCl_2$ and γ-Al_2O_3 for furfural production by dehydration of hemicellulosic carbohydrates. *Appl. Catal. A Gen.* **2019**, *585*, 117188. [CrossRef]
36. Choorefe, M.Y.; Juan, J.C.; Oi, L.E.; Ling, T.C.; Ng, E.P.; Noorsaadah, A.R.; Centi, G.; Lee, K.T. The role of nanosized zeolite y in the H_2 -free catalytic deoxygenation of triolein. *Catal. Sci. Technol.* **2019**, *9*, 772–782.

Article

Biodiesel Production Using Bauxite in Low-Cost Solid Base Catalyst Precursors

Yong-Ming Dai [1], Cheng-Hsuan Hsieh [2], Jia-Hao Lin [3], Fu-Hsuan Chen [4] and Chiing-Chang Chen [3,*]

[1] Department of Chemical and Materials Engineering, National Chin-Yi University of Technology, Taichung 411, Taiwan; forest1105@gmail.com
[2] Department of Materials Science and Engineering, National Tsing-Hua University, Hsinchu 300, Taiwan; frank8771919@gmail.com
[3] Department of Science Education and Application, National Taichung University of Education, Taichung 403, Taiwan; henry811013@yahoo.com.tw
[4] Department of Political Science, National Taiwan University, Taipei 106, Taiwan; fhchen@mail.ntcu.edu.tw
* Correspondence: ccchen@mail.ntcu.edu.tw; Tel.: +886-4-2218-3406; Fax: +886-4-2218-3560

Received: 22 October 2019; Accepted: 11 December 2019; Published: 13 December 2019

Abstract: Investigation was conducted on bauxite mixed with Li_2CO_3 as alkali metal catalysts for biodiesel production. Bauxite contains a high percentage of Si and Al compounds among products. Because of the high expense of commercial materials (SiO_2, Al_2O_3) that makes them not economical, the method was very recently improved by replacing commercial materials with Si and Al from bauxite. This is one of the easiest methods for preparing heterogeneous transesterification catalysts, through one-pot blending, grinding bauxite with Li_2CO_3, and heating at 800 °C for 4 h. The prepared solid-base alkali metal catalyst was characterized in terms of its physical and chemical properties using X-ray powder diffraction and field-emission scanning electron microscopy (FE-SEM). The optimal conditions for the transesterification procedure are to mix methanol oil by molar ratio 9:1, under 65 °C, with catalyst amount 3 wt.%. The procedure is suitable for transesterifying oil to fatty acid methyl ester in the 96% range.

Keywords: bauxite; Li_2CO_3; transesterification; soybean oil

1. Introduction

The need for renewable energy sources to meet ever greater energy requirements is becoming increasingly urgent [1]. Many researchers therefore advocate substituting traditional fuels with renewable alternatives that produce lower emissions of particulate matter and reduce greenhouse effects [2,3]. The current process used for producing fatty acid methyl esters (FAMEs) is the transesterification of vegetable oil catalyzed by alkaline catalysts. In addition, this reaction is associated with some difficulties. First, the catalysts cannot be recovered or reused, therefore they must be neutralized. Second, catalysts should be neutralized and separated from the methyl ester phase after the transesterification reaction, with much wastewater resulting [4,5]. Heterogeneous solid catalysts can be used for overcoming these problems. Heterogeneous solid alkaline catalysts are beneficial because of the good separation of the reaction mixtures and their recyclability. Their low price also decreases the overall production costs [6].

By the combination of silicon and lithium carbonate, our previous research discovered that the lithium silicate compound presented high basicity strength, high chemical stability, and thermo-stability. Such characteristics show extremely high applicability in the catalysis field, especially in transesterification [7]. The $LiAlO_2$ structure has recently been published for transestrification. Under high temperatures, lithium aluminate acquired by calcinating Li_2CO_3 in the waste with aluminum forms high basic solid catalyst [8]. Interestingly, both Li_4SiO_4 and $LiAlO_2$ compounds demonstrate that these two metal oxides are the best solid basic catalyst for transestrification with high stability, lower cost, and solid basicity.

Bauxite contains a large percentage of Si and Al compounds. The expense of pure Si and Al makes the materials uneconomical. The method has been improved by replacing commercial materials with Si and Al from bauxite [9]. In recent years, newer and simpler methods based on the of bauxite containing a large percentage of Si and Al compounds have been introduced [8,9]. One simple method is based on a solid-state method at high temperatures, blending other base metals with bauxite, which contains Si and Al. The concept involves producing new solid basic catalysts, with the goal of reusing these new precursors and offering broad applications in various fields [10–12]. When bauxite is used as a low-cost solid base catalyst, it provides a twofold advantage in relation to environmental pollution. The use of low-cost materials in the manufacturing process is valuable for industrial applications, particularly for producing FAMEs through transesterification of various vegetable oils, where massive quantities of solid-alkali catalyst are processed [13–16]. The cheaper catalyst precursor from bauxite has some advantages, which are mainly economic and environmental. Another study we undertook applied solid base catalysts to transesterification. This indicated that solid-alkali catalysts provided high transesterification efficiency [6,15,16].

In this paper, bauxite is used as a precursor to prepare alkali catalysts. The resulting catalyst was experimented in the transesterification reaction. The influence of various experimental methods—such as the catalyst amount, methanol–vegetable oil proportions, and catalyst reusability—on efficiency was evaluated to detect the optimum conditions for the study of biodiesel production.

2. Results and Discussion

2.1. Characterization of As-Prepared Catalyst

Figures 1 and 2 present the X-ray powder diffraction (XRD) results of catalysts. The structure of the bauxite underwent a phase change to Li_4SiO_4 (JCPDS-742145) and $LiAlO_2$ (JCPDS-0440224) according to the solid-state preparation with Li_2CO_3. Although bauxite was a complex material and the main peaks of its XRD pattern usually overlaid one another, the major phase change in Li_4SiO_4 and $LiAlO_2$ was observed after its calcination. Figure 1 shows different calcination temperatures. At a calcination temperature of 600 °C, Li_2CO_3 (JCPDS-870728) structures had strong intensity. The catalysts can be observed to exhibit the diffraction peak characteristics of Li_4SiO_4, $LiAlO_2$, and Li_2CO_3. For calcination at 800 °C, a stronger intensity was present in the Li_4SiO_4 and $LiAlO_2$ phase [17,18].

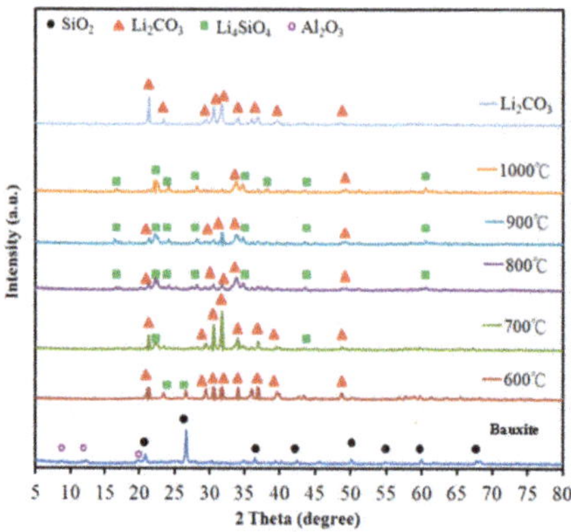

Figure 1. X-ray powder diffraction (XRD) of bauxite mixed with Li_2CO_3 at different calcination temperatures.

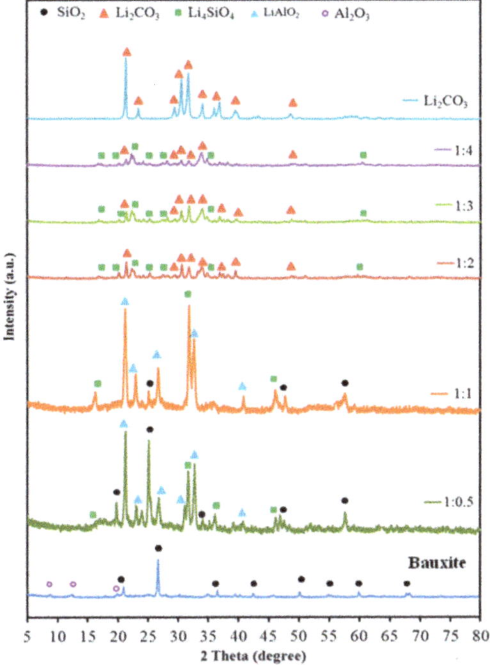

Figure 2. XRD patterns of prepared catalyst under different bauxite and Li_2CO_3 molar ratios.

As result of the above, the XRD patterns of the samples calcined at 600 °C corresponded to the Li_2CO_3 phase. When the calcination temperature reached the melting point of Li_2CO_3 (650 °C), Li_2CO_3 entered a molten state [19]. Subsequently, the catalyst (calcined at 800 °C) exhibits the different XRD diffraction peak due to phase transformation to crystalline Li_4SiO_4 and $LiAlO_2$. It was found that the XRD patterns of catalyst (calcined at 800 °C to 1000 °C) were similar. These diffraction peaks indicated the presence of crystalline Li_4SiO_4 and $LiAlO_2$. The XRD result is attributable to the phenomenon whereby Li_4SiO_4 and $LiAlO_2$ started to form agglomerated blocks in the Li_4SiO_4 and $LiAlO_2$ phase at high calcination temperatures. Therefore, regardless of the calcination temperature, the $LiAlO_2$ and Li_4SiO_4 phases were achieved.

Ordinary bauxite contains two major compounds: Al_2O_3 and SiO_2. The XRD patterns of prepared catalysts for different bauxite and Li_2CO_3 ratios are shown in Figure 2. When the bauxite–Li_2CO_3 molar ratio was 0.5, the diffraction peaks of SiO_2 were observed. With an increase in the bauxite–Li_2CO_3 ratio, the diffraction peaks of Li_4SiO_4 and $LiALO_2$ became increasingly clear. When the bauxite–Li_2CO_3 ratio increased to 2, the heights of the diffraction peaks belonging to Li_2CO_3 increased further. Table 1 presents the efficiency of transesterification to determine the efficiency of all catalysts in the experiment. The reaction test did not provide optimal conditions for the transesterification procedure but provided a method to compare catalytic performance. Results from Table 1 demonstrate that the bauxite exhibited no catalytic performance. When Li_2CO_3 was modified bauxite, the catalysts exhibited catalytic activities. The earlier research of solid base catalysts for transesterification reactions is shown in Table 1. Comparing Li_2CO_3/Bauxite with other solid base catalysts, it could be clearly found that Li_2CO_3/Bauxite showed good catalytic performance for a transesterification reaction.

The structure morphology of catalysts was examined using field-emission scanning electron microscopy (FE-SEM). Figure 3 illustrates morphology noted under FE-SEM for a bauxite-Li_2CO_3 molar ratio of 4 calcined in air with various calcination temperatures. The catalyst has a uniform microscale block and agglomerated block composition with a 20–50 μm size range, as shown in Figure 3.

Small mineral aggregates and agglomerated circular particles were present when the calcination was performed at high temperatures because of the oxide formation. Figure 3 demonstrates the catalyst results, and circular blocks of different sizes with diameters of approximately 50 μm accumulated on the surface. The particles and amorphous silica disappeared.

Table 1. Base strength of bauxite and the prepared catalysts.

Catalyst	Molar Ratio	Basic Strength	* Conversion (%)
Bauxite	0	$7.2 < H^- < 9.8$	0.98
Li_2CO_3/Bauxite	1/1	$9.8 < H^- < 15.0$	95.8
Li_2CO_3/Bauxite	1/2	$9.8 < H^- < 15.0$	95.2
Li_2CO_3/Bauxite	1/3	$15.0 < H^- < 18.4$	96.6
Li_2CO_3/Bauxite	1/4	$15.0 < H^- < 18.4$	96.5
Li_2CO_3	-	$9.8 < H^- < 15.0$	94.2
$LiAlO_2$	-	$9.8 < H^- < 15.0$	96.4
Li_4SiO_4	-	$9.8 < H^- < 15.0$	96.1

* Reaction conditions: 12.5 g soybean oil; methanol/oil molar ratio, 12:1; catalyst amount, 6 wt.%; reaction time, 3 h; methanol reflux temperature and conventional heating method.

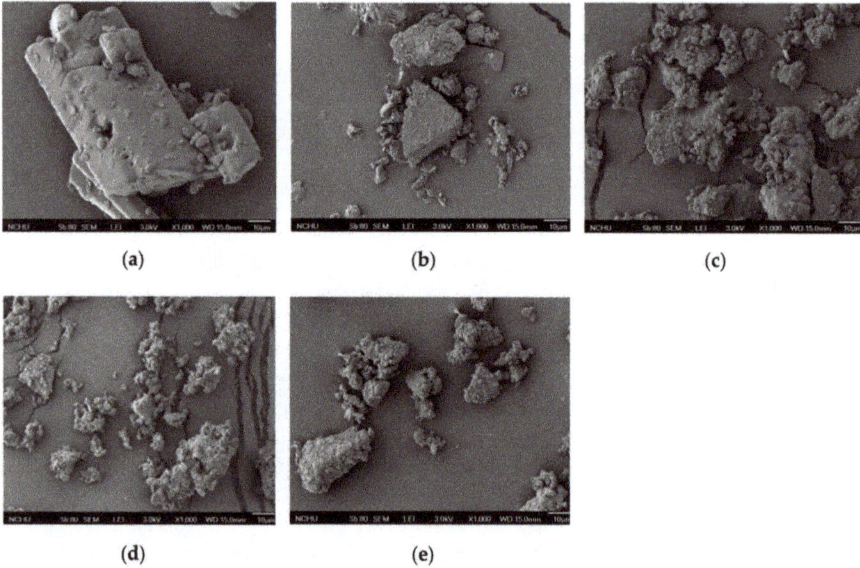

(a) **(b)** **(c)**

(d) **(e)**

Figure 3. Field-emission scanning electron microscopy (FE-SEM) images of bauxite mixed with Li_2CO_3 at different calcination temperatures (**a**) 600 °C, (**b**) 700 °C, (**c**) 800 °C, (**d**) 900 °C, and (**e**) 1000 °C.

2.2. Reaction Studies

Experimental tests on how catalyst amount, reaction time, methanol–vegetable oil, and reusability of the catalysts affect efficiency have been evaluated to detect the optimum conditions for the study of biodiesel production.

Figure 4 shows different calcination temperatures and times, for a bauxite-Li_2CO_3 molar ratio of 4, through the transesterification process. The reaction rate is shown in the results. With the calcination temperature between 600 °C and 1000 °C, the reaction rate of soybean oil increases to the maximum value, and the reaction rate then begins to decrease when the calcination temperature exceeds 800 °C, as shown in Figure 4. Conversion is consequently lower because catalysts begin to form agglomerated blocks at higher calcination temperature. Hence, the optimal condition occurs at 800 °C in this study. Furthermore, the conversion is found to increase with the increase in time from 1 to 5 h and then to decrease as time elapses past 3 h. In Figure 4, the reaction efficiency suggests that FAME production efficiency is lower at lower calcination times and increases when the calcination time reaches the maximum value of 3 h. The conversion rate then decreases, probably due to the formation saponification for longer calcination times. This result is attributed to the fact that Li_4SiO_4 and $LiAlO_2$ start to form agglomerated blocks in the Li_4SiO_4 and $LiAlO_2$ phases for long calcination times. Therefore, regardless of the calcination time, the $LiAlO_2$ and Li_4SiO_4 phase is achieved. A correct calcination time was consequently required to guarantee completion.

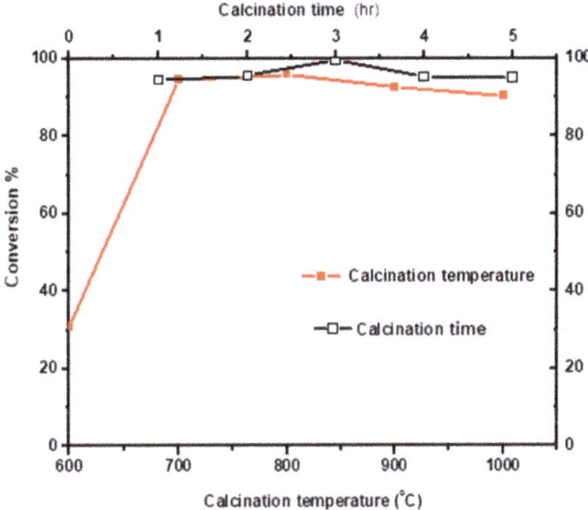

Figure 4. Influence of calcination temperature (°C) and time on the conversion reaction conditions: methanol/oil molar ratio 12:1, with 6 wt.% catalyst loading, reaction time 3 h.

Table 2 presents the efficiency for a bauxite and Li_2CO_3 molar ratio of 4. As shown, the catalyst exhibits a much higher conversion rate than the latter. In addition, increasing the molar ratio of oil to methanol increased the reaction rate (Table 2). Methanol is a reactant for transesterification reactions [8]. Increasing the reactant quantities shifts the equilibrium to the products side [20]. A molar ratio of 6 yielded a lower conversion rate. The highest reaction rate was obtained at a methanol–oil molar ratio of 12. This molar ratio of 12 was considered the optimal condition and was favorable for the transesterification procedure. Throughout the procedure, the catalyst amount represented a crucial parameter for high efficiency. The catalysts possessing strong main activity sites and a high surface area should exhibit higher conversion. Catalytic sites for transesterification are too few when the catalyst–oil ratio is too low. Catalyst amount was varied from 2 to 10 wt.% to oil to evaluate its effect on the conversion rate of the transesterification procedure at 65 °C and a methanol–oil molar ratio of 12. Catalysts loading of 2 wt.% of oil yielded a lower transfer rate. The maximum efficiency was obtained at a catalyst loading of 6 wt.% (Table 2). A catalyst loading of 6 wt.% for oil was considered the optimal condition and favorable for the transesterification process. This may have been due to the formation of resistance to mass transfer [21] for high catalyst quantities. In addition, to determine the catalytic activity of catalysts in the experimental tests, the efficiency of transesterification is presented in Table 2. This indicates that the reaction rate increased with increasing reaction time in the alkali silicate catalysts at a constant reaction temperature. The transesterification comprised three processes: The first process, whereby triglyceride reacted with one molecule of methanol, yielded diglyceride and one molecule of ester. The second process, the reaction of the diglyceride with a second molecule of methanol, yielded monoglyceride and an additional molecule of ester. In the third process, monoglyceride reacted with the third molecule of methanol, yielding glycerin and ester. Consequently, correct transesterification was required to guarantee completion of the reaction. Table 2 demonstrates that the efficiency increases with time and has an optimum value after 3 h.

Table 2. The catalytic performance of catalysts for transesterification of soybean oil with methanol.

Mole Ratio of Methanol/Oil	Catalyst Amount (wt.%)	Reaction Time (h)	Oils	* Conversion (%)
6	6	3	Soybean oil	86.3
12	6	3	Soybean oil	95.9
18	6	3	Soybean oil	94.1
24	6	3	Soybean oil	92.1
12	2	3	Soybean oil	92.4
12	4	3	Soybean oil	95.2
12	6	3	Soybean oil	95.9
12	8	3	Soybean oil	95.0
12	10	3	Soybean oil	94.1
12	6	1	Soybean oil	95.6
12	6	2	Soybean oil	95.8
12	6	3	Soybean oil	95.9
12	6	4	Soybean oil	93.1
12	6	5	Soybean oil	91.1
12	6	3	Coconut oil	89.1
12	6	3	Olive oil	95.2
12	6	3	Castor oil	63.1
12	6	3	Rapeseed oil	95.4
12	6	3	Corn oil	95.1
12	6	3	Waste Cooking oil	92.2

* Reaction conditions: 12.5 g oils; reaction temperature, 65 °C; methanol reflux temperature and conventional heating method.

The efficiencies of oils other than FAME are indicated in Table 2. FAME was successfully synthesized from soybean oil through a transesterification procedure with catalysts using a simple method. The results demonstrate that the efficiency with high free fatty acid concentration was significantly influenced [22].

The reusability for a bauxite–Li$_2$CO$_3$ molar ratio of 4 was evaluated for soybean oil. The catalyst was reusable up to the sixth repetition, with retention of catalyst efficiency, providing a conversion efficiency of >90% and then declining to a conversion efficiency of 80% after the 6th run repetition (Figure 5).

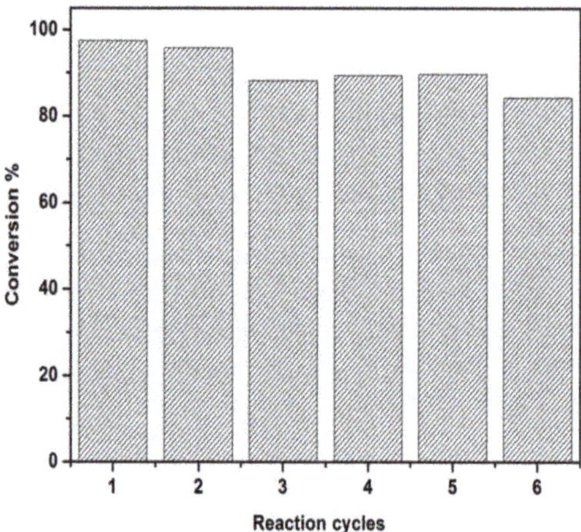

Figure 5. Reusability study after six reaction cycles for catalyst. Reaction conditions: methanol/oil molar ratio 12:1, with 6 wt.% catalyst loading, reaction time 3 h.

3. Experimental Methods and Chemical Materials

3.1. Solid-Alkali Catalyst Preparation

A sample of bauxite was converted into bauxite ash by heat-treating them at 500 °C for 4 h in a muffle furnace. After washing the bauxite ash using deionized water and filtration, the bauxite ash was dried at 120 °C for 16 h. The catalyst was prepared by grinding bauxite ash and Li_2CO_3 (Shimakyu's Pure Chemicals, Osaka, Japan), thoroughly mixing them in a crucible for calcination for 4 h in a muffle furnace, and then cooling them to room temperature.

3.2. Transesterification

Into a cone flask with a cooled condenser, 12.5 g (0.0114 mol) of vegetable oil (Great Wall Business Co., Taiwan) was added, with various methanol (American Chemical Society grade, ECHO Chemical Co., Miaoli, Taiwan) to oil molar ratios (12-24) and catalyst quantities (2-10 wt.%) at 338 K for 4 h of transesterification. All of the experiments were performed under atmospheric pressure. An appropriate amount of deionized water was added into the sample to stop the transesterification reaction. After a period of time, the sample was cooled and separated through a typical filter paper. The extra methanol was removed from filtrate and water prior to the FAME analysis. The FAME composition of the biodiesel was determined by gas chromatography.

3.3. Characterization

The alkali strength analysis of the bauxite used as a precursor to prepare catalyst (H^-) was defined as the acid–base indicator. All alkali catalysts were characterized using an X-ray diffraction instrument with Cu Kα radiation (MAC MXP18, Tokyo, Japan). The structure of the as-prepared alkali was observed employing a field emission scanning electron microscope (FE-SEM, JEOL JSM-6700F). FAME composition in the biodiesel was determined by using a Thermo trace gas chromatograph with a flame ionization detector. A Tr–biodiesel (30 m × 0.25 mm × 0.25 μm film thickness) capillary column was used. Samples were injected under the following conditions: the carrier gas was Nitrogen with a flow rate 2 mL/min., an injector temperature was 200 °C was 90 °C for 0.5 min and increased to 260 °C (programmed temperature) at a rate of 10 °C/min, and the detector temperature was 250 °C. This procedure was developed according to EN 14103.

4. Conclusions

Bauxite was found to contain Si and Al essential compounds suitable as catalysts for biodiesel production. A highly effective catalyst for biodiesel production was synthesized using bauxite and Li_2CO_3 and by calcination at temperatures of 800 °C for 4h. The transesterification procedure of biodiesel reached over 96% and the optimal parameters were 65 °C reaction temperature, 3 h reaction time, 9: 1 methanol-to-oil molar ratio and 3 wt.% catalyst amount. Under the optimal parameters, the catalyst can be effectively reused for 6 runs with a minimal decrease of <10% in the conversion rate. In addition, the results demonstrate that the efficiency with high free fatty acid concentration was significantly influenced. Consequently, the efficiency of the same catalyst under the same reaction conditions differs with different oil. In this work, the application of bauxite as a catalyst for biodiesel production not only provides a cost-effective and environmentally friendly way of using the bauxite, but also reduces the cost of biodiesel production.

Author Contributions: Data conceptualization, Investigation, Writing—original draft, Methodology, Y.-M.D. Methodology, Validation, Investigation, Software, C.-H.H. and J.-H.L. Writing—original draft preparation, F.-H.C. Supervision, Conceptualization, Visualization, Writing—review, Editing, Validation, C.-C.C.

Funding: This research was supported by the Ministry of Science and Technology of the Republic of China (MOST-108-2113-M-142-001).

Acknowledgments: We thanks for the financial support of this research by Ministry of Science and Technology of Taiwan. We also thank for the Center of Expensive Instruments at National Chung Hsing University to perform FE-SEM analysis.

Conflicts of Interest: The authors declare that they have no known competing financial interests or personal relationships that could have appeared to influence the work reported in this paper

References

1. Nigam, P.S.; Singh, A. Production of liquid biofuels from renewable resources. *Prog. Energy Combust. Sci.* **2013**, *37*, 52–68. [CrossRef]
2. Leung, D.Y.C.; Wu, X.; Leung, M.K.H. A review on biodiesel production using catalyzed transesterification. *Appl. Energy* **2010**, *87*, 1083–1095. [CrossRef]
3. Ting, W.J.; Huang, C.M.; Giridhar, N.; Wu, W.T. An enzymatic/acid-catalyzed hybrid process for biodiesel production from soybean oil. *J. Chin. Inst. Chem. Eng.* **2008**, *39*, 203–210. [CrossRef]
4. Kouzu, M.; Umemoto, M.; Kasuno, T.; Tajika, M.; Aihara, Y.; Sugimoto, Y.; Hidaka, J. Biodiesel production from soybean oil using calcium oxide as a heterogeneous catalyst. *J. Jpn. Inst. Energy* **2006**, *85*, 135–141. [CrossRef]
5. Kawashima, A.; Matsubara, K.; Honda, K. Development of heterogeneous base catalysts for biodiesel production. *Bioresour. Technol.* **2008**, *99*, 3439–3443. [CrossRef] [PubMed]
6. Hindryawati, N.; Maniam Rezaul, G.P.; Karim, M.; Chong, K.F. Transesterification of used cooking oil over alkali metal (Li, Na, K) supported rice husk silica as potential solid base catalyst. *JESTECH* **2014**, *17*, 95–103. [CrossRef]
7. Wang, J.X.; Chen, K.T.; Huang, S.T.; Wu, J.S.; Wang, P.H.; Chen, C.C. Production of biodiesel through transesterification of soybean oil using lithium orthosilicate solid catalyst. *Fuel Process. Technol.* **2012**, *104*, 167–173. [CrossRef]
8. Dai, Y.M.; Wu, J.S.; Chen, C.C.; Chen, K.T. Evaluating the optimum operating parameters on transesterification reaction for biodiesel production over a LiAlO$_2$ catalyst. *Chem. Eng. J.* **2015**, *280*, 370–376. [CrossRef]
9. Alves, H.J.; Rocha, A.M.; Monteiro, M.R.; Moretti, C.; Cabrelon, M.D.; Schwengber, C.A.; Milinsk, M.C. Treatment of Clay with KF: New solid catalyst for biodiesel production. *Appl. Clay Sci.* **2014**, *91–92*, 98–104. [CrossRef]
10. Chen, K.T.; Wang, J.X.; Wang, P.H.; Liou, C.Y.; Nien, C.W.; Wu, J.S.; Chen, C.C. Rice husk ash as a catalyst precursor for biodiesel production. *J. Taiwan Inst. Chem. Eng.* **2013**, *44*, 622–629. [CrossRef]
11. Bontempi, E.; Zacco, A.; Borgese, L.; Gianoncelli, A.; Ardesi, R.; Depero, L.E. A new method for municipal solid waste incinerator (MSWI) Fly ash inertization, based on colloidal silica. *J. Environ. Monit.* **2010**, *12*, 2093–2099. [CrossRef] [PubMed]
12. Xie, Z.; Ma, X. The thermal behaviour of the co-combustion between paper sludge and rice straw. *Bioresour. Technol.* **2013**, *146*, 611–618. [CrossRef] [PubMed]
13. Chen, M.Y.; Wang, J.X.; Chen, K.T.; Wen, B.Z.; Lin, W.Y.; Chen, C.C. Transesterification of soybean oil catalyzed by calcium hydroxide which obtained from hydrolysis reaction of calcium carbide. *J. Chin. Chem. Soc.* **2012**, *59*, 170–175. [CrossRef]
14. Klose, F.; Scholz, P.; Kreisel, G.; Ondruschk, B.; Kneise, R.; Knopf, U. Catalysts from waste materials. *Appl. Catal. B. Environ.* **2000**, *28*, 209–221. [CrossRef]
15. Dai, Y.M.; Hsieh, J.H.; Chen, C.C. Transesterification of soybean oil to biodiesel catalyzed by waste silicone solid base catalyst. *J. Chin. Chem. Soc.* **2014**, *61*, 803–805. [CrossRef]
16. Wang, J.X.; Chen, K.T.; Wen, B.Z.; Liao, Y.H.; Chen, C.C. Transesterification of soybean oil to biodiesel using cement as a solid base catalyst. *J. Taiwan Inst. Chem. Eng.* **2012**, *43*, 215–219. [CrossRef]
17. Dai, Y.M.; Wang, Y.F.; Chen, C.C. Synthesis and characterization of magnetic LiFe$_5$O$_8$-LiFeO$_2$ as a solid basic catalyst for biodiesel production. *Catal. Commun.* **2018**, *106*, 20–24. [CrossRef]
18. Wang, J.X.; Chen, K.T.; Huang, S.T.; Chen, C.C. Application of Li$_2$SiO$_3$ as a heterogeneous catalyst in the production of biodiesel from soybean oil. *Chin. Chem. Lett.* **2011**, *22*, 1363–1366. [CrossRef]
19. Wang, S.H.; Wang, Y.B.; Jehng, J.M. Preparation and characterization of hydrotalcite-like compounds containing transition metal as a solid base catalyst for the transesterification. *Appl. Catal. A: Gen.* **2012**, *439–440*, 135–141. [CrossRef]

20. Liao, X.Y.; Zhu, Y.L.; Wang, S.G.; Li, Y.W. Producing triacetylglycerol with glycerol by two steps: Esterification and acetylation. *Fuel Process. Technol.* **2009**, *90*, 988–993. [CrossRef]
21. Ding, Y.; Sun, H.; Duan, J.; Chen, P.; Lou, H.; Zheng, X. Mesoporous Li/ZrO$_2$ as a solid base catalyst for biodiesel production from transesterification of soybean oil with Methanol. *Catal. Commun.* **2011**, *12*, 606–610. [CrossRef]
22. Liu, X.; Piao, X.; Wang, Y.; Zhu, S.; He, H. Calcium methoxide as a solid Base catalyst for the transesterification of soybean oil to biodiesel with methanol. *Fuel* **2008**, *87*, 1076–1082. [CrossRef]

 *catalysts* 

Article

Cleanup and Conversion of Biomass Liquefaction Aqueous Phase to C_3–C_5 Olefins over $Zn_xZr_yO_z$ Catalyst

Stephen D. Davidson [1], Juan A. Lopez-Ruiz [1], Matthew Flake [1], Alan R. Cooper [1], Yaseen Elkasabi [2], Marco Tomasi Morgano [3], Vanessa Lebarbier Dagle [1], Karl O. Albrecht [4] and Robert A. Dagle [1,*]

[1] Energy and Environment Directorate, Institute for Integrated Catalysis Pacific Northwest National Laboratory, Richland, WA 99352, USA; stephen.davidson@pnnl.gov (S.D.D.); juan.lopezruiz@pnnl.gov (J.A.L.-R.); Matthew.Flake@pnnl.gov (M.F.); Alan.Cooper@pnnl.gov (A.R.C.); vanessa.dagle@pnnl.gov (V.L.D.)
[2] USDA-ARS Eastern Regional Research Center, Wyndmoor, PA 19038, USA; yaseen.elkasabi@usda.gov
[3] ARCUS Greencycling Technologies GmbH, 71638 Ludwigsburg, Germany; Marco.TomasiMorgano@arcus-greencycling.technology
[4] Archer Daniels Midland Company, James R. Randall Research Center, Decatur, IL 62521, USA; Karl.Albrecht@adm.com
* Correspondence: robert.dagle@pnnl.gov; Tel.: +1-(509)-371-6264

Received: 14 October 2019; Accepted: 1 November 2019; Published: 6 November 2019

Abstract: The viability of using a $Zn_xZr_yO_z$ mixed oxide catalyst for the direct production of C_4 olefins from the aqueous phase derived from three different bio-oils was explored. The aqueous phases derived from (i) hydrothermal liquefaction of corn stover, (ii) fluidized bed fast pyrolysis of horse litter, and (iii) screw pyrolysis of wood pellets were evaluated as feedstocks. While exact compositions vary, the primary constituents for each feedstock are acetic acid and propionic acid. Continuous processing, based on liquid–liquid extraction, for the cleanup of the inorganic contaminants contained in the aqueous phase was also demonstrated. Complete conversion of the carboxylic acids was achieved over $Zn_xZr_yO_z$ catalyst for all the feedstocks investigated. The main reaction products from each of the feedstocks include isobutene (>30% selectivity) and CO_2 (>23% selectivity). Activity loss from coking was also observed, thereby rendering deactivation of the $Zn_xZr_yO_z$ catalyst, however, complete recovery of catalyst activity was observed following regeneration. Finally, the presence of H_2 in the feed was found to facilitate hydrogenation of intermediate acetone, thereby increasing propene production and, consequently, decreasing isobutene production.

Keywords: biomass-derived aqueous phase upgrading; olefin production; oxide catalyst zinc–zirconia

1. Introduction

Biomass has received much attention as a renewable source for both fuels and chemicals. While there are a wide range of approaches for biomass conversions, a commonality is high cost of processing leading to a high minimum fuel selling price [1–4]. Direct liquefaction of biomass is an appealing approach as much of the hydrocarbon structure can be preserved [5–7]. Processes with direct liquefaction include hydrothermal liquefaction (HTL) and different permutations of pyrolysis that includes fast pyrolysis (FP), catalytic fast pyrolysis (CFP), and intermediate screw pyrolysis (SP) [8–11]. Regardless of approach, aqueous and organic phases can be segregated by gravity separation. While the organic phase has a lower oxygen content, and consists of molecules typically more suited for fuel, the aqueous phase typically comprises a mixture of light oxygenates that are difficult to separate [1,12].

Thus, the aqueous phase is usually considered a waste stream. Several recent works have reported how the aqueous phase can be converted to different co-products, in order to improve the overall carbon efficiency of the liquefaction process [4,12,13]. For example, we recently reported the co-production of either H_2 or propene, from the aqueous phase, and the resulting process economics for producing fuel from the organic phase [13]. From this work, it was found that the largest impact on minimum fuel selling price is increasing the utilization of carbon from the biomass, including as the sale of co-products [13].

With one pathway to co-products utilizing the aqueous phase from biomass liquefaction already demonstrated, the next step is to expand the platform of potential product compounds to make the biorefinery more versatile. Olefins are a promising group of co-products to produce with isobutene being particularly valuable as it is easily converted to fuel additives, solvents, and butyl rubber products [14–18]. Recently, $Zn_xZr_yO_z$ catalysts have been identified as having a unique combination of surface acidic and basic sites, generating a cascade reaction network (Scheme 1) [19–22]. This cascade network allows for the direct conversion of ethanol to isobutene in a single reactor bed [15,22]. Ethanol first undergoes ethanol dehydrogenation and ketonization reactions thus producing acetone. ZnO addition offers the necessary basic sites while also suppressing most of the strong acid sites responsible for undesirable ethanol dehydration. Acetone then undergoes aldol condensation and C–C cleavage over acid sites, while the formation of acetone decomposition products (CH_4 and CO_2) is largely suppressed [22]. Recently, a complete loop starting from syngas and ending with isobutene oligomerization to jet fuel was also demonstrated [15]. In addition, it has been demonstrated that the $Zn_xZr_yO_z$ catalyst is able to convert larger alcohols, and also carboxylic acids and ketones, into mixed olefin streams [22–24]. As previously demonstrated, the reaction mechanism involves the conversion of alcohols into carboxylic acids before undergoing subsequent reaction steps [22]. This makes the $Zn_xZr_yO_z$ catalyst promising for the direct conversion of carboxylic acids present in the aqueous phase from biomass liquefaction to olefin products.

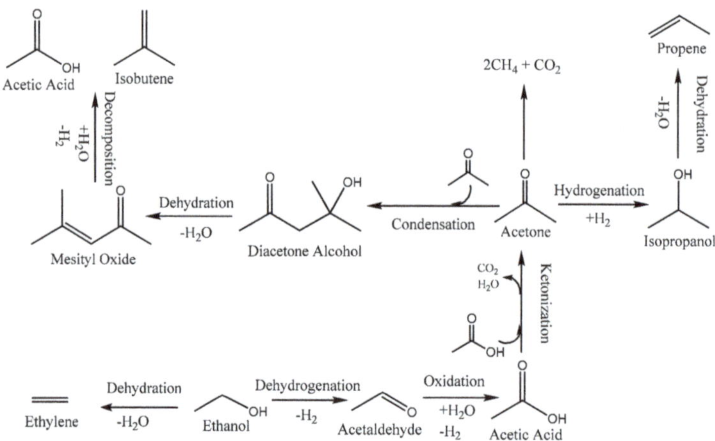

Scheme 1. Reaction network of ethanol to isobutene and major side products adapted from Smith et al. [22].

In this work we examine the feasibility of directly using biomass liquefaction-derived aqueous phases for the direct production of olefins. Three different aqueous phases are studied, derived from hydrothermal liquefaction (supplied by PNNL, PNNL-HTL), a modular fast pyrolysis (supplied by USDA-ARS, USDA-FP), and a screw pyrolysis (supplied by KIT, KIT-SP). The $Zn_xZr_yO_z$ catalyst (specific composition $Zn_1Zr_{2.5}O$) was used for the direct production of C_4+ olefins. Finally, the effect of a gas environment was also investigated to study the effects on product selectivity with a model feedstock of ethanol.

2. Results and Discussion

2.1. Feedstock Cleanup with Continuous Liquid–Liquid Extraction

In our prior work, a batch process was used to separate the target organics (carboxylic acids and alcohols) from the carbon treated stream [13]. The process worked well for high concentrations (~30 wt %) of carboxylic acids, however, we developed a continuous liquid–liquid extraction (LLE) (Figure 1) system to process the aqueous streams with lower concentrations of organics.

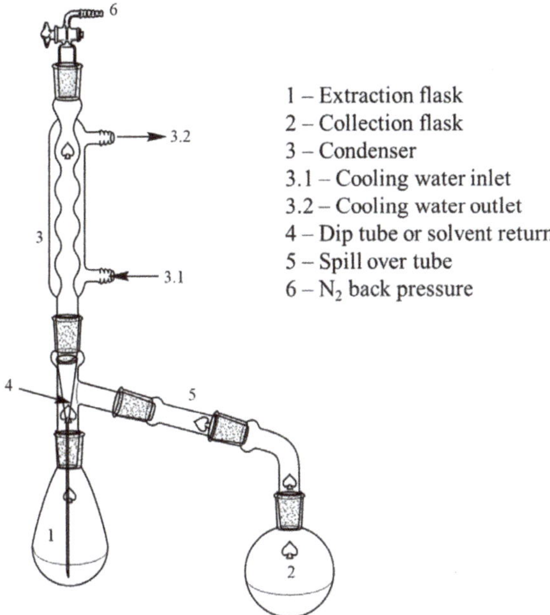

Figure 1. Continuous liquid–liquid extraction apparatus. Operation details can be found in Appendix A.

Details on the preparation of the PNNL-HTL feedstock were previously reported [13]. As received, the feedstocks had a different total organic concentration (based on LC composition), 26.2 wt % for KIT-SP, 14.6 wt % for USDA-FP, and 32.6 wt % for PNNL-HTL. For all feedstocks, acetic acid was the most abundant compound. The composition of the different feedstocks is summarized in Table 1. The full analysis can be found in the Supporting Information. A carbon treatment was performed first to remove color bodies and other compounds from the as received samples as it has been reported to cause deactivation during catalytic upgrading [13]. Following carbon treatment, the organic concentration decreased <10% due to adsorption onto the carbon surface, however, we also observed a loss of aqueous phase as it was retained in the porous structure of the carbon. The overall aqueous loss depended on the number of carbon treatments required for each sample. For example, the USDA-FP feedstock required two rounds of carbon treatment and lost 26.8% of the mass, while the KIT-SP feedstock required five rounds of carbon treatment and lost 69.6% of the mass (~13.6% during each carbon treatment). As previously shown, the carbon treatment removed all the color bodies from the sample and the feedstocks became water clear (see Supporting Figure S1) [13]. However, the carbon treatment increased the inorganic concentration of the feedstock, particularly for K, Mg, Na, and P from ≤100 ppm in all cases to as high as 1560 ppm K in USDA-FP and 4050 ppm K in KIT-SP.

Table 1. Summary of stream composition at the different stages of the clean-up process. PNNL-HTL is derived from corn stover, processed using hydrothermal liquefaction (HTL) and originally published elsewhere [13]. USDA-FP is derived from horse litter, processed using fast pyrolysis (FP). KIT-SP is derived from beech wood chips, processed using intermediate screw pyrolysis (SP). More information about the composition can be found in the Supporting Information.

	PNNL-HTL					USDA-FP					KIT-SP				
	Initial	Carbon Treated	Raffinate	Refined Extract	Final Feedstock	Initial	Carbon Treated	Raffinate	Refined Extract	Final Feedstock	Initial	Carbon Treated	Raffinate	Refined Extract	Final Feedstock
%C Retained	100	65.5	26.0	29.0	29.0	100	73.2	15.1	21.7	15.1	100	30.4	11.7	14.9	11.8
%Carboxylic Acid Retained	100	62.6	18.8	34.9	34.9	100	88.9	24.8	53.5	45.9	100	33.5	6.55	22.2	22.4
Acetic Acid [wt.%]	22.3	19.4	7.65	63.4	29.1	4.78	3.07	1.45	39.6	9.24	9.68	6.62	1.80	34.3	20.1
Propionic Acid [wt.%]	3.93	2.33	0.241	11.2	4.79	2.89	1.95	1.47	9.23	1.68	2.72	2.07	1.39	5.42	2.98
Non-Participating Carbon [wt.%]	6.40	6.09	7.89	5.77	0.16	5.95	3.00	5.34	27.33	1.55	13.8	8.03	12.2	41.1	2.57
Total C [wt.%]	32.6	27.8	15.8	80.4	34.0	8.35	8.02	8.26	76.16	12.5	26.2	16.7	15.4	80.8	25.6
S/C [a]					2.8					11.2					4.7
Na [ppm]	7040	7320	8580	25.2	22.9	9.30	202	147	18.3	12.8	28.2	446	365	7.96	9.64
K [ppm]	746	1960	2350	11.6	3.97	12.3	1560	1450	85.2	17.2	15.7	4050	1420	20.5	12.8
S [ppm]	BDL[b]	BDL	BDL	BDL	BDL	10.6	65.1	44.9	BDL	BDL	56.5	68.5	65.8	BDL	BDL

[a] S/C = molar ratio of steam-to-carbon. [b] BDL = below detection limit.

The carbon-treated aqueous phases were then placed inside a continuous LLE to remove and concentrate organic molecules from the aqueous phase. Addition of methyl tert-butyl ether (MTBE) to the carbon-treated aqueous samples created two phases, an aqueous phase (the raffinate) and an organic phase (the extract). Under batch LLE, this process was found to increase the total organic concentration from 34–80 wt %. The continuous LLE used here was found to generate extract (i.e., organic molecules extracted from aqueous phase) with the same concentration as the one obtained in the batch LLE. Further, both LLE systems generated the same amount of extract (e.g., ~0.60 $g_{carboxylic\ acids}$ $g_{aqueous\ phase\ fed}^{-1}$). However, the removal of MTBE was less effective in the continuous LLE than in the rotary-vaporization used in the batch LLE. For example, the extract from the continuous LLE had about 10 times higher concentration of MTBE (20.9 and 35.2 wt % for USDA-FP and KIT-SP respectively) than the extract obtained in batch LLE (2.91 wt % for HTL). Therefore, an additional distillation step was necessary to remove the remaining MTBE from the extract generated in the continuous LLE to produce the final feedstock. The composition of the main constituents found in the streams at the different steps of the clean-up process are summarized in Table 1. A full analysis of the stream can be found in the Supporting Information.

2.2. Catalytic Upgrading of Cleaned Up Aqueous Phase to Isobutene

The final PNN-HTL and KIT-SP feedstocks were selected to explore the catalytic upgrading of carboxylic acids into olefins using the $Zn_1Zr_{2.5}O$ catalyst. The product selectivity and catalyst stability were evaluated as a function of gas hour space velocity (GHSV). The final PNNL-HTL feedstock had a lower concentration of non-participating carbon compared to the final KIT-SP feedstock, 0.16 and 2.57 wt % respectively.

From these GHSV screening (Figure 2), the product distribution obtained with both feedstocks appeared similar to that previously observed with ethanol and propanol [22]. As depicted in Scheme 1 and in our previous work, [22] acetone and iso- and n-butene (i.e., iso-$C_4^=$ and n-$C_4^=$) are direct products of acetic acid. However, methyl ethyl ketone (MEK) and methyl butane ($C_5^=$) are produced from the combination of propanoic acid with acetic acid as depicted in Scheme 2. This mechanism also agrees with our recent study for conversion of MEK to olefins over $Zn_xZr_yO_z$ catalysts [23]. As shown in Figure 2, the ketone and olefin selectivity are directly affected by the GHSV, emphasizing that the ketones are indeed reaction intermediates for the olefin production as depicted in Schemes 1 and 2. For example, at the highest GHSV (i.e., lowest contact time) there was already complete conversion of the carboxylic acids but we only observed ketone intermediates (e.g., acetone from acetic acid-acetic acid self-ketonization, MEK from acetic acid-propionic acid cross-ketonization, and 3-pentanone from propionic acid self-ketonization). However, the selectivity towards ketone formation decreased as the GHSV decreased (i.e., higher contact time). For example, at ~1200 h^{-1} (i.e., the lowest SV studied) there was almost complete conversion of the ketone intermediates to their respective olefins, as illustrated in Schemes 1 and 2, with a combined ketone selectivity of 8.6 and 3.9% for the PNNL-HTL and KIT-SP feedstocks respectively. This suggested that $Zn_xZr_yO_z$ is an effective catalyst for the direct production of olefins from carboxylic acids as well as alcohols.

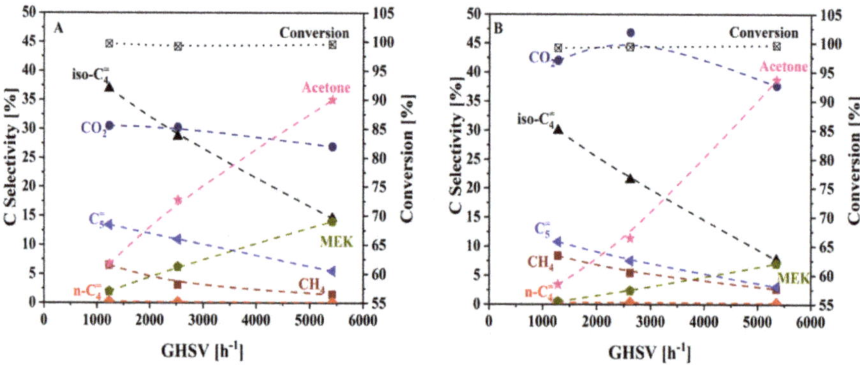

Figure 2. Gas hour space velocity (GHSV) profile of (**A**) PNNL-HTL feedstock, S/C = 2.8, and (**B**) KIT-SP feedstock, S/C = 4.7. Temperature: 450 °C, catalyst loading 0.7 g, N_2 50 vol%. Complete conversion of carboxylic acids was observed at all GHSV's studied. 3-pentanone was observed in trace (0.21% for PNNL-HTL and 0.10% for KIT-SP) amounts at ~5400 h^{-1}, but was otherwise not observed.

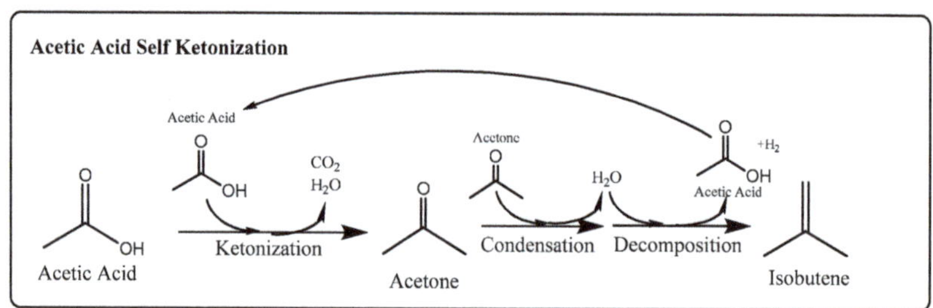

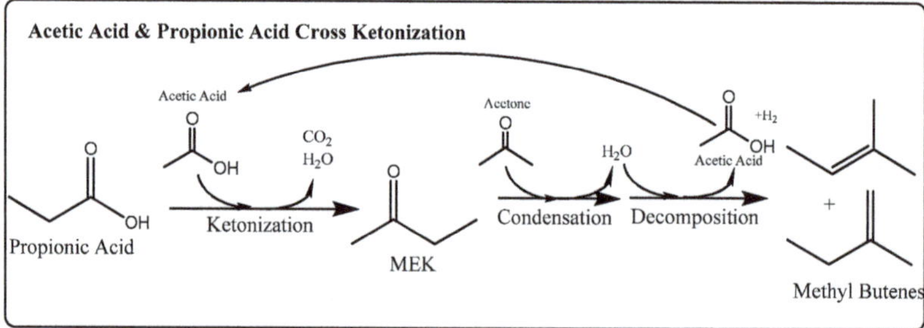

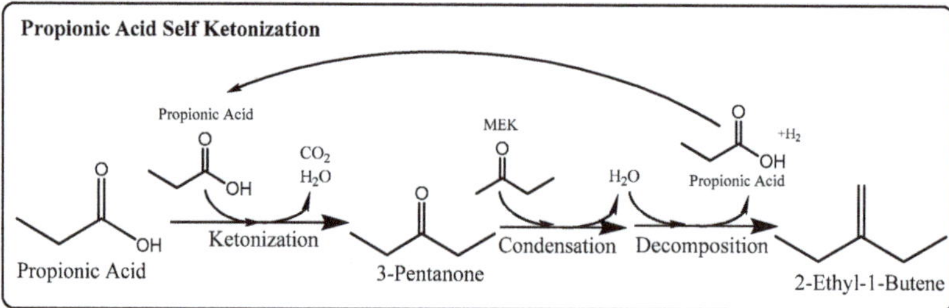

Scheme 2. Reaction network of carboxylic acid to olefin ethanol to isobutene and major side products.

As shown in Figure 3, the overall product selectivity obtained with the three feedstocks favors olefin over ketone formation when operated at low GHSV and same reaction conditions, however, the individual olefin selectivity differs. For example, the PNNL-HTL and HIT-SP feedstocks were ~15% selective towards C_5^+ production, the USDA-FP sample was ≥5%. We hypothesize this difference in C_5^+ selectivity is due to the differences in acetic acid and propanoic acid concentration. For example, PNNL-HTL feedstocks had the highest overall concentration of carboxylic acids (33.9 wt %) and propanoic acid (4.79 wt %), while the USDA-FP feedstock had the lowest overall concentration of carboxylic concentration (10.9 wt %) and propanoic acid (1.68 wt %).

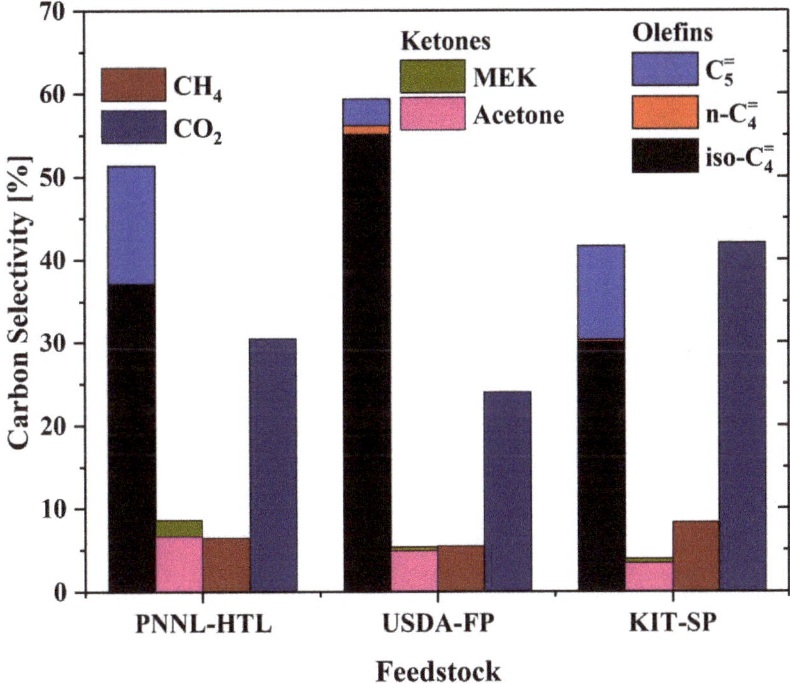

Figure 3. Major product selectivity of all feedstocks. Temperature: 450 °C, GHSV: 1200 h^{-1}, catalyst loading 0.7 g, N_2 50 vol%, S/C = 2.8, 11.2, 4.7 for PNNL-HTL, USDA-FP, and KIT-SP, respectively. The product selectivity was measured at 100% conversion of carboxylic acids.

Figure 4 depicts the stability of the $Zn_1Zr_{2.5}O$ catalyst when tested with the PNNL-HTL and KIT-SP feedstocks containing different concentrations of non-participating carbon species, 0.16 and 2.57 wt %, respectively. The initial and final conversion and primary product selectivities are summarized in Table 2. The PNNL-HTL feedstock (with the lowest non-participating carbon concentrations) showed a stability profile like the one previously observed with model ethanol feedstocks [19,22]. While the higher GHSV did increase selectivity to the ketone products (~23.5 and ~10.0% for acetone and MEK respectively) overall selectivity was relatively stable for ~20 h time on stream (TOS), after which the ketone selectivity began increasing significantly while olefin selectivity decreased. This further supported our speculation that the ketones are the intermediate of the olefins. In contrast, the KIT-SP feedstock started with higher ketone selectivity, ~49% selectivity at 10 h TOS, and continued to increase for the duration of the test, ~65% selectivity by 22 h TOS. While the conversion was constant with the PNNL-HTL feedstock at 100% for the duration of the experiment, it decreased with the KIT-SP feedstock. As shown in Figure 4B, the conversion started to decrease at 26 h TOS and dropped to 82% by the end of the experiment. This difference in stability with the two feedstocks indicated that, while the $Zn_xZr_yO_z$ catalyst can accommodate a range of compounds, the catalyst will deactivate faster in the presence of more non-participating compounds in the feedstock [13].

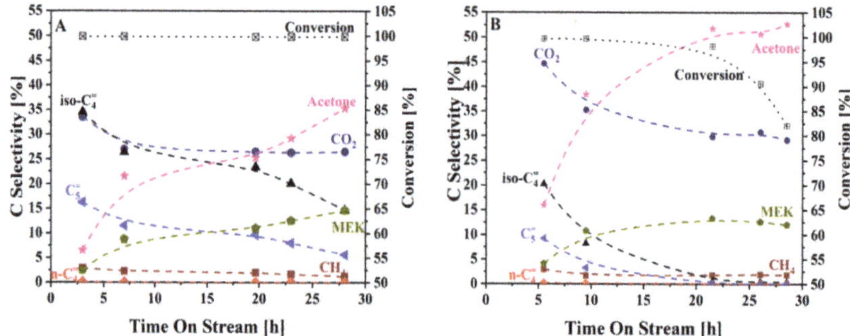

Figure 4. High GHSV stability of (**A**) PNNL-HTL feedstock, S/C = 2.8, and (**B**) KIT-SP feedstock, S/C = 4.7. Temperature: 450 °C, GHSV: 5500 h^{-1}, catalyst loading 0.7 g, N$_2$ 50 vol%.

Table 2. Comparison of initial and final activity and selectivity of PNNL-HTL and KIT-SP feedstocks shown in Figure 4.

	PNNL-HTL		KIT-SP	
	Initial	Final	Initial	Final
Conversion (%)	99.8	99.8	99.7	82.1
Selectivity (%)				
All Olefins	53.2	21.5	30.8	0.77
All Ketones	9.11	50.0	20.1	64.8
Isobutene	34.5	14.7	20.2	0.39
Acetone	6.60	35.3	16.0	52.7

The long-term stability of the $Zn_xZr_yO_z$ catalyst (Figure 5) was tested with the USDA-FP feedstock as it had the lowest total carboxylic acid content and some non-participating carbon compounds, 10.9 and 1.55 wt %, respectively. We chose a low SV (1200 h^{-1}) to maximize olefin production which is more representative of industrial targets. The $Zn_xZr_yO_z$ catalyst was relatively stable for close to 100 h, and by 95 h TOS the isobutene selectivity was stable >50% (net olefins ~60%). The ketone selectivity, particularly acetone, slowly increased over this period from 3.4% at 47 h TOS to 8.9% at 95 h TOS. At 120 h TOS the $Zn_xZr_yO_z$ catalyst showed catalytic deactivation as ketone breakthrough was observed, with ketone selectivity at 25.3% (23.9% and 1.37% for acetone and MEK respectively). Following in situ catalyst regeneration, the $Zn_xZr_yO_z$ showed complete recovery of activity with isobutene selectivity increasing to 58.5% and acetone selectivity dropping back to 2.70%. The catalysts showed catalyst deactivation after another 50 h of reaction as acetones selectivity increased to 7.49%. A second in situ catalyst regeneration completely regenerated the $Zn_xZr_yO_z$ activity.

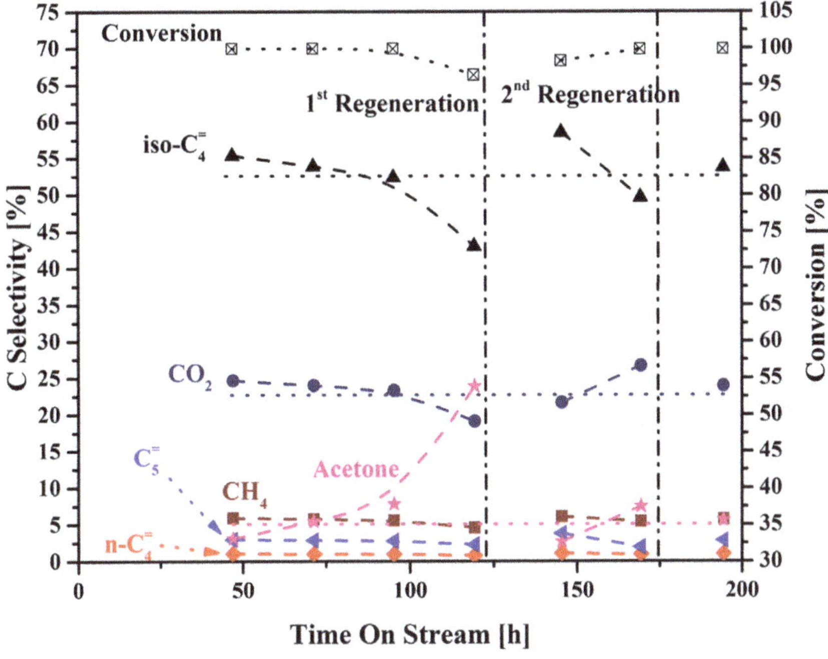

Figure 5. Stability and regeneration of USDA-FP feedstock. Temperature: 450 °C, S/C = 11.2, GHSV: 1200 h^{-1}, catalyst loading 0.7 g, N$_2$ 50 vol%. The catalyst regeneration was done by calcination in 5 vol% O$_2$ in N$_2$ at 500 °C for 4 h.

2.3. Catalytic Upgrading of Ethanol to Propene

H$_2$ is often used to improve catalyst stability, however the role of H$_2$ on the olefin formation reactions has not been widely studied [1,12,22]. Whereas H$_2$ could hydrogenate olefins into paraffins, it could also hydrogenate reaction intermediates and shift the product distribution towards different olefin products. As shown in Scheme 1, acetone can be hydrogenated to isopropanol which then can dehydrate to propene. To simplify this portion of our study, a 20 wt % ethanol in water feed was used rather than the more complex aqueous phase feedstocks discussed earlier. Propene production in the absence of H$_2$ from the complex feedstocks (i.e., PNNL-HTL, USDA-FP, and KIT-SP) was negligible (<1% selectivity, typically ~0.2% selectivity).

As shown in Figure 6, the propene selectivity was enhanced by cofeeding H$_2$. For example, when the N$_2$ carrier gas was replaced by H$_2$ in in 25 vol% increments, the propene (C$_3$H$_6$) selectivity increased roughly linearly with H$_2$ composition from 5.40 to 18.6%, while the isobutene selectivity decreased from 43.1 to 32.6%. The change in product distribution is because the hydrogenation of acetone (and subsequent dehydration to produce propene) is enhanced in the presence of H$_2$ while acetone self-condensation (to produce isobutene) is inhibited as depicted in Scheme 1 and described in our previous work [22]. Further, the CO$_2$ selectivity also decreased as a result of inhibiting the isobutene formation as expected from the reaction network described in Scheme 1. The CH$_4$ selectivity remained relatively constant at ~4%, indicating that changing the gas environment does not significantly impact decomposition or methanation reactions. We speculate that cofeeding H$_2$ during the upgrading of the feedstocks explored in this work over the Zn$_x$Zr$_y$O$_z$ would cause a similar enhancement in propene production and shift the C$_5^=$ production more towards C$_4^=$ [23]. The mixed olefin product stream could be then oligomerized to higher value products as demonstrated by Saavedra Lopez et al. [14].

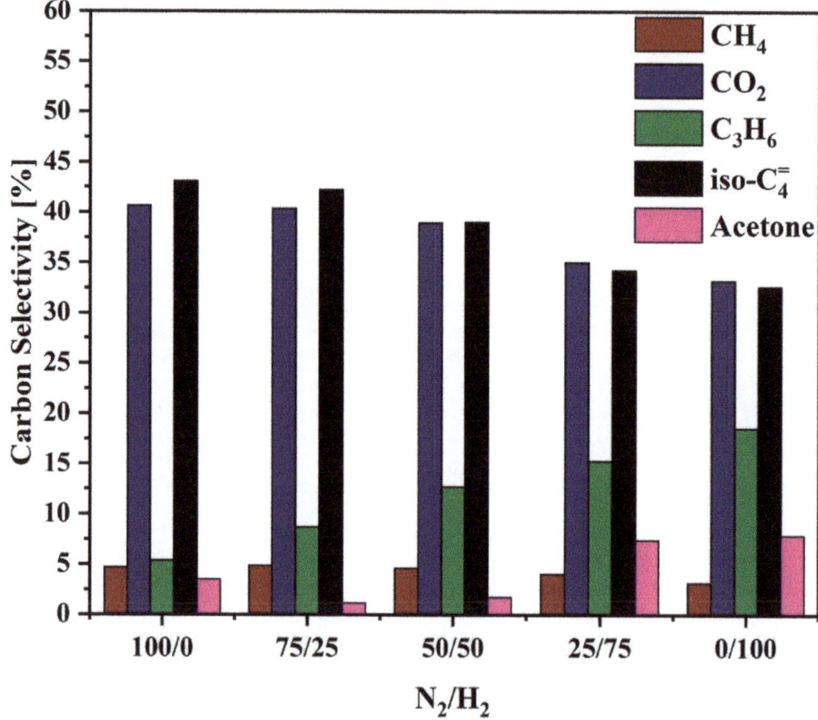

Figure 6. Effect of gas environment on product selectivity. Feedstock: 20 wt % ethanol in water, S/C = 5.1, Temperature: 450 °C, GHSV: 1200 h^{-1}, catalyst loading 0.7 g, N$_2$/H$_2$ composition was constant at 50 vol%.

3. Materials and Methods

3.1. Feedstock Sources

Three different feedstocks were investigated during this study. The first feedstock, provided by PNNL and designated PNNL-HLT, was the same as used in a prior study [13]. Briefly, the aqueous feed produced from HTL of corn stover was used and modified to simulate ~100 recycles in the HTL system to reach the desired concentration of organic compounds in the aqueous phase [4,13,25]. The second feedstock, provided by USDA and designated USDA-FP, was produced using FP of horse litter. Details on the FP system can be found elsewhere; briefly, the horse litter was fed via hopper to a novel FP system where a fluidized bed is used to perform the fast pyrolysis, recycled process gas can also be used to decrease the oxygen content of the final oil, the effluent from the fluidized bed is then passed through a cyclone separator for solid removal, and a condenser train was used for sample collection [11,26,27]. The third feedstock, provided by KIT and designated KIT-SP, was produced using STYX screw pyrolysis of bark-free beech wood pellets. Details on the STYX screw pyrolysis system can be found elsewhere; briefly, the wood pellets were fed via hopper to an integrated pyrolysis reactor with a screw mechanism to drive material through the reactor, product gasses are removed at various points along the integrated reactor and sent to a condenser train for sample collection, filters in the reactor prevent solids from being entrained in the gas flows, and all solids are removed at the end to the of the reactor at the screw outlet [10,28,29].

3.2. Feedstock Clean Up

A multi-step process was used to clean up the various bio-oil aqueous phases shown in Figure 7. The PNNL-HTL feedstock is the same as reported previously, cleaned using the batch process first reported [13]. The USDA-FP and KIT-SP feedstocks were both cleaned using continuous liquid–liquid extraction. First the as received feedstocks were treated with activated carbon (AC) obtained from Pacific Activated Carbon Co. Inc. (*PACCO*, CTC, coconut shell, lot# 2007-2-22AC) to remove trace contaminants from the blended HTL-derived aqueous feed at an AC/feedstock ratio of $^1/_{10}$ by weight. The slurry was capped to avoid evaporation of H_2O and light organics and agitated at 250 rpm at ambient temperature for 24 h. The AC was then removed via vacuum filtration. For USDA-FP, two carbon treatments were sufficient to achieve the desired color change of the feedstock. For KIT-SP, five carbon treatments were required to achieve the desired color change of the feedstock (Supporting Figure S1). Methyl tert-butyl ether (MTBE; Sigma-Aldrich, anhydrous 99.8%) was then used as a solvent to extract the desired carboxylic acid molecules (acetic acid and propanoic acid) from the carbon-treated mixture at a mixture/MTBE ratio of $\sim^1/_1$ by weight. The carbon-treated mixture and the MTBE were charged into a continuous liquid–liquid extraction vessel (Figure 1) and MTBE circulated internally for ~48 h. After separation, the MTBE and raffinate from the extractor were placed in a cone-shaped separatory funnel to separate the MTBE (top) and the raffinate (bottom) (~10 min). The MTBE was combined with the collection pot to form the extract. The extract was then boiled to remove the majority of MTBE. This MTBE was then recycled for subsequent extractions. The refined extract, containing mostly acetic acid, propanoic acid, and residual MTBE was diluted with deionized (DI) water to have enough material for benchtop distillation. The distillation resulted in two cuts and the pot residual. The first cut was the majority of the residual MTBE, while the second cut had the lightest components from the refined extract. The final feedstock was made by combining the second cut from distillation, the pot residual, and DI H_2O to adjust the acetic acid concentration to the desired amount. For easy comparison of the concentration of organics across feedstocks, the molar ratio of steam-to-carbon (S/C) was used, that is, the molar ratio of water to the total number of moles of carbon in the feedstock. Non-participating carbon refers to any compound detected via LC that does not appear in the reaction network shown in Scheme 1, that is, non-alcohol, non-ketone, and non-carboxylic acid compounds.

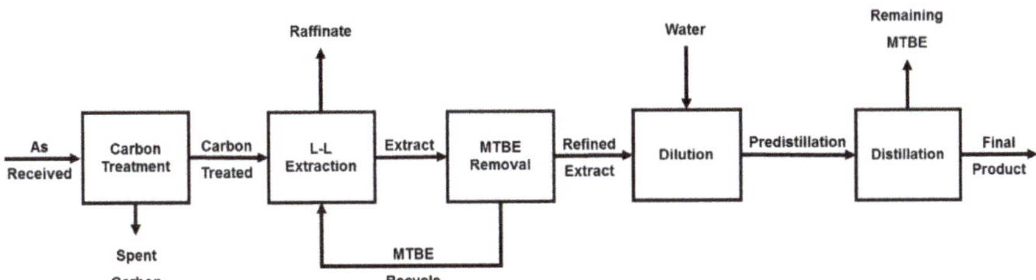

Figure 7. Clean up protocol for bio-oil aqueous phase prior to testing adapted from Davidson et al. [13].

3.3. Catalyst Preparation

Mixed oxide catalysts were synthesized via wet impregnation of a $Zn(NO_3)_2 \cdot 6H_2O$ solution on $Zr(OH)_4$ as described elsewhere [22]. The $Zr(OH)_4$ was initially dried overnight at 105 °C to remove any excess water on the surface before impregnation. After impregnation, the catalysts were dried overnight at room temperature and then for 4 h at 105 °C prior to calcination. The catalysts then were calcined via a 3 °C min^{-1} ramp to 400 °C for 2 h, followed by a 5 °C min^{-1} ramp to the final calcination temperature of 550 °C for 3 h. Extensive details of the catalyst characteristics have been published previously and are not reproduced here [19–22].

3.4. Reactor and Reaction Conditions

Olefin production reactions were performed in a $\frac{1}{2}$″ inch outer-diameter, fixed-bed alumina reactor. Alumina reactors were used to minimize potential side reactions that might occur with stainless steel reactors. Typically, 0.70 g of $Zn_1Zr_{2.5}O$ was loaded undiluted. Prior to testing, catalysts were degassed under N_2 at 450 °C for 8 h.

The catalysts were tested at atmospheric pressure and at a reaction temperature of 450 °C. The liquid feed was fed via HPLC pump to a vaporizer held at 150 °C. A N_2 gas flow rate (typically ~50 mol% in the feed) was introduced into the system to serve as the carrier gas and internal reference standard. Both liquid and gas feed were adjusted accordingly to achieve desired GHSV. On-line gas products were tracked via four-channel MicroGC® and condensable products were analyzed off line via LC analysis, as described previously [13]. The reported conversion is based on the ratio of feed carboxylic acids to carboxylic acids remaining after reaction. For all reactions total carbon balance was >95% and measurement errors in composition are <1%.

4. Conclusions

In this study we demonstrated a cleanup process, previously reported using batch processing, here adapted to continuous flow liquid–liquid extraction. Aqueous phases, rich in carboxylic acids, were obtained from several liquefaction processes, including HTL, FP, and SP, and evaluated for cleanup and catalytic upgrading. Future advances to the cleanup process are required in order to improve hydrocarbon retention. Cleaned feedstocks were converted to primarily C_4 and C_5 olefins, and CO_2 over a $Zn_xZr_yO_z$ mixed oxide catalyst. The reaction network appears consistent with previous reports using model ethanol and acetic acid feedstocks. Carboxylic acids are converted primarily to iso-olefins over $Zn_xZr_yO_z$ catalyst. Specifically, iso-butene is produced from acetic acid, a primary constituent from the biomass-derived aqueous phase. Conversion of acetic acid and propionic acid over $Zn_xZr_yO_z$ involve cross-coupling reactions and produce branched C_5 olefins. Non-participating hydrocarbons in the feedstock (e.g., sugars, polyols) are not believed to participate in the reaction mechanism through ketone intermediates and contribute to catalyst deactivation. Finally, the presence of H_2 in the feed was found to facilitate hydrogenation of intermediate acetone, thereby increasing propene production and, consequently, decreasing isobutene production.

Supplementary Materials: The following are available online at http://www.mdpi.com/2073-4344/9/11/923/s1, Table S1: LC analysis of USDA-FP feedstock after key steps of clean up protocol, Table S2: LC analysis of KIT-SP feedstock after key steps of clean up protocol, Table S3: ICP analysis of USDA-FP feedstock after key steps of clean up protocol, Table S4: ICP analysis of KIT-SP feedstock after key steps of clean up protocol, Figure S1: KIT-SP feedstock A) Initial B) During Carbon Treatment, After Third Carbon Treatment ~72 h C) After All Carbon Treatment, Five Carbon Treatments ~120 h, Table S5: Summary of dilute feedstock cleanup.

Author Contributions: S.D.D. lead most experiments, data analysis, and operations. J.A.L.-R. contributed to the development of the cleanup protocol. M.F. synthesized and prepared the $Zn_xZr_yO_z$ catalyst. A.R.C., Y.E., and M.T.M. provided the major feedstocks studied in this work (PNNL-HTL, USDA-FP, and KIT-SP respectively). V.L.D., K.O.A., and R.A.D. lead conceptualization, project administration, and funding acquisition.

Funding: This work was financially supported by the U.S. Department of Energy's (DOE) Bioenergy Technologies Office under Contract DE-AC05-76RL01830 and in collaboration with the Chemical Catalysis for Bioenergy Consortium (ChemCatBio), a member of the Energy Materials Network (EMN), and was performed at Pacific Northwest National Laboratory (PNNL). PNNL is a multi-program national laboratory operated for the DOE by Battelle Memorial Institute.

Acknowledgments: The authors thank Teresa Lemmon for analytical support of this work and collaborators at NREL for providing dilute feedstocks. ARS acknowledges the help of Mark Schaffer and Neil Goldberg for operation of the pyrolysis system. The views and opinions of the authors expressed herein do not necessarily state or reflect those of the United States Government or any agency thereof. Neither the United States Government nor any agency thereof, nor any of their employees, makes any warranty, expressed or implied, or assumes any legal liability or responsibility for the accuracy, completeness, or usefulness of any information, apparatus, product, or process disclosed, or represents that its use would not infringe privately owned rights.

Conflicts of Interest: The authors declare no conflict of interest. The funders had no role in the design of the study; in the collection, analyses, or interpretation of data; in the writing of the manuscript; or in the decision to publish the results.

Abbreviations

AC	Activated carbon
BDL	Below detection limit
DI	Deionized
FP	Fast pyrolysis
GHSV	Gas hourly space velocity
HTL	Hydrothermal liquefaction
KIT	Karlsruhe Institute of Technology
LLE	Liquid–liquid extraction
MEK	Methyl ethyl ketone (butanone)
MTBE	Methyl Tert-butyl ether
PNNL	Pacific Northwest National Lab
S/C	Steam-to-carbon ratio (molar ratio)
SP	Screw pyrolysis
SV	Space velocity
TOS	Time on stream
USDA	United States Department of Agriculture

Appendix A

Operating conditions of the LLE proved critical to optimizing organic extraction. For these extractions the heavy solvent valve was closed. Carbon-treated feed was first charged into the extraction column. Then the remaining height of the extraction column was filled with MTBE. To control the ratio of solvent to feedstock, the remaining quantity of MTBE was then charged into the collection flask. Once charged, the glass fittings were all sealed with vacuum grease and clamps. The circulator for the condenser was set to 5 °C and allowed to equilibrate before starting heating. A stir bar and plate with no heating provided mixing to the extraction column, while a stir bar and plate with heating via mineral oil bath provided heat and mixing to the collection flask. A low flow of N_2 was applied to the head space above the condenser to minimize loss of material during extraction. Once equilibrated at ambient temperature the mineral oil bath was heated to 80 °C; this vaporized the MTBE from the collection flask to the condenser, where it condensed and ran down a drip tube to the bottom of the extraction flask. As the MTBE rose through the extraction flask it extracted the target organics from the feedstock. Upon reaching the top of the extraction flask, the organic rich MTBE then flowed back to the collection flask where the MTBE vaporized again but the other organics did not. This stable cycle of the LLE was then continued for ~48 h.

References

1. Huber, G.W.; Iborra, S.; Corma, A. Synthesis of Transportation Fuels from Biomass: Chemistry, Catalysts, and Engineering. *Chem. Rev.* **2006**, *106*, 4044–4098. [CrossRef] [PubMed]
2. Bond, J.Q.; Upadhye, A.A.; Olcay, H.; Tompsett, G.A.; Jae, J.; Xing, R.; Alonso, D.M.; Wang, D.; Zhang, T.; Kumar, R.; et al. Production of renewable jet fuel range alkanes and commodity chemicals from integrated catalytic processing of biomass. *Energy Environ. Sci.* **2014**, *7*, 1500–1523. [CrossRef]
3. Huber, G.W.; Corma, A. Synergies between Bio- and Oil Refineries for the Production of Fuels from Biomass. *Angew. Chem. Int. Ed.* **2007**, *46*, 7184–7201. [CrossRef] [PubMed]

4.   Zhu, Y.; Biddy, M.J.; Jones, S.B.; Elliott, D.C.; Schmidt, A.J. Techno-economic analysis of liquid fuel production from woody biomass via hydrothermal liquefaction (HTL) and upgrading. *Appl. Energy* **2014**, *129*, 384–394. [CrossRef]

5.   Chheda, J.N.; Huber, G.W.; Dumesic, J.A. Liquid-phase catalytic processing of biomass-derived oxygenated hydrocarbons to fuels and chemicals. *Angew. Chem. Int. Ed.* **2007**, *46*, 7164–7183. [CrossRef] [PubMed]

6.   Panisko, E.; Wietsma, T.; Lemmon, T.; Albrecht, K.; Howe, D. Characterization of the aqueous fractions from hydrotreatment and hydrothermal liquefaction of lignocellulosic feedstocks. *Biomass Bioenergy* **2015**, *74*, 162–171. [CrossRef]

7.   Zhang, Q.; Chang, J.; Wang, T.; Xu, Y. Review of biomass pyrolysis oil properties and upgrading research. *Energy Convers. Manag.* **2007**, *48*, 87–92. [CrossRef]

8.   Elliott, D.C.; Beckman, D.; Bridgwater, A.V.; Diebold, J.P.; Gevert, S.B.; Solantausta, Y. Developments in direct thermochemical liquefaction of biomass: 1983–1990. *Energy Fuels* **1991**, *5*, 399–410. [CrossRef]

9.   Elliott, D.C.; Biller, P.; Ross, A.B.; Schmidt, A.J.; Jones, S.B. Hydrothermal liquefaction of biomass: Developments from batch to continuous process. *Bioresour. Technol.* **2015**, *178*, 147–156. [CrossRef]

10.   Funke, A.; Tomasi Morgano, M.; Dahmen, N.; Leibold, H. Experimental comparison of two bench scale units for fast and intermediate pyrolysis. *J. Anal. Appl. Pyrolysis* **2017**, *124*, 504–514. [CrossRef]

11.   Boateng, A.A.; Daugaard, D.E.; Goldberg, N.M.; Hicks, K.B. Bench-Scale Fluidized-Bed Pyrolysis of Switchgrass for Bio-Oil Production. *Ind. Eng. Chem. Res.* **2007**, *46*, 1891–1897. [CrossRef]

12.   Huber, G.W.; Dumesic, J.A. An overview of aqueous-phase catalytic processes for production of hydrogen and alkanes in a biorefinery. *Catal. Today* **2006**, *111*, 119–132. [CrossRef]

13.   Davidson, S.D.; Lopez-Ruiz, J.A.; Zhu, Y.; Cooper, A.R.; Albrecht, K.O.; Dagle, R.A. Clean up and Catalytic Upgrading of Hydrothermal Liquefaction-Derived Aqueous Phase into Fuels and Chemicals via Ketone Intermediates. *ACS Sustain. Chem. Eng.* **2019**. submitted.

14.   Saavedra, L.J.; Dagle, R.A.; Dagle, V.L.; Smith, C.; Albrecht, K.O. Oligomerization of ethanol-derived propene and isobutene mixtures to transportation fuels: Catalyst and process considerations. *Catal. Sci. Technol.* **2019**, *9*, 1117–1131. [CrossRef]

15.   Dagle, V.L.; Smith, C.; Flake, M.; Albrecht, K.O.; Gray, M.J.; Ramasamy, K.K.; Dagle, R.A. Integrated process for the catalytic conversion of biomass-derived syngas into transportation fuels. *Green Chem.* **2016**, *18*, 1880–1891. [CrossRef]

16.   Yoon, J.W.; Chang, J.-S.; Lee, H.-D.; Kim, T.-J.; Jhung, S.H. Trimerization of isobutene over a zeolite beta catalyst. *J. Catal.* **2007**, *245*, 253–256. [CrossRef]

17.   Marchionna, M.; Di Girolamo, M.; Patrini, R. Light olefins dimerization to high quality gasoline components. *Catal. Today* **2001**, *65*, 397–403. [CrossRef]

18.   Carr, A.G.; Dawson, D.M.; Bochmann, M. Zirconocenes as Initiators for Carbocationic Isobutene Homo- and Copolymerizations. *Macromolecules* **1998**, *31*, 2035–2040. [CrossRef]

19.   Liu, C.; Sun, J.; Smith, C.; Wang, Y. A study of ZnxZryOz mixed oxides for direct conversion of ethanol to isobutene. *Appl. Catal. A Gen.* **2013**, *467*, 91–97. [CrossRef]

20.   Sun, J.; Baylon, R.A.L.; Liu, C.; Mei, D.; Martin, K.J.; Venkitasubramanian, P.; Wang, Y. Key Roles of Lewis Acid–Base Pairs on ZnxZryOz in Direct Ethanol/Acetone to Isobutene Conversion. *J. Am. Chem. Soc.* **2015**, *138*, 507–517. [CrossRef]

21.   Sun, J.; Zhu, K.; Gao, F.; Wang, C.; Liu, J.; Peden, C.H.F.; Wang, Y. Direct Conversion of Bio-ethanol to Isobutene on Nanosized ZnxZryOz Mixed Oxides with Balanced Acid–Base Sites. *J. Am. Chem. Soc.* **2011**, *133*, 11096–11099. [CrossRef] [PubMed]

22.   Smith, C.; Dagle, V.; Flake, M.; Ramasamy, K.; Kovarik, L.; Bowden, M.; Onfroy, T.; Dagle, R. Conversion of Syngas-Derived C2+ Mixed Oxygenates to C3-C5 Olefins over ZnxZryOz Mixed Oxides Catalysts. *Catal. Sci. Technol.* **2016**, *6*, 2325–2336. [CrossRef]

23.   Dagle, V.L.; Dagle, R.A.; Kovarik, L.; Baddour, F.; Habas, S.E.; Elander, R. Single-step Conversion of Methyl Ethyl Ketone to Olefins over ZnxZryOz Catalysts in Water. *ChemCatChem* **2019**, *11*, 3393–3400. [CrossRef]

24.   Crisci, A.J.; Dou, H.; Prasomsri, T.; Román-Leshkov, Y. Cascade Reactions for the Continuous and Selective Production of Isobutene from Bioderived Acetic Acid Over Zinc-Zirconia Catalysts. *ACS Catal.* **2014**, *4*, 4196–4200. [CrossRef]

25. Schmidt, A.J.; Lindstorm, T.; Talmadge, M.; Zhu, Y. Mid Stage 2 Report on Hydrothermal Liquefaction Strategy for the NABC Leadership Team. In *PNNL-21768*; Energy, U.S.D.O., Ed.; Pacific Northwest National Laboratory: Richland, WA, USA, 2012.
26. Mullen, C.A.; Boateng, A.A.; Goldberg, N.M. Production of Deoxygenated Biomass Fast Pyrolysis Oils via Product Gas Recycling. *Energy Fuels* **2013**, *27*, 3867–3874. [CrossRef]
27. Boateng, A.A.; Schaffer, M.A.; Mullen, C.A.; Goldberg, N.M. Mobile demonstration unit for fast- and catalytic pyrolysis: The combustion reduction integrated pyrolysis system (CRIPS). *J. Anal. Appl. Pyrolysis* **2019**, *137*, 185–194. [CrossRef]
28. Tomasi Morgano, M.; Leibold, H.; Richter, F.; Seifert, H. Screw pyrolysis with integrated sequential hot gas filtration. *J. Anal. Appl Pyrolysis* **2015**, *113*, 216–224. [CrossRef]
29. Tomasi Morgano, M.; Leibold, H.; Richter, F.; Stapf, D.; Seifert, H. Screw pyrolysis technology for sewage sludge treatment. *Waste Manag.* **2018**, *73*, 487–495. [CrossRef] [PubMed]

Article

Ga/HZSM-5 Catalysed Acetic Acid Ketonisation for Upgrading of Biomass Pyrolysis Vapours

Hessam Jahangiri [1,2,3], Amin Osatiashtiani [3], Miloud Ouadi [1], Andreas Hornung [1,4,5], Adam F. Lee [6,*] and Karen Wilson [6,*]

1   School of Chemical Engineering, University of Birmingham, Birmingham B15 2TT, UK;
    h.jahangiri@bham.ac.uk (H.J.); m.ouadi@bham.ac.uk (M.O.); andreas.hornung@umsicht.fraunhofer.de (A.H.)
2   School of Water, Energy and Environment, Cranfield University, Cranfield MK43 0AL, UK
3   European Bioenergy Research Institute, Aston University, Birmingham B4 7ET, UK;
    a.osatiashtiani@aston.ac.uk
4   Fraunhofer UMSICHT, Fraunhofer Institute for Environmental, Safety and Energy Technology, An der
    Maxhütte 1, 92237 Sulzbach-Rosenberg, Germany
5   Friedrich-Alexander University Erlangen-Nuremberg, Schlossplatz 4, 91054 Erlangen, Germany
6   Applied Chemistry & Environmental Science, RMIT University, Melbourne, VIC 3000, Australia
*   Correspondence: adam.lee2@rmit.edu.au (A.F.L.); karen.wilson2@rmit.edu.au (K.W.);
    Tel.: +61-399-252-623 (A.F.L.); +61-(03)-9925-2122 (K.W.)

Received: 19 September 2019; Accepted: 5 October 2019; Published: 11 October 2019

Abstract: Pyrolysis bio-oils contain significant amounts of carboxylic acids which limit their utility as biofuels. Ketonisation of carboxylic acids within biomass pyrolysis vapours is a potential route to upgrade the energy content and stability of the resulting bio-oil condensate, but requires active, selective and coke-resistant solid acid catalysts. Here we explore the vapour phase ketonisation of acetic acid over Ga-doped HZSM-5. Weak Lewis acid sites were identified as the active species responsible for acetic acid ketonisation to acetone at 350 °C and 400 °C. Turnover frequencies were proportional to Ga loading, reaching ~6 min^{-1} at 400 °C for 10Ga/HZSM-5. Selectivity to the desired acetone product correlated with the weak:strong acid site ratio, being favoured over weak Lewis acid sites and reaching 30% for 10Ga/HZSM-5. Strong Brønsted acidity promoted competing unselective reactions and carbon laydown. 10Ga/HZSM-5 exhibited good stability for over 5 h on-stream acetic acid ketonisation.

Keywords: pyrolysis; ketonisation; bio-oil; turnover frequencies (TOFs)

---

1. Introduction

Hydrocarbons sourced from non-edible or waste lignocellulosic or algal biomass are an attractive source of sustainable liquid transportation fuels to mitigate current dependence on fossil fuels and associated anthropogenic climate change [1,2]. However, biomass-derived fuels are incompatible with existing distribution infrastructure and vehicle engines without (catalytic) upgrading to improve their physicochemical properties [2,3]. A range of thermochemical technologies exist for bio-oil production, including hydrothermal liquefaction [4,5] and pyrolysis [6,7], or gasification [8,9] and subsequent Fischer–Tropsch synthesis [10,11]. Pyrolysis routes have gained particular attention over the past 30 years, offering a high liquid (bio-oil) yield which can be used directly as a drop-in fuel, blended with conventional diesel, or as an efficient energy vector [12,13]. Pyrolysis bio-oils are mixtures of oxygenated compounds which typically comprise phenolics, furanics, carboxylic acids and other small oxygenates whose composition varies with biomass source and processing [6,14,15]. The high oxygen content of crude bio-oils results in a heating value half that of petroleum-derived fuels, while the presence of carboxylic acids renders the oils corrosive (pH 2–3) [15] and chemically unstable due to presence of small reactive oxygenates (e.g., unsaturated aldehydes) which may undergo acid-catalysed polymerisation [2]. Crude bio-oils must therefore be upgraded to remove corrosive components and improve stability prior to subsequent hydrodeoxygenation to improve their calorific value.

A range of catalytic routes exist for upgrading pyrolysis bio-oils, including esterification [16], hydrodeoxygenation (HDO) [17], aldol condensation [18] and ketonisation [19]. Each route has advantages and disadvantages. For example, esterification operates at low temperature in the liquid phase but requires an external source of short-chain alcohols, and produces water by-product which must be separated [20]. HDO is effective for the production of cyclic and aliphatic alkanes as liquid fuels, but requires a renewable H_2 input and precious metal catalysts that are stable in (often acidic) bio-oils and coke-resistant [21]. Aldol condensation stabilises bio-oils by converting some reactive oxygenates over solid base catalysts, but does not address the intrinsic acidity of bio-oils that can deactivate HDO catalysts [22]. Ketonisation affords an intermediate deoxygenation step that can be close-coupled to a pyrolysis reactor to upgrade vapours before then condense into a bio-oil, thereby improving bio-oil acidity, and achieving partial deoxygenation [19], although it is also accompanied by a small loss of carbon as CO_2. Ketonisation [22–24], takes place through the condensation of two carboxylic (e.g., acetic) acid molecules to form a ketone (e.g., acetone), CO_2 and H_2O (Figure 1). An important advantage of ketonisation over esterification for acid neutralisation is that the former can be performed in the vapour phase without additional reactants, thus enabling close-coupling to a pyrolysis reactor to upgrade bio-oil vapours prior to their condensation as a bio-oil [24,25]. Ketonisation also facilitates bio-oil deoxygenation and concomitant hydrocarbon chain growth [26], and hence improves the calorific value of the resulting condensate (in addition to its pH and stability).

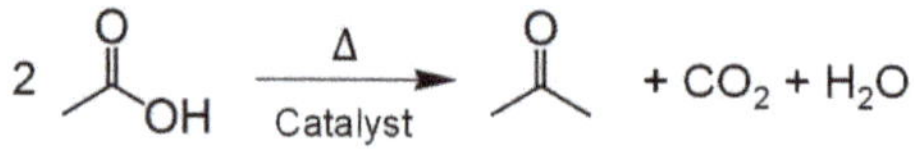

Figure 1. Acetic acid ketonisation.

Ketonisation is widely studied in organic synthesis [25,27], being catalysed by diverse heterogeneous catalysts including alkaline earth metal oxides such as BaO, MgO [28,29], transition metal oxides including MnO_2 [24,30,31], TiO_2 [32,33], Fe_3O_4 [22,24,34], CeO_2 [35,36] and ZrO_2 [37,38] and actinide oxides such as ThO_2 [39]. The mechanism of ketonisation and corresponding sensitivity to catalyst properties remains the subject of ongoing debate [25,40]. Ketonisation over basic and reducible oxide catalysts proceeds via two distinct pathways depending on the lattice energy of the metal oxide, with lower energy lattice (stronger bases) forming stable carboxylates that thermally decompose at elevated temperature (>420 °C) [41] to yield ketones, whereas higher energy lattices favour a lower temperature surface catalysed route [25]. However, there are fewer reports of carboxylic acid ketonisation over zeolites, being limited to HZSM-5, HZSM-11, HZSM-34, HZSM-35, mordenite, erionite and zeolite Beta. Of these, HZSM-5(100) is the most favourable for acetic acid ketonisation to acetone [42], being very selective to xylenols and acetone at moderate reaction temperatures (320 °C) and forming acetone as the major product >350 °C [42]. Zeolite modification by transition metals and lanthanides such as Ce, Co, Ni and Ga increases aromatic product yields during catalytic pyrolysis [43–45], however to our knowledge Ga-promoted zeolites have never been investigated for carboxylic acid ketonisation. Gallium can be introduced into zeolites by incipient wetness impregnation and ion exchange [46], although the choice of preparation method had little impact on aromatic products from Ga/HZSM-5-catalysed fast pyrolysis [46].

HZSM-5 is an attractive catalyst for acetic acid ketonisation, owing to its corrosion resistance, high surface area, commercial availability and ability to stabilise undercoordinated cations and hence tune surface composition and resulting catalytic performance [25]. The presence of strongly acidic protons in zeolites reportedly promotes the formation of surface acyl species, rather than carboxylates, following acid adsorption. Subsequent coupling of a carboxylic acid and surface acyl yields an acid anhydride, which in turn dissociates to liberate CO_2 and a ketone as illustrated in Figure 2 [25,47], although Chang et al propose that ketonisation proceeds by nucleophilic attack of an acylium ion by adsorbed carboxylate [48]. The latter is similar to a ketene intermediate pathway, in with the acylium ion is directly formed by acid protonation and water loss [48]. The relationship between Lewis/Brønsted acid character and ketonisation activity/selectivity over zeolites remains poorly established.

1. Adsorption of carboxylic acid with elimination of water,
2. Surface acyl formation,
3. Surface adsorption of second carboxylic acid,
4. Coupling of adsorbed carboxylic acid with a surface acyl group to form an acid anhydride,
5. Adsorbed anhydride decomposition to form CO_2 and acetone.

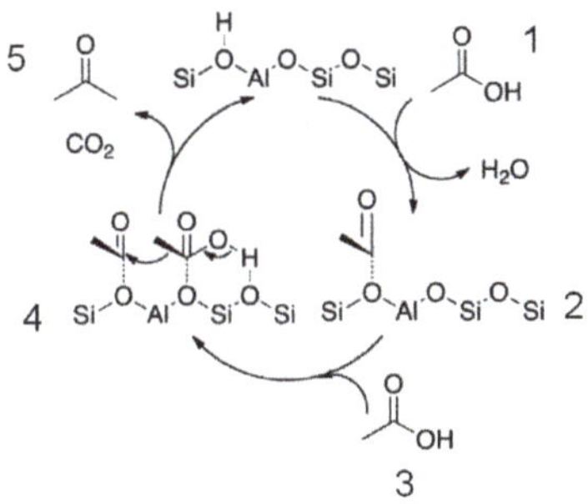

Figure 2. Proposed mechanism for acetic acid ketonisation to acetone over zeolites. Reproduced with permission from reference [47], Copyright © 2016, Elsevier.

Herein, we report the impact of Ga doping on the surface acidity of HZSM-5 and associated reactivity for the continuous vapour phase ketonisation of acetic acid to acetone. Activity and selectivity to acetone were proportional to Ga loading, reflecting the formation of weak Lewis acid sites and suppressed coking.

2. Results and Discussion

2.1. Catalyst Characterisation

Crystalline phases were characterised by powder X-ray diffraction (XRD). Figure S1 shows diffraction patterns for HZSM-5 as a function of Ga loadings to HZSM-5, and pure bulk Ga_2O_3. No reflections associated with gallium oxide phases were observed for any loadings, indicating either the presence of highly dispersed Ga_2O_3 nanoparticles throughout the HZSM-5 pore network, or the exchange of Ga^{3+} with Al^{3+} ions in the framework (or protons at the surface of the zeolite). It is well documented that protons play an important role in regulating the interaction of metal oxides with zeolite surfaces [49,50]. Fang and co-workers report that impregnation favours the formation of Ga_2O_3 and small amounts of GaO^+ at surface of ZSM-5, thereby introducing weak Lewis acid sites [51]. In contrast ion-exchanged Ga/HZSM-5 prepared by refluxing the zeolite in aqueous $Ga(NO_3)_3$ at 70–100 °C [46,52] appears to favour framework dealumination through Ga ion-exchange. Although the latter syntheses resemble our impregnating conditions we cannot conclude whether Ga resides as surface GaO^+ clusters or within the zeolite framework. The reference gallium oxide was phase-pure monoclinic (m-Ga_2O_3) with reflections at $2\theta = 19.0°$, $30.4°$, $30.5°$, $31.8°$, $33.5°$, $35.2°$, $37.4°$, $38.5°$, $42.8°$, $45.9°$, $48.7°$ and $57.5°$ [53,54]. Crystallite sizes of the parent HZSM-5 (Table 1) were independent of Ga loading, and significantly smaller than the zeolite. Nitrogen porosimetry revealed type IV isotherms for xGa/HZSM-5 (Figure S2), with the observed mesoporosity attributed to interparticle voids [55]. Corresponding Brunauer-Emmett-Teller (BET) surface areas, total pore volumes and micropore volumes continuously decreased with increased Ga loading (Table 1), attributed to partial pore blockage, possibly as a result of extra-framework gallium deposition within the micropores [56]. Bulk Ga_2O_3 exhibited a very low surface area <10 $m^2.g^{-1}$.

Table 1. Elemental analysis and physicochemical properties of catalysts.

Catalysts	Ga Loading [a] /wt%	S_{BET} /$m^2 \cdot g^{-1}$	V_{Total} [b] /$mL \cdot g^{-1}$	V_{micro} [c] /$mL \cdot g^{-1}$	Crystallite Size [d] /nm	Acid Loading [e] /$mmol.g^{-1}$	Weak:Strong Acid Site Ratio [f]
HZSM-5	0	427	0.294	0.141	65.7	1.13	0.29
0.5 Ga/HZSM-5	0.3	420	0.290	0.140	55.2	1.09	0.38
3 Ga/HZSM-5	3.0	338	0.249	0.106	50.3	1.03	0.53
10 Ga/HZSM-5	9.0	313	0.210	0.099	62.6	0.80	0.83
Ga_2O_3	75	7.6	0.08	–	26.4	0.14	0.40

[a] ICP-OES, [b] total pore volume at P/Po = 0.98, [c] t-plot method, [d] XRD, [e] propylamine desorption, [f] XPS.

Elemental analysis revealed the surface Ga content was consistently lower than the bulk determined by XPS and ICP-AES analysis respectively (Table S1), consistent with the selective incorporation of gallium inside the HZSM-5 pore network. The formation of large Ga_2O_3 particles on the external surface of zeolite crystallites can be discounted due to the absence of associated XRD patterns. Note that the lower Ga surface versus bulk content for the m-Ga_2O_3 reference reflects oxygen termination of gallium surfaces [57]. O 1s XP spectra of HZSM-5 revealed a single broad peak with a 533 eV binding energy associated with Si–O–Si and Si–O–Al environments [58,59] (Figure 3a), which was unaffected by low levels of Ga doping, but shifted to lower binding energy for 10Ga/HZSM-5, approaching that of Ga_2O_3 at 530.7 eV [60,61]. A similar trend was observed for the Ga $2p_{3/2}$ XP spectra (Figure 3b), which exhibited a single broad peak at 1119.0 eV for low Ga loadings, whose binding energy decreased towards that of m-Ga_2O_3 at 1117.9 eV for 10Ga/HZSM-5 [62]. These data demonstrate that the local environment of gallium in HZSM-5 is chemically distinct from that in bulk Ga_2O_3, consistent with either highly dispersed Ga_2O_3 nanoparticles, or ion-exchange of Ga^{3+} into the zeolite framework [62,63]. Corresponding Al and Si 2p XP spectra of xGa/HZSM-5 (Figure S3a,b) each evidenced a single chemical environment with respective binding energies of approximately 75.1 eV and 103.8 eV, consistent with the literature for HZSM-5 [58,59]. Al and Si 2p peaks shifted to lower binding energy for 10Ga/HZSM-5 indicative of significant ion-exchange and concomitant formation of extra-framework alumina.

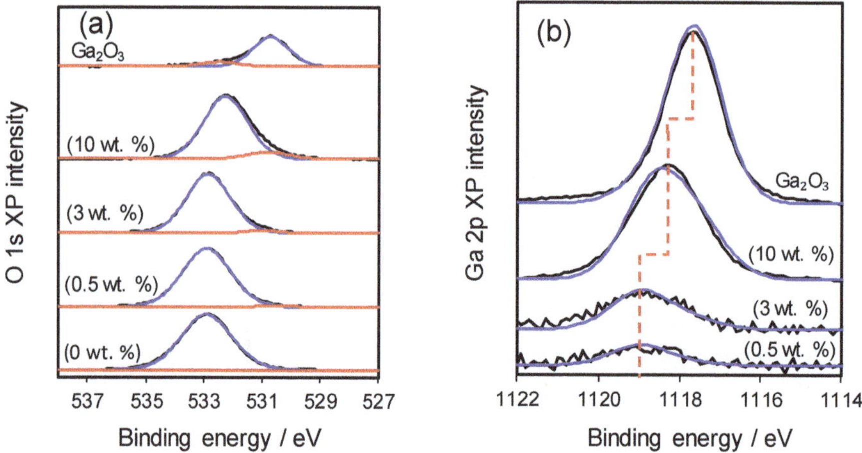

Figure 3. (a) O 1s and (b) Ga 2p XP spectra of xGa/HZSM-5 and Ga_2O_3.

The acid properties of xGa/HZSM-5 and m-Ga_2O_3 were first investigated by diffuse reflectance Fourier transform infrared spectroscopy (DRIFTS) following pyridine adsorption (Figure S4a).

Strong bands at 1444 cm^{-1} and 1545 cm^{-1} were assigned to pyridine bound to Lewis and Brønsted acid sites respectively, the intense band at 1490 cm^{-1} to pyridine bound to both acid sites and the weak 1600 cm^{-1} band to pyridine bound to Lewis acid sites [56]. The relative Lewis/Brønsted acid character was quantified from the ratio of 1444 cm^{-1} and 1545 cm^{-1} band intensities (Figure S4b). The Lewis:Brønsted ratio exhibited a small increase for 10Ga/HZSM-5, in accordance with literature reports [56,64,65]. Corresponding DRIFTS for pyridine on m-Ga$_2$O$_3$ revealed two weak bands at 1452 cm^{-1} and 1614 cm^{-1} indicative of pure Lewis acid character as previously reported [66,67]. Acid strength was subsequently probed by temperature-programmed reaction spectroscopy (TPRS) of propylamine. Reactively-formed propene (arising from propylamine decomposition over acid sites) was evolved in two desorptions at ~480 °C and ~540–555 °C associated with strong and weak acid sites respectively (Figure 4); the former possibly arising from high-index facets or defects [68,69]. The desorption temperature of both peaks was independent of Ga loading, however the ratio of weak:strong acid sites decreased monotonically reaching ~0.83 for 10Ga/HZSM-5. The decreased acid strength was consistent with ion-exchange of less electronegative Ga^{3+} for Al^{3+} into the zeolite surface, which is expected to decrease hydroxyl polarisation and hence Brønsted acid strength [70]. Acid site loadings and weak:strong acid site ratio respectively decreased and increased with Ga loading (Table 1 and Figure S5), however the total acid site density was approximately constant at ~2.6 μmol·m^{-2}. The acid site density of m-Ga$_2$O$_3$ was significantly higher at 18.4 μmol·m^{-2}, with a weak:strong acid site ratio of 0.40 akin to 0.5Ga/HZSM-5, however the absolute Ga loading was far lower than any of the xGa/HZSM-5 materials.

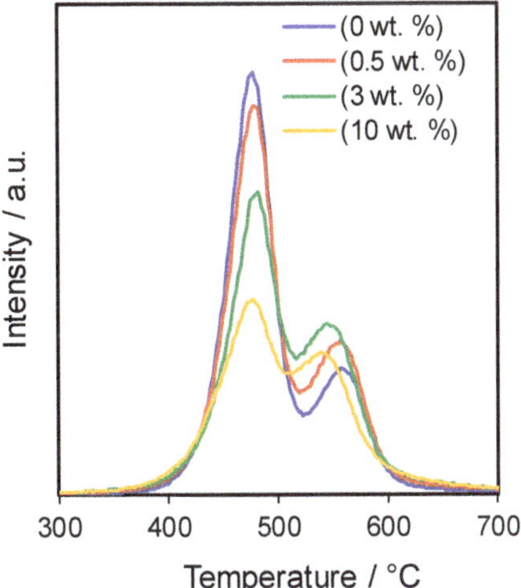

Figure 4. Reactively-formed propene from propylamine temperature-programmed reaction spectroscopy (TPRS) over xGa/HZSM-5.

2.2. Catalytic Activity in Ketonisation

Vapour phase acetic acid ketonisation was subsequently studied over xGa/HZSM-5 in a fixed-bed continuous flow reactor. Turnover frequencies (TOFs) were derived by normalising the steady state rate of acetic acid conversion to the acid site loadings from Table 1. At 350 °C, TOFs were almost independent of Ga loading, exhibiting only a small increase for 10Ga/HZSM-5. Increasing the reaction temperature to 400 °C increased TOFs for all catalysts as previously reported [19,71], with a monotonic rise with Ga loading now apparent (Figure 5). Catalytic reactivity mirrored the weak:strong acid site ratio for both reaction temperatures, indicating that ketonisation preferentially occurs over weak acid sites within xGa/HZSM-5. Limited deactivation (<15%) was observed for 5 h on-stream for all xGa/HZSM-5 catalysts (Figure S6), attributed to pore/site-blocking by coke or strongly bound bidentate carboxylate species [72], or structural changes, whereas the Ga_2O_3 reference exhibited minimal deactivation. Powder XRD revealed negligible change zeolite structure following the reaction (Figure S7), however elemental analysis confirmed the presence of surface carbon post-reaction for all xGa/HZSM-5 catalysts (falling from 12 wt% for the parent HZSM-5 and xGa/HZSM-5 samples to only 1 wt% for Ga_2O_3, Table S2).

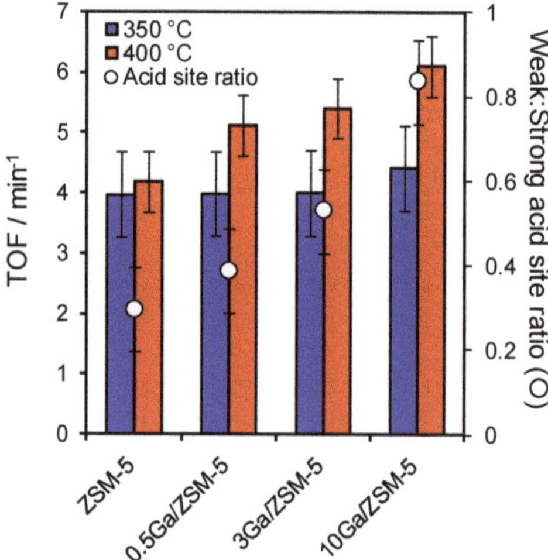

Figure 5. Turnover frequencies (TOFs) for acetic acid ketonisation over xGa/HZSM-5 and corresponding weak:strong acid site ratio. Reaction conditions: 200 mg catalyst, 0.2 mL·min^{-1} acetic acid, 50 mL·min^{-1} N_2, 1 bar.

Acetone selectivity at iso-conversion increased with Ga loading at both 350 °C and 400 °C (Figure 6), concomitant with the rise in weak:strong acid site ratio and Lewis acidity [73]. Vervecken also reported an increase in acetone selectivity >350 °C for acetic acid ketonisation over HZSM-5(100) [42], attributed to a higher activation energy for ketonisation that competing aromatisation (which forms xylenols, phenolics and other aromatics). The maximum acetone selectivity for 10Ga/HZSM-5 was 30%; the principal by-products were CO_2, xylenol, phenol and hydrocarbons [42]. The observation that weak Lewis acid sites and/or related acid-base pairs are the active species for vapour phase acetic acid ketonisation (Figure S8) is consistent with previous experimental [74–77] and computational studies [72]. As discussed in the Introduction, acidic protons in zeolites promote the formation of surface acyl species, which may couple with carboxylate species formed over weaker acid sites to yield an acid anhydride intermediate which in turn decomposes to liberate CO_2 and acetone [25]. However, Chang et al report that ketonisation over HZSM-5 occurs via nucleophilic attack of an acylium ion by carboxylate species [48]; the acylium ion being formed by acid protonation and dehydration [48]. In the case of xGa/HZSM-5, Ga loadings >10 wt% may further increase acetone productivity (and to a lesser extent selectivity) at lower reaction temperature. Although all xGa/HZSM-5 catalysts were stable for 5 h on-stream at 400 °C, future extended ageing and recycling tests are required to optimise formulation and performance.

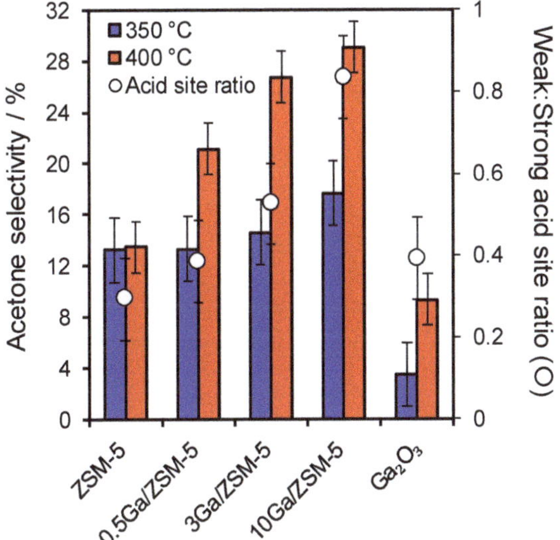

Figure 6. Correlation between acetone selectivity from acetic acid ketonisation at iso-conversion (23% and 29% at 350 °C and 400 °C, respectively) and weak:strong acid site ratio for xGa/HZSM-5. Reaction conditions: 200 mg catalyst, 0.2 mL·min^{-1} acetic acid, 50 mL·min^{-1} N$_2$, 1 bar.

3. Materials and Methods

3.1. Catalyst Synthesis

A commercial HZSM-5 (SiO$_2$:Al$_2$O$_3$ = 30, Zeolyst International, CBV 3024E) was first calcined in air at 550 °C for 4 h to remove any surface residues. Then 3 g of calcined HZSM-5 was subsequently wet-impregnated with appropriate concentrations of a 20 mL aqueous solution of Ga(NO$_3$)$_3$·xH$_2$O (crystalline, 99.9% trace metals basis, Sigma-Aldrich, UK) to prepare Ga gallium zeolites with nominal loadings of 0.5 wt% (0.005 M), 3 wt% (0.03 M) and 10 wt% (0.1 M). In each case the resulting slurry was stirred for 6 h at ambient temperature, dried overnight at 90 °C, and finally calcined at 500 °C under static for 4 h. The resulting catalysts were designated xGa/HZSM-5 where x is the nominal Ga loading. Bulk Ga$_2$O$_3$ (≥99.99% trace metals basis, Sigma-Aldrich, UK) was also calcined at 500 °C for 4 h as a reference material.

3.2. Catalyst Characterisation

The bulk Ga loading was determined by elemental analysis using a Thermofisher iCAP 7000 ICP-OES (Thermofisher, UK). Identification of crystalline phases was performed using a Bruker D8 Advance powder X-ray diffractometer (Bruker, UK) with Cu K$_\alpha$ radiation for angles between 2θ = 10–80° with a step size of 0.04°. Volume averaged particle sizes were estimated from the Scherrer equation using the peak width of characteristic HZSM-5 and Ga$_2$O$_3$ reflections at 2θ = 14.8° and 35.2° respectively. Surface areas, pore size distributions and mesopore volumes were determined by N$_2$ porosimetry using a Quantasorb Nova 4000 e porosimeter and Novawin 11.03 software (Quantachrome, UK). Samples were outgassed in vacuo at 300 °C for 18 h according to Quantachrome recommendations for microporous zeolites prior to analysis, with specific surface areas calculated by applying the Brunauer–Emmet–Teller (BET) model over the range P/P$_0$ = 0.02–0.07 of the adsorption isotherm. Micropore volumes were determined using the t-plot method developed by Lippens and de Boer over the range P/P$_0$ = 0.2–0.5. X-ray photoelectron spectroscopy (XPS) measurements were performed using a Kratos Axis HSi photoelectron spectrometer (Kratos Analytical, UK) equipped with a charge neutraliser and a monochromated Al K$_\alpha$ X-ray source (hν = 1486.7 eV). Spectra were recorded using an analyser pass energy of 20 eV and X-ray power of 225 W at a normal emission. Spectral fitting was performed using CasaXPS version 2.3.14 (Casa Software Ltd, UK), with binding energies corrected to the C 1s peak at 284.6 eV; high resolution C and O 1s, Ga, Al and Si 2p XP spectra fitted using a common Gaussian/Lorentzian line shape. Spectra were Shirley background-subtracted and surface compositions quantified by application of element- and instrument-specific response factors. Errors in surface composition were estimated by varying the background subtraction procedure across reasonable limits and re-calculating fits. The carbon content of spent catalysts was measured using a Thermo Scientific Flash 2000 organic elemental analyser (Thermofisher, UK) calibrated to sulfanilamide, fitted with a Cu/CuO CHNS quartz tube and a thermal conductivity detector. Samples were prepared by adding ~10 mg catalyst and ~2 mg V_2O_5 to tin crucibles.

Acid site loadings and strength were determined by n-propylamine (Sigma Aldrich, UK, ≥99%) temperature-programmed reaction spectroscopy (TPRS) using a Mettler Toledo TGA/DSC 2 STARe system (Mettler Toledo, UK) connected to a Pfeiffer Vacuum ThermoStar GSD 301 T3 (Pfeiffer, UK) benchtop mass spectrometer (MS). Propylamine adsorption was performed by adding liquid n-propylamine to pre-weighed samples (1 mL per 20 mg) and placing in an alumina crucible. Excess physisorbed propylamine was removed by drying in vacuo at 30 °C for 1 h prior to analysis. Samples were heated in the thermogravimetric analyser (TGA) from 40 °C to 800 °C at a ramp rate of 10 °C.min^{-1} under flowing N_2 (40 mL·min^{-1}), with evolved gases analysed by MS to monitor reactively formed propene. Lewis/Brønsted character was determined by diffuse reflectance infrared Fourier transform spectroscopy (DRIFTS) of samples following pyridine adsorption over diluted samples (10 wt% in KBr). Excess physisorbed pyridine was removed in vacuo at 30 °C overnight prior to room temperature measurement using a Nicolet Avatar 370 MCT (Thermofisher, UK) with Smart Collector accessory and liquid nitrogen cooled mercury cadmium telluride (MCT-A) detector. DRIFTS spectra were background-subtracted, and the ratio of the transmitted intensities of the 1450 cm^{-1} and 1540 cm^{-1} peaks used to quantify the ratio between Lewis and Brønsted acid sites.

3.3. Catalytic Ketonisation

Acetic acid ketonisation was performed in a bespoke continuous flow packed-bed reactor with online gas chromatography (GC) analysis. The reactor comprised a 1 cm o.d., quartz tube, within which the catalyst bed was placed centrally and retained by quartz wool plugs. A constant catalyst bed volume of 4 cm^3 was used in all experiments, comprised of 200 mg of catalyst diluted with fused silica granules. The reactor tube was positioned in a temperature-programmable furnace with a thermocouple placed in contact with the catalyst bed. Acetic acid (Sigma-Aldrich, UK, ACS reagent ≥ 99.7%) was fed in a down-flow fashion into the reactor using an Agilent 1260 Infinity Isocratic Pump (Agilent, UK) and N_2 as the carrier gas (50 mL·min^{-1}). All reactor lines were heated to 130 °C to prevent condensation, and a 1 cm diameter metal tube packed with fused silica granules was used to ensure acetic acid vaporisation before the reactor. Products were analysed online by a Varian 3800 GC (Varian, UK) with heated gas-sampling valve, equipped with a BR-Q PLOT column (30 m × 0.53 mm i.d.,). Acetone and acetic acid were detected using a flame ionisation detector (FID). The GC was calibrated for acetic acid and acetone by triplicate injections of 50 µl standard solutions through a split/splitless injector. Acetic acid conversion and selectivity were calculated according to Equations (1) and (2).

$$\text{Conversion} = \frac{n_{AcOH_0} - n_{AcOH}}{n_{AcOH_0}} \times 100 \tag{1}$$

$$\text{Selectivity} = \frac{2 \times n_{Acetone}}{n_{AcOH_0} - n_{AcOH}} \times 100 \tag{2}$$

where n_{AcOH_0} is the initial moles of acetic acid, n_{AcOH} is the final moles acetic acid and $n_{Acetone}$ represents the moles of produced as acetone. Acetone was the primary product over all catalysts. No acetic acid conversion was observed in the absence of a catalyst.

4. Conclusions

A family of Ga-doped HZSM-5 materials were synthesis by wet impregnation as solid acid catalysts for the vapour phase ketonisation of acetic acid, a potential route to upgrading pyrolysis bio-oil vapours. XRD indicates that Ga is either incorporated into the parent zeolite framework or highly dispersed across the zeolite surface as GaO^+ clusters for loadings spanning 0.3–10 wt% Ga doping has little impact on the zeolite textural properties but increased the Lewis acid character concomitant with a decrease in acid strength relative to HZSM-5. Turnover frequencies for acetic acid ketonisation, and acetone selectivity at iso-conversion, were both proportional to the weak:strong acid site ratio, evidencing that ketonisation over xGa/HZSM-5 preferentially occurs over weak (Lewis) acid sites. The most active catalyst was 10 wt% Ga/HZSM-5, which was stable for 5 h on-stream despite significant carbon laydown, with an acetone selectivity of 30%. Acetic acid ketonisation is an attractive route to upgrading biomass pyrolysis vapours through close-coupling with Ga/HZSM-5 catalysts derived from earth abundant elements.

Supplementary Materials: The following are available online at http://www.mdpi.com/2073-4344/9/10/841/s1, Figure S1: XRD patterns of xGa/HZSM-5 and bulk Ga_2O_3; Figure S2: N_2 adsorption-desorption isotherms of xGa/HZSM-5 and Ga_2O_3; Figure S3: (a) Al 2p and (b) Si 2p XP spectra of xGa/HZSM-5; Figure S4: (a) DRIFT spectra of pyridine-saturated xGa/HZSM-5 and Ga_2O_3 and (b) corresponding Lewis:Brønsted acid site ratio (1444 cm^{-1}: 1545 cm^{-1} bands) for xGa/HZSM-5; Figure S5: Density of strong and weak acid sites for xGa/HZSM-5 from propylamine TPRS; Figure S6: Acetic acid conversion over xGa/HZSM-5, and Ga_2O_3 vs time on stream. Reaction conditions: 200 mg catalyst, at 400 °C, 0.2 mL·min^{-1} acetic acid, 50 mL·min^{-1} N_2, 1 bar; Figure S7: XRD patterns of (a) fresh and (b) post-reaction xGa/HZSM-5 and Ga_2O_3; Figure S8: Correlation between acetone selectivity from acetic acid ketonisation at iso-conversion (23% and 29% at 350 °C and 400 °C, respectively) and acid strength from propylamine temperature-programmed reaction spectroscopy (higher temperatures indicate weaker acidity) over xGa/HZSM-5, and Ga_2O_3; Table S1: Surface and bulk composition of xGa/HZSM-5 and Ga_2O_3; Table S2: Carbon content of used xGa/HZSM-5 and Ga_2O_3 after 5 h reaction.

Author Contributions: Conceptualisation, A.H., A.F.L. and K.W.; Formal analysis, H.J., A.O., A.F.L. and K.W.; Investigation, H.J.; Methodology, H.J.; Supervision, M.O., A.H., A.F.L., K.W.; Writing-original draft, H.J., A.O. and M.O.; Writing-review & editing, A.F.L. and K.W.

Funding: We thank the EPSRC (EP/K036548/2, EP/K014676/1, EP/N009924/1) for financial support. K.W. thanks the Royal Society for the award of an Industry Fellowship.

Conflicts of Interest: The authors declare no conflict of interest.

References

1. Wilson, K.; Lee, A.F. Catalyst design for biorefining. *Philos. Trans. R. Soc. A* **2016**, *374*, 20150081. [CrossRef]
2. Alonso, D.M.; Bond, J.Q.; Dumesic, J.A. Catalytic conversion of biomass to biofuels. *Green Chem.* **2010**, *12*, 1493–1513. [CrossRef]
3. Leitner, W.; Klankermayer, J.; Pischinger, S.; Pitsch, H.; Kohse-Höinghaus, K. Advanced Biofuels and Beyond: Chemistry Solutions for Propulsion and Production. *Angew. Chem. Int. Ed.* **2017**, *56*, 5412–5452. [CrossRef]
4. Toor, S.S.; Rosendahl, L.; Rudolf, A. Hydrothermal liquefaction of biomass: A review of subcritical water technologies. *Energy* **2011**, *36*, 2328–2342. [CrossRef]
5. Elliott, D.C.; Biller, P.; Ross, A.B.; Schmidt, A.J.; Jones, S.B. Hydrothermal liquefaction of biomass: Developments from batch to continuous process. *Bioresour. Technol.* **2015**, *178*, 147–156. [CrossRef]
6. Bridgwater, A.V. Review of fast pyrolysis of biomass and product upgrading. *Biomass Bioenergy* **2012**, *38*, 68–94. [CrossRef]
7. Santos, J.; Ouadi, M.; Jahangiri, H.; Hornung, A. Integrated intermediate catalytic pyrolysis of wheat husk. *Food Bioprod. Process.* **2019**, *114*, 23–30. [CrossRef]
8. Yung, M.M.; Jablonski, W.S.; Magrini-Bair, K.A. Review of Catalytic Conditioning of Biomass-Derived Syngas. *Energy Fuels* **2009**, *23*, 1874–1887. [CrossRef]
9. Ouadi, M.; Fivga, A.; Jahangiri, H.; Saghir, M.; Hornung, A. A Review of the Valorization of Paper Industry Wastes by Thermochemical Conversion. *Ind. Eng. Chem. Res.* **2019**, *58*, 15914–15929. [CrossRef]

10. Jahangiri, H.; Bennett, J.; Mahjoubi, P.; Wilson, K.; Gu, S. A review of advanced catalyst development for Fischer-Tropsch synthesis of hydrocarbons from biomass derived syn-gas. *Catal. Sci. Technol.* **2014**, *4*, 2210–2229. [CrossRef]

11. Sartipi, S.; Makkee, M.; Kapteijn, F.; Gascon, J. Catalysis engineering of bifunctional solids for the one-step synthesis of liquid fuels from syngas: A review. *Catal. Sci. Technol.* **2014**, *4*, 893–907. [CrossRef]

12. Hassan, H.; Lim, J.K.; Hameed, B.H. Recent progress on biomass co-pyrolysis conversion into high-quality bio-oil. *Bioresour. Technol.* **2016**, *221*, 645–655. [CrossRef] [PubMed]

13. Papari, S.; Hawboldt, K. A review on the pyrolysis of woody biomass to bio-oil: Focus on kinetic models. *Renew. Sustain. Energy Rev.* **2015**, *52*, 1580–1595. [CrossRef]

14. Sfetsas, T.; Michailof, C.; Lappas, A.; Li, Q.; Kneale, B. Qualitative and quantitative analysis of pyrolysis oil by gas chromatography with flame ionization detection and comprehensive two-dimensional gas chromatography with time-of-flight mass spectrometry. *J. Chromatogr. A* **2011**, *1218*, 3317–3325. [CrossRef]

15. Ciddor, L.; Bennett, J.A.; Hunns, J.A.; Wilson, K.; Lee, A.F. Catalytic upgrading of bio-oils by esterification. *J. Chem. Technol. Biotechnol.* **2015**, *90*, 780–795. [CrossRef]

16. Pirez, C.; Caderon, J.-M.; Dacquin, J.-P.; Lee, A.F.; Wilson, K. Tunable KIT-6 Mesoporous Sulfonic Acid Catalysts for Fatty Acid Esterification. *ACS Catal.* **2012**, *2*, 1607–1614. [CrossRef]

17. Zacher, A.H.; Olarte, M.V.; Santosa, D.M.; Elliott, D.C.; Jones, S.B. A review and perspective of recent bio-oil hydrotreating research. *Green Chem.* **2014**, *16*, 491–515. [CrossRef]

18. Snell, R.W.; Combs, E.; Shanks, B.H. Aldol Condensations Using Bio-oil Model Compounds: The Role of Acid–Base Bi-functionality. *Top. Catal.* **2010**, *53*, 1248–1253. [CrossRef]

19. Jahangiri, H.; Osatiashtiani, A.; Bennett, J.A.; Isaacs, M.A.; Gu, S.; Lee, A.F.; Wilson, K. Zirconia catalysed acetic acid ketonisation for pre-treatment of biomass fast pyrolysis vapours. *Catal. Sci. Technol.* **2018**, *8*, 1134–1141. [CrossRef]

20. Manayil, J.C.; Inocencio, C.V.; Lee, A.F.; Wilson, K. Mesoporous sulfonic acid silicas for pyrolysis bio-oil upgrading via acetic acid esterification. *Green Chem.* **2016**, *18*, 1387–1394. [CrossRef]

21. Marker, T.L.; Felix, L.G.; Linck, M.B.; Roberts, M.J. Integrated hydropyrolysis and hydroconversion (IH 2) for the direct production of gasoline and diesel fuels or blending components from biomass, part 1: Proof of principle testing. *Environ. Prog. Sustain. Energy* **2012**, *31*, 191–199. [CrossRef]

22. Bennett, J.A.; Parlett, C.M.A.; Isaacs, M.A.; Durndell, L.J.; Olivi, L.; Lee, A.F.; Wilson, K. Acetic Acid Ketonization over Fe$_3$O$_4$/SiO$_2$ for Pyrolysis Bio-Oil Upgrading. *ChemCatChem* **2017**, *9*, 1648–1654. [CrossRef]

23. Pham, T.N.; Shi, D.; Resasco, D.E. Kinetics and Mechanism of Ketonization of Acetic Acid on Ru/TiO2 Catalyst. *Top. Catal.* **2014**, *57*, 706–714. [CrossRef]

24. Heracleous, E.; Gu, D.; Schüth, F.; Bennett, J.A.; Isaacs, M.A.; Lee, A.F.; Wilson, K.; Lappas, A.A. Bio-oil upgrading via vapor-phase ketonization over nanostructured FeO x and MnO x: Catalytic performance and mechanistic insight. *Biomass Convers. Biorefinery* **2017**, *7*, 319–329. [CrossRef]

25. Pham, T.N.; Sooknoi, T.; Crossley, S.P.; Resasco, D.E. Ketonization of Carboxylic Acids: Mechanisms, Catalysts, and Implications for Biomass Conversion. *ACS Catal.* **2013**, *3*, 2456–2473. [CrossRef]

26. Gaertner, C.A.; Serrano-Ruiz, J.C.; Braden, D.J.; Dumesic, J.A. Ketonization Reactions of Carboxylic Acids and Esters over Ceria–Zirconia as Biomass-Upgrading Processes. *Ind. Eng. Chem. Res.* **2010**, *49*, 6027–6033. [CrossRef]

27. Wortz, C.G. Process for the Production of Ketones. U.S. Patent 2,108,156, 15 February 1938.

28. Nagashima, O.; Sato, S.; Takahashi, R.; Sodesawa, T. Ketonization of carboxylic acids over CeO2-based composite oxides. *J. Mol. Catal. A Chem.* **2005**, *227*, 231–239. [CrossRef]

29. Pestman, R.; Koster, R.M.; van Duijne, A.; Pieterse, J.A.Z.; Ponec, V. Reactions of carboxylic acids on oxides. 2. Bimolecular reaction of aliphatic acids to ketones. *J. Catal.* **1997**, *168*, 265–272. [CrossRef]

30. Gliński, M.; Kijeński, J. Catalytic Ketonization of Carboxylic Acids Synthesis of Saturated and Unsaturated Ketones. *React. Kinet. Catal. Lett.* **2000**, *69*, 123–128. [CrossRef]

31. Parida, K.M.; Samal, A.; Das, N.N. Catalytic ketonization of monocarboxylic acids over Indian Ocean manganese nodules. *Appl. Catal. A Gen.* **1998**, *166*, 201–205. [CrossRef]

32. Martinez, R.; Huff, M.C.; Barteau, M.A. Ketonization of acetic acid on titania-functionalized silica monoliths. *J. Catal.* **2004**, *222*, 404–409. [CrossRef]

33. Kim, K.S.; Barteau, M.A. Structure and Composition Requirements for Deoxygenation, Dehydration, and Ketonization Reactions of Carboxylic-Acids on Tio2 (001) Single-Crystal Surfaces. *J. Catal.* **1990**, *125*, 353–375. [CrossRef]

34. Pestman, R.; Koster, R.M.; Pieterse, J.A.Z.; Ponec, V. Reactions of carboxylic acids on oxides. 1. Selective hydrogenation of acetic acid to acetaldehyde. *J. Catal.* **1997**, *168*, 255–264. [CrossRef]

35. Randery, S.D.; Warren, J.S.; Dooley, K.M. Cerium oxide-based catalysts for production of ketones by acid condensation. *Appl. Catal. A Gen.* **2002**, *226*, 265–280. [CrossRef]

36. Dooley, K.M.; Bhat, A.K.; Plaisance, C.P.; Roy, A.D. Ketones from acid condensation using supported CeO2 catalysts: Effect of additives. *Appl. Catal. A Gen.* **2007**, *320*, 122–133. [CrossRef]

37. Gliński, M.; Kijeński, J. Decarboxylative coupling of heptanoic acid. Manganese, cerium and zirconium oxides as catalysts. *Appl. Catal. A Gen.* **2000**, *190*, 87–91.

38. Okumura, K.; Iwasawa, Y. Zirconium Oxides Dispersed on Silica Derived from Cp2ZrCl2, [(i-PrCp)2ZrH(μ-H)]2, and Zr(OEt)4Characterized by X-Ray Absorption Fine Structure and Catalytic Ketonization of Acetic Acid. *J. Catal.* **1996**, *164*, 440–448. [CrossRef]

39. Kistler, S.S.; Swann, S.; Appel, E.G. Aërogel Catalysts—Thoria: Preparation of Catalyst and Conversions of Organic Acids to Ketones. *Ind. Eng. Chem.* **1934**, *26*, 388–391. [CrossRef]

40. Pacchioni, G. Ketonization of Carboxylic Acids in Biomass Conversion over TiO2 and ZrO2 Surfaces: A DFT Perspective. *ACS Catal.* **2014**, *4*, 2874–2888. [CrossRef]

41. Patil, K.C.; Chandrashekhar, G.V.; George, M.V.; Rao, C.N.R. Infrared spectra and thermal decompositions of metal acetates and dicarboxylates. *Can. J. Chem.* **1968**, *46*, 257–265. [CrossRef]

42. Vervecken, M.; Servotte, Y.; Wydoodt, M.; Jacobs, L.; Martens, J.A.; Jacobs, P.A. Zeolite-Induced Selectivity in the Conversion of the Lower Aliphatic Carboxylic Acids. In *Chemical Reactions in Organic and Inorganic Constrained Systems*; Setton, R., Ed.; Springer: Dordrecht, The Netherlands, 1986; pp. 95–114.

43. Iliopoulou, E.F.; Stefanidis, S.D.; Kalogiannis, K.G.; Delimitis, A.; Lappas, A.A.; Triantafyllidis, K.S. Catalytic upgrading of biomass pyrolysis vapors using transition metal-modified ZSM-5 zeolite. *Appl. Catal. B-Environ.* **2012**, *127*, 281–290. [CrossRef]

44. French, R.; Czernik, S. Catalytic pyrolysis of biomass for biofuels production. *Fuel Process. Technol.* **2010**, *91*, 25–32. [CrossRef]

45. Neumann, G.T.; Hicks, J.C. Effects of Cerium and Aluminum in Cerium-Containing Hierarchical HZSM-5 Catalysts for Biomass Upgrading. *Top. Catal.* **2012**, *55*, 196–208. [CrossRef]

46. Cheng, Y.T.; Jae, J.; Shi, J.; Fan, W.; Huber, G.W. Production of Renewable Aromatic Compounds by Catalytic Fast Pyrolysis of Lignocellulosic Biomass with Bifunctional Ga/ZSM-5 Catalysts. *Angew. Chem. Int. Ed.* **2012**, *51*, 1387–1390. [CrossRef] [PubMed]

47. Gumidyala, A.; Sooknoi, T.; Crossley, S. Selective ketonization of acetic acid over HZSM-5: The importance of acyl species and the influence of water. *J. Catal.* **2016**, *340*, 76–84. [CrossRef]

48. Chang, C.D.; Chen, N.Y.; Koenig, L.R.; Walsh, D.E. Synergism in Acetic-Acid Methanol Reactions over Zsm-5 Zeolites. *Abstr. Pap. Am. Chem. Soc.* **1983**, *185*, 49-Fuel.

49. Tessonnier, J.-P.; Louis, B.; Walspurger, S.; Sommer, J.; Ledoux, M.-J.; Pham-Huu, C. Quantitative Measurement of the Brönsted Acid Sites in Solid Acids: Toward a Single-Site Design of Mo-Modified ZSM-5 Zeolite. *J. Phys. Chem. B* **2006**, *110*, 10390–10395. [CrossRef]

50. Li, B.; Li, S.; Li, N.; Chen, H.; Zhang, W.; Bao, X.; Lin, B. Structure and acidity of Mo/ZSM-5 synthesized by solid state reaction for methane dehydrogenation and aromatization. *Microporous Mesoporous Mater.* **2006**, *88*, 244–253. [CrossRef]

51. Fang, Y.; Su, X.; Bai, X.; Wu, W.; Wang, G.; Xiao, L.; Yu, A. Aromatization over nanosized Ga-containing ZSM-5 zeolites prepared by different methods: Effect of acidity of active Ga species on the catalytic performance. *J. Energy Chem.* **2017**, *26*, 768–775. [CrossRef]

52. Amin, N.A.S.; Ali, A. Characterization of Modified HZSM–5 with Gallium and its Reactivity in Direct Conversion of Methane to Liquid Hydrocarbons. *J. Teknol.* **2001**, *35*, 21–30.

53. Li, J.Y.; Chen, X.L.; Qiao, Z.Y.; He, M.; Li, H. Large-scale synthesis of single-crystalline beta-Ga2O3 nanoribbons, nanosheets and nanowires. *J. Phys. Condens. Matter* **2001**, *13*, L937–L941. [CrossRef]

54. Huang, C.-C.; Yeh, C.-S. GaOOH, and [small beta]- and [gamma]-Ga2O3 nanowires: Preparation and photoluminescence. *New J. Chem.* **2010**, *34*, 103–107. [CrossRef]

55. Wang, S.; Yin, Q.; Guo, J.; Ru, B.; Zhu, L. Improved Fischer—Tropsch synthesis for gasoline over Ru, Ni promoted Co/HZSM-5 catalysts. *Fuel* **2013**, *108*, 597–603. [CrossRef]

56. Rodrigues, V.D.; Eon, J.G.; Faro, A.C. Correlations between Dispersion, Acidity, Reducibility, and Propane Aromatization Activity of Gallium Species Supported on HZSM5 Zeolites. *J. Phys. Chem. C* **2010**, *114*, 4557–4567. [CrossRef]

57. Tamba, D.; Kubo, O.; Oda, M.; Osaka, S.; Takahashi, K.; Tabata, H.; Kaneko, K.; Fujita, S.; Katayama, M. Surface termination structure of α-Ga2O3 film grown by mist chemical vapor deposition. *Appl. Phys. Lett.* **2016**, *108*, 251602. [CrossRef]

58. Grunert, W.; Muhler, M.; Schroder, K.P.; Sauer, J.; Schlogl, R. Investigations of Zeolites by Photoelectron and Ion-Scattering Spectroscopy.2. A New Interpretation of Xps Binding-Energy Shifts in Zeolites. *J. Phys. Chem.* **1994**, *98*, 10920–10929. [CrossRef]

59. Borade, R.B.; Clearfield, A. Characterization of Acid Sites in Beta and Zsm-20 Zeolites. *J. Phys. Chem.* **1992**, *96*, 6729–6737. [CrossRef]

60. Guo, D.; Wu, Z.; An, Y.; Li, P.; Wang, P.; Chu, X.; Guo, X.; Zhi, Y.; Lei, M.; Li, L. Unipolar resistive switching behavior of amorphous gallium oxide thin films for nonvolatile memory applications. *Appl. Phys. Lett.* **2015**, *106*, 042105. [CrossRef]

61. Wei, W.; Qin, Z.; Fan, S.; Li, Z.; Shi, K.; Zhu, Q.; Zhang, G. Valence band offset of β-Ga2O3/wurtzite GaN heterostructure measured by X-ray photoelectron spectroscopy. *Nanoscale Res. Lett.* **2012**, *7*, 1–5. [CrossRef]

62. Xiao, H.; Zhang, J.F.; Wang, X.X.; Zhang, Q.D.; Xie, H.J.; Han, Y.Z.; Tan, Y.S. A highly efficient Ga/ZSM-5 catalyst prepared by formic acid impregnation and in situ treatment for propane aromatization. *Catal. Sci. Technol.* **2015**, *5*, 4081–4090. [CrossRef]

63. Carli, R.; Bianchi, C.L. Xps Analysis of Gallium Oxides. *Appl. Surf. Sci.* **1994**, *74*, 99–102. [CrossRef]

64. Altwasser, S.; Raichle, A.; Traa, Y.; Weitkamp, J. Preparation of gallium-containing catalysts by solid-state reaction of acidic Zeolites with elemental gallium. *Chem. Eng. Technol.* **2004**, *27*, 1262–1265. [CrossRef]

65. Anunziata, O.A.; Pierella, L.B. Nature of the Active-Sites in H-Zsm-11 Zeolite Modified with $Zn^{(2+)}$ and $Ga^{(3+)}$. *Catal. Lett.* **1993**, *19*, 143–151. [CrossRef]

66. Lavalley, J.C.; Daturi, M.; Montouillout, V.; Clet, G.; Arean, C.O.; Delgado, M.R.; Sahibed-dine, A. Unexpected similarities between the surface chemistry of cubic and hexagonal gallia polymorphs. *Phys. Chem. Chem. Phys.* **2003**, *5*, 1301–1305. [CrossRef]

67. Vimont, A.; Lavalley, J.C.; Sahibed-Dine, A.; Arean, C.O.; Delgado, M.R.; Daturi, M. Infrared spectroscopic study on the surface properties of gamma-gallium oxide as compared to those of gamma-alumina. *J. Phys. Chem. B* **2005**, *109*, 9656–9664. [CrossRef] [PubMed]

68. Phumman, P.; Niamlang, S.; Sirivat, A. Fabrication of Poly(p-Phenylene)/Zeolite Composites and Their Responses Towards Ammonia. *Sensors* **2009**, *9*, 8031–8046. [CrossRef]

69. Morterra, C.; Cerrato, G.; Ferroni, L.; Negro, A.; Montanaro, L. Surface characterization of tetragonal ZrO2. *Appl. Surf. Sci.* **1993**, *65*, 257–264. [CrossRef]

70. Dompas, D.H.; Mortier, W.J.; Kenter, O.C.H.; Janssen, M.J.G.; Verduijn, J.P. The influence of framework-gallium in zeolites: Electronegativity and infrared spectroscopic study. *J. Catal.* **1991**, *129*, 19–24. [CrossRef]

71. Pulido, A.; Oliver-Tomas, B.; Renz, M.; Boronat, M.; Corma, A. Ketonic Decarboxylation Reaction Mechanism: A Combined Experimental and DFT Study. *ChemSusChem* **2013**, *6*, 141–151. [CrossRef]

72. Wang, S.; Iglesia, E. Experimental and Theoretical Evidence for the Reactivity of Bound Intermediates in Ketonization of Carboxylic Acids and Consequences of Acid–Base Properties of Oxide Catalysts. *J. Phys. Chem. C* **2017**, *121*, 18030–18046. [CrossRef]

73. Schreiber, M.W.; Plaisance, C.P.; Baumgartl, M.; Reuter, K.; Jentys, A.; Bermejo-Deval, R.; Lercher, J.A. Lewis-Bronsted Acid Pairs in Ga/H-ZSM-5 To Catalyze Dehydrogenation of Light Alkanes. *J. Am. Chem. Soc.* **2018**, *140*, 4849–4859. [CrossRef] [PubMed]

74. Mattsson, A.; Österlund, L. Adsorption and Photoinduced Decomposition of Acetone and Acetic Acid on Anatase, Brookite, and Rutile TiO2 Nanoparticles. *J. Phys. Chem. C* **2010**, *114*, 14121–14132. [CrossRef]

75. Ma, Q.; Liu, Y.; Liu, C.; He, H. Heterogeneous reaction of acetic acid on MgO, [small alpha]-Al_2O_3, and $CaCO_3$ and the effect on the hygroscopic behaviour of these particles. *Phys. Chem. Chem. Phys.* **2012**, *14*, 8403–8409. [CrossRef] [PubMed]

76.   Finocchio, E.; Willey, R.J.; Busca, G.; Lorenzelli, V. FTIR studies on the selective oxidation and combustion of light hydrocarbons at metal oxide surfaces Part 3.-Comparison of the oxidation of C3 organic compounds over Co3O4, MgCr2O4 and CuO, Journal of the Chemical Society. *Faraday Trans.* **1997**, *93*, 175–180. [CrossRef]
77.   Geiculescu, A.C.; Spencer, H.G. Thermal Decomposition and Crystallization of Aqueous Sol-Gel Derived Zirconium Acetate Gels: Effects of the Additive Anions. *J. Sol-Gel Sci. Technol.* **2000**, *17*, 25–35. [CrossRef]

 *catalysts*

Article

Ru-Catalyzed Oxidative Cleavage of Guaiacyl Glycerol-*β*-Guaiacyl Ether-a Representative *β*-O-4 Lignin Model Compound

Mayra Melián-Rodríguez [1], **Shunmugavel Saravanamurugan** [1,2], **Sebastian Meier** [1], **Søren Kegnæs** [1] **and Anders Riisager** [1,*]

[1]   Centre for Catalysis and Sustainable Chemistry, Department of Chemistry, Building 207, Technical University of Denmark, DK-2800 Kgs. Lyngby, Denmark; mayra.melian@gmail.com (M.M.-R.); srsaravana78@gmail.com (S.S.); semei@kemi.dtu.dk (S.M.); skk@kemi.dtu.dk (S.K.)

[2]   Laboratory of Bioproduct Chemistry, Center of Innovative and Applied Bioprocessing (CIAB), Sector-81 (Knowledge City), Mohali-140 306 Punjab, India

*   Correspondence: ar@kemi.dtu.dk; Tel.: +45-4525-2233

Received: 19 September 2019; Accepted: 30 September 2019; Published: 3 October 2019

Abstract: The introduction of efficient and selective catalytic methods for aerobic oxidation of lignin and lignin model compounds to aromatics can extend the role of lignin applications in biorefineries. The current study focussed on the catalytic oxidative transformation of guaiacyl glycerol-β-guaiacyl ether (GGGE)–a β-O-4 lignin model compound to produce basic aromatic compounds (guaiacol, vanillin and vanillic acid) using metal-supported catalysts. Ru/Al_2O_3, prepared with ruthenium(IV) oxide hydrate, showed the highest yields of the desired products (~60%) in acetonitrile in a batch reactor at 160 °C and 5-bar of 20% oxygen in argon. Alternative catalysts containing other transition metals (Ag, Fe, Mn, Co and Cu) supported on alumina, and ruthenium catalysts based on alternative supports (silica, spinel, HY zeolite and zirconia) gave significantly lower activities compared to Ru/Al_2O_3 at identical reaction conditions. Moreover, the Ru/Al_2O_3 catalyst was successfully reused in five consecutive reaction runs with only a minor decrease in catalytic performance.

Keywords: aerobic oxidation; ruthenium; heterogeneous catalysis; lignin valorization; guaiacyl glycerol-β-guaiacyl ether

---

1. Introduction

Lignocellulosic biomass, predominantly comprised of cellulose, hemicellulose and lignin, is an abundant and renewable carbon-based alternative to fossil resources [1]. Several applications with nanocomposites of cellulose and hemicelluloses have been reported, for example water-purification [2], (bio)sensing [3] and anti-microbial treatment [4]. However, lately, the transformation of materials, as well as their monomeric C_6 and C_5 carbohydrates, to value-added chemicals and fuels have been studied extensively [5–15]. In comparison, lignin has received much less attention as feedstock, possibly due to its complex polymeric structure and lower reactivity, even though it is a major part of lignocellulosic biomass, typically 30% by weight and 40% by energy content. However, recent developments have demonstrated lignin to be a potentially important feedstock for producing chemicals, especially aromatic compounds [16–23]. This progress is important for the future of biorefineries as valorization of the entire biomass substrate improves economic viability.

The direct transformation of lignin usually requires mechanical pretreatment and harsh reaction conditions, due to its poor solubility and complex heterogeneous structure. Thus, in order to understand the reactivity of lignin, in general, various lignin model compounds containing different structural linkages, such as α-O-4 and β-O-4, have been widely used as substrates. Among the different linkages, the most abundant structural unit in lignin is the β-O-4 (Figure 1), representing approximately 60% of hardwood and 45–50% of softwood [24].

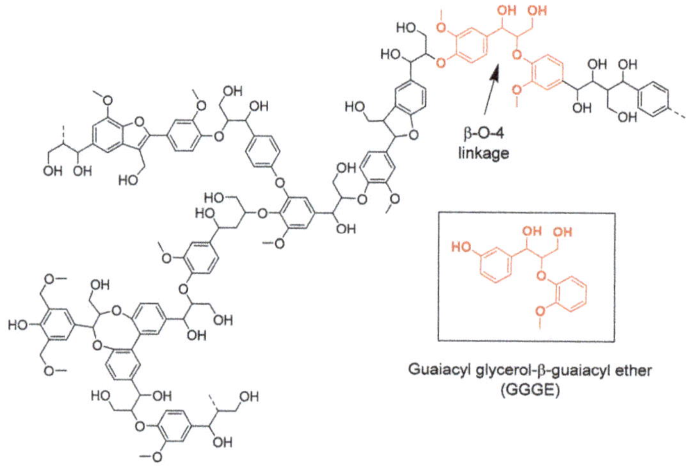

Figure 1. Schematic representation of typical lignin fragments and the corresponding β-O-4 lignin model compound guaiacyl glycerol-β-guaiacyl ether (GGGE).

Several studies have converted lignin and simple aromatic model compounds by catalytic oxidation under typically harsh reaction conditions and afforded low yield and/or selectivity to the targeted products, whereas dimeric lignin model compounds containing β-O-4 linkages have only been scarcely studied [25,26]. In this context, the exploration of bulky β-O-4 lignin model compounds may provide valuable insight that can be transferred to the reactivity of the complex lignin molecule, especially on cleavage of β-O-4 linkages and further reactivity of the formed monomers [18].

Aerobic oxidation of the bulky lignin model compound guaiacyl glycerol-β-guaiacyl ether (GGGE) has primarily been examined with vanadium-based homogeneous catalyst systems. Hence, Son et al. reported a vanadium-based catalyst for the non-oxidative C-O bond cleavage of dimeric lignin model compounds with a conversion of 80% [27]. Furthermore, vanadium complexes showed promising catalytic activity for oxidative C–C bond cleavage of GGGE compounds, and promoted multistep reactions affording C–C and C–O cleavage products from alternative dimeric β-O-4 lignin model compounds [28–30]. Alternatively, Rahimi et al. introduced a two-step, metal-free organocatalytic method using first 4-acetamido-TEMPO as the catalyst for chemoselective aerobic oxidation of the secondary benzylic alcohols in GGGE followed by C–C cleavage using H_2O_2 [23]. In addition, Leitner et al. more recently introduced a highly active and selective ruthenium-complex catalyst system for C–C bond cleavage of β-O-4 lignin linkages involving a dehydrogenation-initiated retro-aldol reaction [31]. Despite these promising homogeneous catalyst systems for selective cleavage of C–C and C–O bonds, separation and recyclability of the catalysts remains cumbersome for such catalytic systems [30,32]. In contrast, solid catalysts with supported metals/metal oxides can easily be recovered from liquid reaction mixtures, and can often be recycled multiple times with preservation of the catalytic performance.

Supported ruthenium catalysts such as Ru/alumina, are an effective and reusable heterogeneous catalyst system for aerobic oxidation of both activated and non-activated alcohols in the presence of sulfur, nitrogen and carbon-carbon double bonds [33], and they are therefore interesting in the context of oxidative lignin valorization. We previously examined such catalysts for the aerobic oxidation of the lignin model compound veratryl alcohol to veratraldehyde in water and methanol with good results [34]. In the present study, analogous ruthenium supported catalysts with the different supports γ-alumina (Ru/Al$_2$O$_3$), silica (Ru/SiO$_2$), zirconia (Ru/ZrO$_2$), spinel (Ru/MgAl$_2$O$_4$) and USY zeolite (Ru/HY), which were prepared, characterized and applied for aerobic oxidative cleavage of GGGE in acetonitrile to produce guaiacol, vanillin and vanillic acid under mild reaction conditions (Scheme 1). Acetonitrile was preferred as the reaction solvent to ensure dissolution of GGGE and the products, and reaction parameters such as temperature and time were optimized for the promising catalyst Ru/Al$_2$O$_3$ to increase the selectivity for the desirable products, and the recyclability of the catalyst examined by performing consecutive reaction runs. For comparison, other alumina-supported metal catalysts M/Al$_2$O$_3$ (M = Mn, Ag, Cu and Fe) were prepared and evaluated.

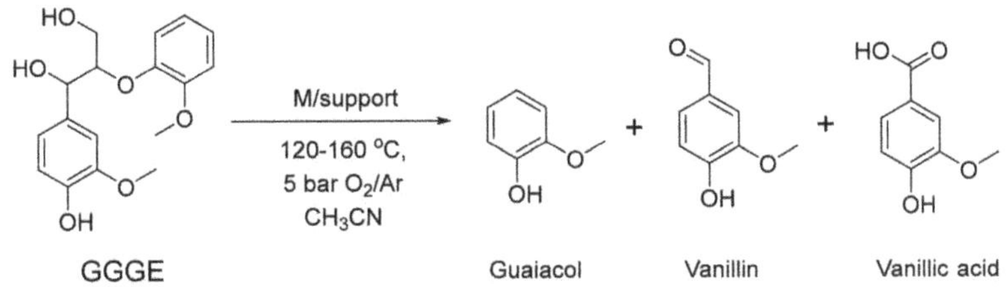

Scheme 1. Catalytic aerobic oxidation of guaiacyl glycerol-β-guaiacyl ether (GGGE) to guaiacol, vanillin and vanillic acid with supported metal catalysts.

2. Results and Discussion

2.1. Catalyst Screening

Catalysts with 5 wt.% M/Al$_2$O$_3$ (M = Ru, Ag, Fe, Mn, Cu) were initially tested for the aerobic oxidation of GGGE into guaiacol, vanillin and vanillic acid in acetonitrile at 160 °C with 5 bar 20% oxygen in argon for 20 h. The results are presented in Table 1.

Table 1. Catalytic oxidation of guaiacyl glycerol-β-guaiacyl ether (GGGE) over different metal/alumina catalysts [a].

Entry	Catalyst	BET Surface Area (m^2/g)	GGGE Conv. (%)	Product Yield (%)		
				Guaiacol	Vanillin	Vanillic Acid
1	-	–	70	11	<1	<1
2	Al$_2$O$_3$	204	73	15	<1	<1
3	5 wt.% Fe/Al$_2$O$_3$	154	92	23	7	4
4	5 wt.% Mn/Al$_2$O$_3$	152	>99	21	8	9
5	5 wt.% Cu/Al$_2$O$_3$	158	>99	9	3	<1
6	5 wt.% Ag/Al$_2$O$_3$	164	>99	27	10	8
7	5 wt.% Ru/Al$_2$O$_3$ (1) [b]	148	>99	28	11	11
8	5 wt.% Ru/Al$_2$O$_3$ (1) [b,c]	148	-	3	42	6
9	3 wt.% Ru/Al$_2$O$_3$ (1) [b]	160	>99	20	8	6
10	1 wt.% Ru/Al$_2$O$_3$ (1) [b]	157	>99	18	8	6
11	5 wt.% Ru/Al$_2$O$_3$ (2) [d]	152	>99	24	9	3
12	5 wt.% Ru/Al$_2$O$_3$ (3) [e]	166	>99	34	13	11

[a] Reaction conditions: 10 mL 0.017 M GGGE in acetonitrile, 40 mg catalyst, 160 °C, 5 bar (20% oxygen in argon), 20 h. [b] Catalyst prepared using ruthenium(III) chloride. [c] 100 mg vanillin used as a substrate in 10 mL acetonitrile. [d] Catalyst prepared using ruthenium(III) acetylacetonate. [e] Catalyst prepared using ruthenium(IV) oxide.

In blank experiments with alumina support alone or without the catalyst about 70% of GGGE was converted, however, as expected only low yields of the desired product guaiacol (<15%) and traces of vanillin and vanillic acid (<1%) were obtained, suggesting that GGGE was possibly transformed into (unidentified) byproducts, e.g., polymers (Table 1, entries 1 and 2). In contrast, when employing the Ru/Al$_2$O$_3$ and Ag/Al$_2$O$_3$ catalysts the yield of guaiacol improved to 28 and 27%, respectively, with the former catalyst performing the best and providing the highest yields of both vanillin (11%) and vanillic acid (11%) (Table 1, entries 6 and 7). When applying the other 5 wt.% metal/alumina catalysts the GGGE conversion remained also close to quantitative, but lower yields of the targeted products were obtained (Table 1, entries 3–5). Notably, in the case of 5 wt.% Cu/Al_2O_3 the transformation of GGGE to (unidentified) byproducts was even promoted compared to the blank experiments. With the 5 wt.% Ru/Al_2O_3 (1) catalyst, an additional experiment was performed using vanillin as the starting substrate instead of GGGE (under similar reaction conditions) to examine whether guaiacol was partly formed from vanillin by consecutive decarbonylation of vanillin (Table 1, entry 8). The obtained results showed a poor yield of guaiacol (3%) and vanillic acid (6%) along with a moderate conversion of vanillin (58%), inferring that guaiacol predominantly formed from the cleaving of the β-O-4 linkage in GGGE and vanillic acid predominantly formed from oxidation of vanillin.

2.2. Effect of Ru Precursor, Ru Loading and Catalyst Support

For metal/metal oxide catalysts it is typically found that the metal loading, metal precursor and support material influence the catalytic performance through changes in the physical- and structural properties [35–39]. Accordingly, Ru/Al_2O_3 catalysts were prepared with different metal loadings and supports using ruthenium(III) chloride precursor and the resultant catalysts were tested for the GGGE oxidation. Similarly, catalysts with 5 wt.% Ru were prepared with the three different precursors ruthenium(III) chloride, ruthenium(III) acetylacetonate and ruthenium(IV) oxide (Ru/Al_2O_3 (1), Ru/Al_2O_3 (2) and Ru/Al_2O_3 (3), respectively) and the resultant catalysts were tested (Tables 1 and 2).

For the catalysts with 1 and 3 wt.% Ru loading, the yields of guaiacol (18–20%) as well as vanillin (8%) and vanillic acid (6%) were lower than for 5 wt.% Ru/Al_2O_3 (1), confirming that the catalytic activity was dependent on the amount of the metal inventory (Table 1, entries 9 and 10). TEM images of the catalysts further showed that the former catalysts possessed relatively large Ru-particles of sizes 80–100 nm, while the 5 wt.% catalyst had particles with sizes of 40–60 nm, implying that smaller particles improved the catalytic conversion of GGGE to guaiacol, vanillin and vanillic acid (Figure S1). With the preferred 5 wt.% Ru metal loading, the yields of guaiacol, vanillin and vanillic acid were lower when ruthenium(III) acetylacetonate (i.e. Ru/Al_2O_3 (2)) was used as the precursor instead of ruthenium(III) chloride (Table 1, entry 11). In contrast, a slightly improved catalytic activity in terms of vanillin (34%) and guaiacol yields (13%) was observed when employing ruthenium(IV) oxide precursor (i.e. Ru/Al_2O_3 (3)) compared to Ru/Al_2O_3 (1) (Table 1, entry 12), whereas the yield of vanillic acid remained unchanged (11%). TEM images revealed that the Ru-particle sizes of the catalysts decreased in the order Ru/Al_2O_3 (2) (100–200 nm) > Ru/Al_2O_3 (1) (40–60 nm) > Ru/Al_2O_3 (3) (10–40 nm) (Figure 2). This size order followed the order of catalytic performance toward formation of guaiacol, vanillin and vanillic acid, corroborating that the metal precursor influenced particle formation, and the corresponding catalytic performance, as also observed previously when catalysts were prepared with different metal precursors [38].

The BET surface areas of the Ru catalysts as well as the other metal-based catalysts (148–166 m^2/g) were significantly lower than the alumina support alone (204 m^2/g), indicating some pore blocking in the catalysts by metal oxide particles and therefore likely also some change in pore size distributions for the different catalysts. For the 5 wt.% Ru/Al_2O_3 (1) catalyst with the lowest surface area, the increased acidity of the ruthenium(III) chloride precursor solution may have also possibly contributed in part to lowering the surface area by alteration of the Al_2O_3 support surface. Notably, for analogous Ru/Al_2O_3 catalysts prepared by similar methods a comparable relative decrease (20–30%) in surface area has also been found [37].

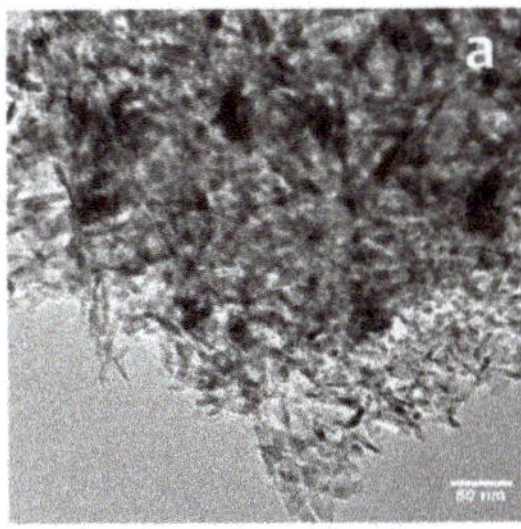
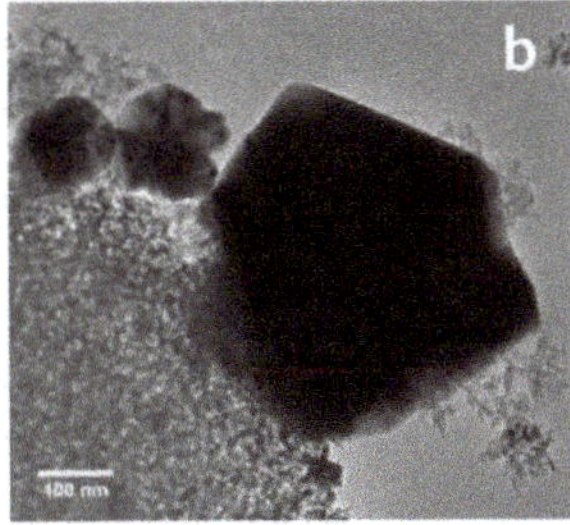
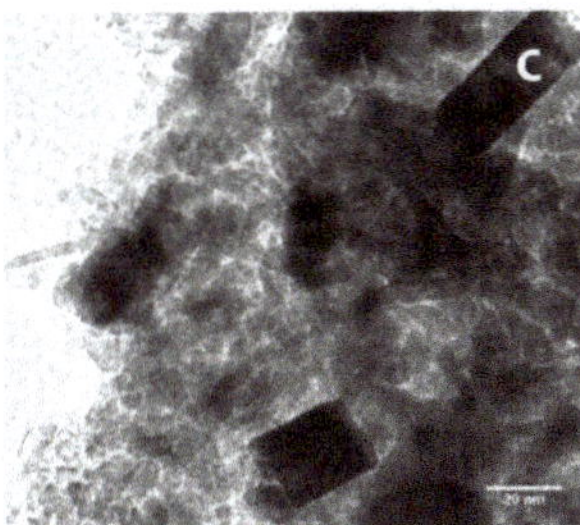

Figure 2. High-resolution TEM images of 5 wt.% Ru/Al$_2$O$_3$ (1) (**a**), 5 wt.% Ru/Al$_2$O$_3$ (2) (**b**) and 5 wt.% Ru/Al$_2$O$_3$ (3) (**c**) catalysts.

The influence of the catalyst support was examined for catalysts containing 5 wt.% Ru prepared using ruthenium(III) chloride precursor and conventional supports such as SiO$_2$, MgAl$_2$O$_4$ (spinel), HY (Si/Al ~ 6) and ZrO$_2$ (Table 2). All the catalysts based on the alternative supports gave full GGGC conversion with product yields of guaiacol (15–22%), vanillin (7–12%) and vanillic acid (8–10%) (Table 2, entries 1–4), which were comparable to the analogous 5 wt.% Ru/Al$_2$O$_3$ (1) (Table 1, entry 7). This finding suggested that the characteristics of the support materials was of minor importance for the catalytic performance under the applied conditions.

Table 2. Catalytic oxidation of GGGE with alternative Ru/support catalysts [a].

Entry	Catalyst	BET Surface Area [b] (m^2/g)	GGGEconv. (%)	Product Yield (%)		
				Guaiacol	Vanillin	Vanillic Acid
1	5 wt.% Ru/SiO$_2$	278	>99	22	7	10
2	5 wt.% Ru/spinel	63	>99	20	12	8
3	5 wt.% Ru/HY(6) [c]	698	>99	15	8	9
4	5 wt.% Ru/ZrO$_2$	97	>99	20	9	8

[a] Reaction conditions: 10 ml 0.017 M GGGE in acetonitrile, 40 mg catalyst (prepared using ruthenium (III) chloride), 160 °C, 5 bar (20% oxygen in argon), 20 h. [b] BET surface areas of support materials. [c] The number in parenthesis corresponds to the Si/Al ratio.

2.3. Effect of Reaction Time, Reaction Temperature and Oxygen Pressure

In order to optimize the reaction towards formation of guaiacol/vanillin/vanillic acid, the influence of reaction temperature and reaction time were examined using the 5 wt.% Ru/Al$_2$O$_3$ (3) catalyst and the results are illustrated in Figures 3 and 4, respectively. When the aerobic oxidation of GGGE was performed for 20 h at a relatively low temperature (120 °C), a low yield of guaiacol (8%) was obtained along with 8% vanillin and <2% vanillic acid with 52% conversion (Figure 3). At 140 °C, GGGE conversion (85%) product yields improved slightly, while full substrate conversion (>99%) and maximum product yields were found at 160 °C (see also Table 1, entry 12). With reaction times of less than 20 h at 160 °C, the GGGE conversion as well as the product formation was significantly lower and vanillic acid formed only after 3 h of reaction (Figure 4). Similarly, at a prolonged reaction time of 30 h the product yields decreased noticeably (21% guaiacol, 11% vanillin and 7% vanillic acid), indicating the formation of byproducts both by side reactions as well as product degradation. At even higher reaction temperatures (180 and 200 °C) the quantitative conversion was maintained but the yield of the products decreased significantly. This was likely due to deactivation of the Ru/Al$_2$O$_3$ (3) catalyst by Ru particle aggregation and formation of byproducts (unidentified) by consecutive reactions of the products, thus confirming 160 °C and 20 h to be the optimal conditions for obtaining the highest product yields.

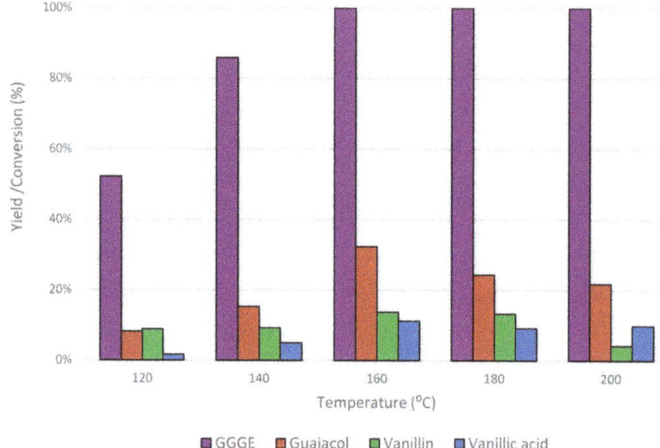

Figure 3. Temperature study for the GGGE oxidation with 5 wt.% Ru/Al$_2$O$_3$ (3) catalyst. Reaction conditions: 10 mL 0.017 M GGGE in acetonitrile, 40 mg catalyst, 20 h, 5 bar (20% oxygen in argon).

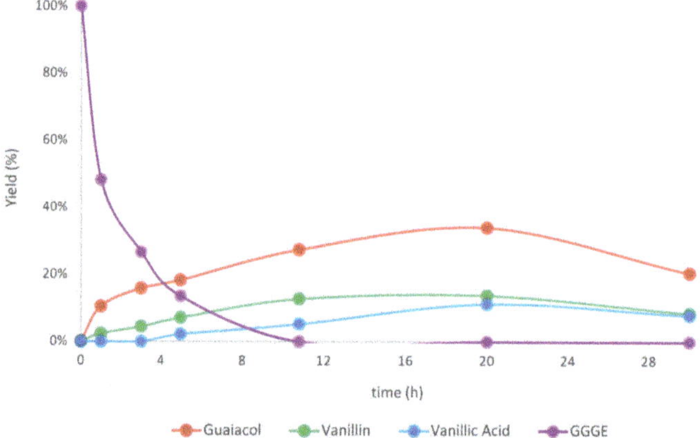

Figure 4. Time-course study for the GGGE oxidation with 5 wt.% Ru/Al$_2$O$_3$ (3) catalyst. Reaction conditions: 10 mL 0.017 M GGGE in acetonitrile, 40 mg catalyst, 160 °C, 5 bar (20% oxygen in argon). All data were obtained from individual experiments.

The importance of oxygen being present for the product formation was further evaluated by performing a catalytic reaction under optimized conditions (160 °C, 20 h) with pure argon atmosphere (20 bar). The GGGE was quantitatively converted (>99%) as was found previously using 5 bar of 20% oxygen in argon (see Table 1, entry 12). However, only a moderate yield of guaiacol (24%) and very poor yields of vanillin (3%) and vanillic acid (<1%) were formed under argon atmosphere, thus confirming that oxygen promoted guaiacol formation and was a prerequisite for the production of vanillin and vanillic acid, as was also expected. Notably, full GGGE conversion and very similar product yields (32% guaiacol, 11% vanillin, 11% vanillic acid) were obtained using 5 bar of air instead of 5 bar of 20% oxygen in argon as also anticipated since both had similar oxygen content (i.e., P$_{O2}$ ≈ 1 bar). In contrast, <1% yield of the desired oxidation products were obtained using water as the solvent under similar reaction conditions (results not shown), possibly due to low oxygen solubility at the reaction conditions.

High-resolution NMR analysis of the post-reaction mixture obtained at the optimal reaction conditions (160 °C and 20 h) was performed in order to validate the reaction products, and to obtain insight into byproduct formation with the aim of understanding the loss of carbon from the overall carbon balance of the process. Guaiacol, vanillin and vanillic acid were confirmed to be the predominant reaction products, while a variety of minor aromatic byproducts was formed (Figure 5). These byproducts included 2-methoxy-1,4-benzoquinone as the main aromatic byproduct (2.5% yield) and benzoic acid alongside its derivatives (2% yield), as well as a plethora of additional, unidentified aromatic byproducts contributing to the loss of carbon. The 2-methoxy-1,4-benzoquinone was identified using in situ spectroscopy on crude post-reaction material (Figure S2), and has previously been described as a degradation product in the manganese peroxidase-catalyzed oxidation of guaiacol [40]. Hence, its presence indicated that overoxidation of the main reaction products occured at reaction conditions that were more severe than the optimum conditions, thus rationalizing the decline especially in guaiacol and vanillin yields at longer times or higher temperatures (Figures 3 and 4).

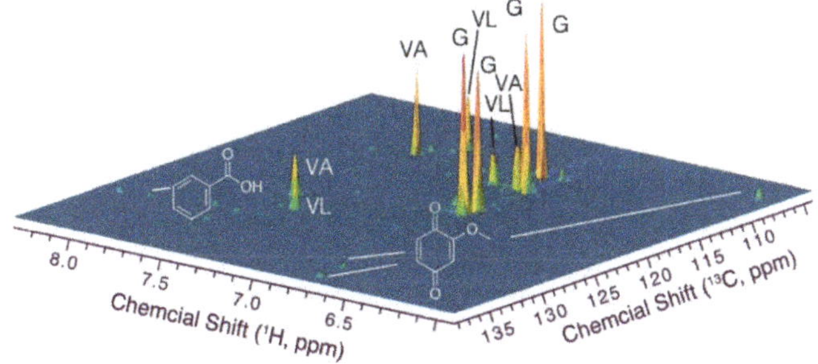

Figure 5. ^{1}H-^{13}C HSQC spectrum of post-reaction mixture displayed showing the main products guaiacol (G), vanillin (VL) and vanillinc acid (VA) alongside a variety of minor byproducts, including benzoic acid and 2-methoxy-1,4-benzoquinone.

2.4. Catalyst Reuse

To examine catalyst viability, the 5 wt.% Ru/Al$_2$O$_3$ (3) catalyst was finally subjected to reuse in five consecutive oxidation reactions of GGGE under optimized reaction conditions, i.e., 160 °C and 20 h (Figure 6). The results revealed that the catalyst was recyclable and maintained good catalytic performance with high GGGE conversion during the five reaction runs. However, some decrease in the yields of guaiacol (34 to 30%), vanillin (13 to 8%) and vanillic acid (11 to 6%) occurred over the five recycles. This decline suggested that part of the catalytically active metal sites of the Ru/Al$_2$O$_3$ (3) catalyst was gradually lost during the reaction sequence. In this connection, TEM analysis of the catalyst after the five-time use confirmed that the spent catalyst contained Ru-based nanoparticles that were more uniformly shaped and larger (30–100 nm) than the fresh catalyst (7–60 nm) (Figure S3). The larger particles most likely formed during the intermediate catalyst calcinations, as also previously observed for Ru supported catalysts [41–43], and were expected to contain less catalytically active sites and therefore be (comparably) less active than smaller-sized particles.

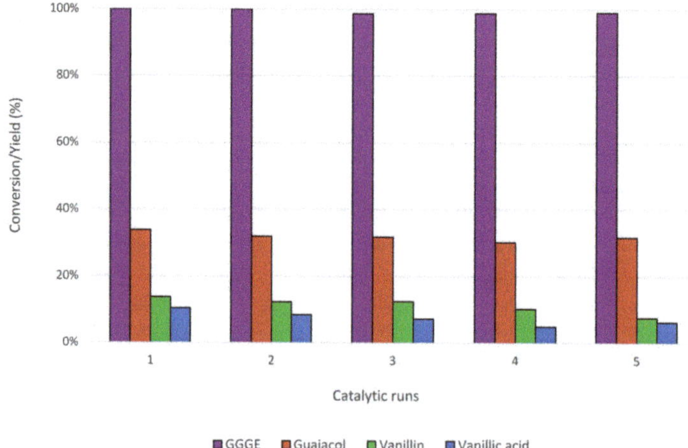

Figure 6. Reuse of 5 wt.% Ru/Al$_2$O$_3$ (3) catalyst in five consecutive GGGE oxidation reactions. Reaction conditions: GGGE to catalyst mass ratio = 1.30, 160 °C, 5 bar (20% oxygen in argon).

3. Materials and Methods

3.1. General

Guaiacyl glycerol-β-guaiacyl ether (>99%, GGGE) was prepared from acetovanillone through a multiple step synthesis route using a reported procedure [44]. Vanillin (99%), vanillic acid (99%) manganese(II) acetate tetrahydrate (>99%), copper(II) acetate monohydrate (p.a.), silver(I) nitrate (p.a.), iron(III) nitrate nonahydrate (p.a.), ruthenium(IV) oxide hydrate (>99.9%), ruthenium(III) acetylacetonate (>97%), ruthenium(III) chloride hydrate (>99%), zirconium(IV) oxide (99%), magnesium aluminate (spinel) (99%), silica gel 60 (high-purity grade) and acetonitrile (99.8%) were purchased from Sigma Aldrich. γ-Aluminium oxide (>99%) was provided by Saint Gobain. Compressed oxygen (>99.99%) and argon (>99.99%) were purchased from Air Liquide, Denmark. All chemicals and gases were used as received.

3.2. Catalyst Preparation

The supported catalysts were prepared by wet impregnation where the appropriate amount of the metal precursor (Ru, Mn, Fe, Cu or Ag) was dissolved in water, the corresponding support (alumina, silica, zirconia, spinel or HY zeolite) slowly added and the resulting suspension stirred for 3 h. Then the mixture was dried overnight at 80 °C followed by calcination at 450 °C for 6 h in static air to obtain the corresponding supported catalyst with a metal content of 1–5 wt.% corresponding to the amount of added metal precursor. All catalysts were stored in a desiccator and used without further activation.

3.3. Catalyst Characterization

TEM images of the Ru supported catalysts were recorded on a FEI Tecnai Transmission Electron Microscope at 200 kV with samples deposited on a carbon support. EDS analysis was performed with an Oxford INCA system. Surface areas of the supported catalysts were determined by nitrogen sorption measurements at liquid nitrogen temperature on a Micromeritics ASAP 2020. Samples were outgassed in a vacuum at 150 °C for 4 h prior to the measurements, and the total surface areas were calculated according to the BET method.

3.4. Catalytic Oxidation Reactions

Catalytic experiments were performed in an autoclave (Microclave 50 mL reactor, Autoclave Engineers) charged with an appropriate amount of guaiacyl glycerol-β-guaiacyl ether (GGGE), catalyst (40 mg) and acetonitrile (10 mL) as the solvent. The reactor was pressurized with 5 bar of 20% oxygen in argon, and heated to the desired reaction temperature. Mechanical stirring of the reactor (300 rpm) was started once a temperature of 20 °C below the set point was reached. After the reaction, the autoclave was quenched in cold water, the catalyst removed by filtration and the reaction mixture subjected to analysis.

In the catalyst recyclability study, the catalyst was recovered by filtration after each reaction run, thoroughly washed with acetonitrile, dried overnight at 60 °C and calcined at 450 °C for 6 h before being used in the next reaction run as described above.

3.5. Product Analysis

Aliquots of the reaction mixture were subjected to HPLC analysis (Agilent 1200 series instrument, Agilent C-18 column, 15 cm length) in order to quantify the yield and conversion. The eluent was in all cases 60 vol.% aqueous acetonitrile solution. The conversions of GGGE and yields of the products (guaiacol, vanillin, vanillic acid) were determined from individual standard solutions with products identified by GC-MS analysis.

3.6. NMR Spectroscopy

NMR analyses were conducted to validate the presence and amount of the main reaction products and to obtain insight into the nature of major byproducts. For NMR analyses, samples were condensed on a rotary evaporator at 40 °C and re-dissolved in deuterated acetonitrile (>99.8 atom% deuterium). NMR spectra were recorded on an 800 MHz Bruker Avance II spectrometer equipped with a 18.7 T magnet (Oxford, United Kingdom) and a TCI z-gradient cryoprobe (Bruker, Karlsruhe, Germany) at 25 °C. The 2D NMR spectra included TOCSY with a 10 kHz spin lock field that was applied for 60 ms (2048 × 256 complex data points with 640 ms and 80 ms acquisition times), a sensitivity enhanced ^{1}H-^{13}C HSQC (1024 × 1024 complex points, sampling 128 ms and 73 ms acquisition times), ^{1}H-^{13}C HMBC (2048 × 256 complex data points, sampling 213 and 11 ms acquisition times) and a ^{1}H-^{13}C HSQC TOCSY (1024 × 512 complex points with 107 ms and 42 ms acquisition times). Identification of byproducts in reaction mixtures were done by comparison to pure reference standards (e.g., acetic and benzoic acid) or by de novo structure determination (e.g., 2-methoxy-1,4-benzoquinone). All spectra were acquired in Bruker Topspin and processed in the same software with extensive zero filling in all dimensions.

4. Conclusions

An optimized 5 wt.% Ru/Al$_2$O$_3$ catalyst prepared with ruthenium(IV) oxide hydrate precursor was found to give superior yield of the monomeric aromatics guaiacol (34%), vanillin (13%) and vanillic acid (11%) in the aerobic oxidation of the lignin-model compound GGGE in acetonitrile at 5 bar (20% oxygen in argon) under optimized reaction conditions (160 °C, 20 h). In comparison, Ru/Al$_2$O$_3$ catalysts prepared with other ruthenium precursors were found to give lower yields of the desired products, which could be correlated to the particle sizes of the Ru-species measured by TEM. Furthermore, catalysts containing other transition metals (Ag, Fe, Mn, Co and Cu) supported on alumina, and ruthenium catalysts based on alternative supports (silica, spinel, HY and zirconia) were also significantly less active compared to the Ru/Al$_2$O$_3$ catalysts for the GGGE oxidation when using identical reaction conditions. Notably, the optimized Ru/Al$_2$O$_3$ catalyst proved robust for recycling in five consecutive reaction runs with only minor activity loss corresponding to thermal regeneration between the runs.

Improved performance of Ru/Al$_2$O$_3$ catalysts was obtained by the optimization of catalyst preparation and reaction conditions for the oxidation of a lignin-model compound to monomeric aromatics. However, the reactivity of lignin-model systems may not be directly transferable to a complex system with native lignin, where more stable catalysts displaying higher activity and selectivity may be required in order to extend the role of lignin applications in biorefineries.

Supplementary Materials: The following are available online at http://www.mdpi.com/2073-4344/9/10/832/s1, Figure S1: High-resolution TEM images of (A) fresh 1 wt.% Ru/Al$_2$O$_3$ (1) catalyst, (B) used 1 wt.% Ru/Al$_2$O$_3$ (1) catalyst, (C) fresh 3 wt.% Ru/Al$_2$O$_3$ (1) catalyst, (D) used 3 wt.% Ru/Al$_2$O$_3$ (1) catalyst, (E) fresh 5 wt.% Ru/Al$_2$O$_3$ (1) catalyst and (F) used 5 wt.% Ru/Al$_2$O$_3$ (1) catalyst, Figure S2: Overlay of ^{1}H-^{13}C HSQC and ^{1}H-^{13}C HMBC NMR spectra of post-reaction material displayed as contour plots, showing the single and multiple-bond correlations in 2-methoxy-1,4-benzoquinone, with the inset displaying the full chemical shift assignment of the compound. Figure S3: High-resolution TEM catalysts images of (left) fresh 5 wt.% Ru/Al$_2$O$_3$ (3) catalyst, and (right) 5 wt.% Ru/Al$_2$O$_3$ (3) catalyst after five reaction runs.

Author Contributions: Data curation and analysis, M.M.-R., S.M. and S.S.; Writing of manuscript, M.M.-R., S.M., S.S., S.K. and A.R.; Management, A.R.

Funding: This research received no external funding.

Acknowledgments: The Danish Agency for Science, Technology and Innovation (International Network Programme, 12-132649), Haldor Topsøe A/S and the Technical University of Denmark. The 800 MHz NMR spectra were recorded using the spectrometer at the DTU NMR Center, supported by the Villum Foundation.

Conflicts of Interest: The authors declare no conflict of interest.

References

1. Vlachos, D.G.; Caratzoulas, S. The roles of catalysis and reaction engineering in overcoming the energy and the environment crisis. *Chem. Eng. Sci.* **2010**, *65*, 18–29. [CrossRef]

2. Sharma, P.R.; Chattopadhyay, A.; Sharma, S.K.; Geng, L.; Amiralian, N.; Martin, D.; Hsiao, B.S. Nanocellulose from Spinifex as an Effective Adsorbent to Remove Cadmium (II) from Water. *ACS Sustain. Chem. Eng.* **2018**, *6*, 3279–3290. [CrossRef]

3. Golmohammadi, H.; Morales-Narváez, E.; Naghdi, T.; Merkoçi, A. Nanocellulose in Sensing and Biosensing. *Chem. Mater.* **2017**, *29*, 5426–5446. [CrossRef]

4. Sharma, P.R.; Varma, A.J. Functional nanoparticles obtained from cellulose: Engineering the shape and size of 6-carboxycellulose. *Chem. Commun.* **2013**, *49*, 8818–8820. [CrossRef] [PubMed]

5. Larsen, J.; Haven, M.Ø.; Thirup, L. Inbicon makes lignocellulosic ethanol a commercial reality. *Biomass Bioenergy* **2012**, *46*, 36–45. [CrossRef]

6. Chen, K.; Xu, L.J.; Yi, J. Bioconversion of Lignocellulose to Ethanol: A Review of Production Process. *Adv. Mater. Res.* **2011**, *280*, 246–249. [CrossRef]

7. Lindedam, J.; Bruun, S.; Jørgensen, H.; Felby, C.; Magid, J. Cellulosic ethanol: Interactions between cultivar and enzyme loading in wheat straw processing. *Biotechnol. Biofuels* **2010**, *3*, 25. [CrossRef]

8. Paniagua, M.; Saravanamurugan, S.; Melián-Rodríguez, M.; Melero, J.A.; Riisager, A. Xylose Isomerization with Zeolites in a Two-Step Alcohol-Water Process. *ChemSusChem* **2015**, *8*, 1088–1094. [CrossRef] [PubMed]

9. Vidal, B.C.; Dien, B.S.; Ting, K.C.; Singh, V. Influence of Feedstock Particle Size on Lignocellulose Conversion—A Review. *Appl. Biochem. Biotechnol.* **2011**, *164*, 1405–1421. [CrossRef]

10. Alonso, D.M.; Bonda, J.Q.; Dumesic, J.A. Catalytic conversion of biomass to biofuels. *Green Chem.* **2010**, *12*, 1493–1513. [CrossRef]

11. Chamnankid, B.; Ratanatawanate, C.; Faungnawakij, K. Conversion of xylose to levulinic acid over modified acid functions of alkaline-treated zeolite Y in hot-compressed water. *Chem. Eng. J.* **2014**, *258*, 341–347. [CrossRef]

12. Shi, X.; Wu, Y.; Li, P.; Yi, H.; Yang, M.; Wang, G. Catalytic conversion of xylose to furfural over the solid acid SO$_4$$^{2-}$-/ZrO$_2$-Al$_2O_3$/SBA-15 catalysts. *Carbohydr. Res.* **2011**, *346*, 480–487. [CrossRef] [PubMed]

13. Klemm, D.; Heublein, B.; Fink, H.P.; Bohn, A. Cellulose: Fascinating biopolymer and sustainable raw material. *Angew. Chem. Int. Ed.* **2005**, *44*, 3358–3393. [CrossRef] [PubMed]

14. Klemm, D.; Cranston, E.D.; Fischer, D.; Gama, M.; Kedzior, S.A.; Kralisch, D.; Kramer, F.; Kondo, T.; Lindström, T.; Nietzsche, S.; et al. Nanocellulose as a natural source for groundbreaking applications in materials science: Today's state. *Mater. Today* **2018**, *21*, 720–748. [CrossRef]

15. Sharma, P.R.; Joshi, R.; Sharma, S.K.; Hsiao, B.S. A Simple Approach to Prepare Carboxycellulose Nanofibers from Untreated Biomass. *Biomacromolecules* **2017**, *188*, 2333–2342. [CrossRef] [PubMed]

16. Zakzeski, J.; Bruijnincx, P.C.; Jongerius, A.L.; Weckhuysen, M.B. The catalytic valorization of lignin for the production of renewable chemicals. *Chem. Rev.* **2010**, *110*, 3552–3599. [CrossRef] [PubMed]

17. Li, C.; Zhao, X.; Wang, A.; Huber, G.W.; Zhang, T. Catalytic Transformation of Lignin for the Production of Chemicals and Fuels. *Chem. Rev.* **2015**, *115*, 11559–11624. [CrossRef] [PubMed]

18. Deng, W.; Zhang, H.; Wu, X.; Li, R.; Zhang, Q.; Wang, Y. Oxidative conversion of lignin and lignin model compounds catalyzed by CeO_2-supported Pd nanoparticles. *Green Chem.* **2015**, *17*, 5009–5018. [CrossRef]

19. Wahyudiono, T.; Kanetake, M.; Sasaki, M.; Goto, M. Decomposition of a Lignin Model Compound under Hydrothermal Conditions. *Chem. Eng. Technol.* **2007**, *30*, 1113–1122. [CrossRef]

20. Sedai, B.; Díaz-Urrutia, C.; Baker, R.T.; Wu, R.; Silks, L.P.; Hanson, S.K. Aerobic oxidation of β-1 lignin model compounds with copper and oxovanadium catalysts. *ACS Catal.* **2013**, *3*, 3111–3122. [CrossRef]

21. Nguyen, J.D.; Matsuura, B.S.; Stephenson, C.R.J. A photochemical strategy for lignin degradation at room temperature. *J. Am. Chem. Soc.* **2014**, *136*, 1218–1221. [CrossRef] [PubMed]

22. Rahimi, A.; Azarpira, A.; Kim, H.; Ralph, J.; Stahl, S.S. Chemoselective metal-free aerobic alcohol oxidation in lignin. *J. Am. Chem. Soc.* **2013**, *135*, 6415–6418. [CrossRef] [PubMed]

23. Wang, S.; Gao, W.; Xiao, L.-P.; Shi, J.; Sun, R.-C.; Song, G. Hydrogenolysis of biorefinery corncob lignin into aromatic phenols over activated carbon-supported nickel. *Sustain. Energy Fuels* **2019**, *3*, 401–408. [CrossRef]

24. Sedai, B.; Baker, R.T. Copper Catalysts for Selective C-C Bond Cleavage of β-O-4 Lignin Model Compounds. *Adv. Synth. Catal.* **2014**, *356*, 3563–3574. [CrossRef]

25. Cheng, C.; Wang, J.; Shen, D.; Xue, J.; Guan, S.; Gu, S.; Luo, K.H. Catalytic Oxidation of Lignin in Solvent Systems for Production of Renewable Chemicals: A Review. *Polymers* **2017**, *9*, 240. [CrossRef]

26. Guadix-Montero, S.; Sankar, M. Review on Catalytic Cleavage of C–C Inter-unit Linkages in Lignin Model Compounds: Towards Lignin Depolymerisation. *Top. Catal.* **2018**, *61*, 183–198. [CrossRef]

27. Son, S.; Toste, F.D. Non-Oxidative Vanadium-Catalyzed C-O Bond Cleavage: Application to Degradation of Lignin Model Compounds. *Angew. Chem. Int. Ed.* **2010**, *49*, 3791–3794. [CrossRef]

28. Hanson, S.K.; Baker, R.T.; Gordon, J.C.; Scott, B.L.; Thorn, D.L. Aerobic oxidation of lignin models using a base metal vanadium catalyst. *Inorg. Chem.* **2010**, *49*, 5611–5618. [CrossRef]

29. Hanson, S.K.; Wu, R.; Silks, L.A.P. C-C or C-O bond cleavage in a phenolic lignin model compound: Selectivity depends on vanadium catalyst. *Angew. Chem. Int. Ed.* **2012**, *51*, 3410–3413. [CrossRef]

30. Zhang, G.; Scott, B.L.; Wu, R.; Silks, L.A.P.; Hanson, S.K. Aerobic oxidation reactions catalyzed by vanadium complexes of bis(phenolate) ligands. *Inorg. Chem.* **2012**, *51*, 7354–7361. [CrossRef]

31. Vom Stein, T.; den Hartog, T.; Buendia, J.; Stoychev, S.; Mottweiler, J.; Bolm, C.; Klankermayer, J.; Leitner, W. Ruthenium-Catalyzed C-C Bond Cleavage in Lignin Model Substrates. *Angew. Chem. Int. Ed.* **2015**, *54*, 5859–5863. [CrossRef] [PubMed]

32. Stahl, S.S.; Madison, A.R. Selective Aerobic Alcohol Oxidation Method For Conversion Of Lignin Into Simple Aromatic Compounds. U.S. Patent 8969534 B2, 3 March 2015.

33. Yamaguchi, K.; Mizuno, N. Supported ruthenium catalyst for the heterogeneous oxidation of alcohols with molecular oxygen. *Angew. Chem. Int. Ed.* **2002**, *41*, 4538–4542. [CrossRef]

34. Melián-Rodríguez, M.; Saravanamurugan, S.; Kegnæs, S.; Riisager, A. Aerobic Oxidation of Veratryl Alcohol to Veratraldehyde with Heterogeneous Ruthenium Catalysts. *Top. Catal.* **2015**, *58*, 1036–1042. [CrossRef]

35. Chang, Y.F.; McCarty, J.G.; Wachsman, E.D. Effect of ruthenium-loading on the catalytic activity of Ru-NaZSM-5 zeolites for nitrous oxide decomposition. *Appl. Catal. B Environ.* **1995**, *6*, 21–33. [CrossRef]

36. Harvey, T.G.; Pratt, K.C. The effect of ruthenium loading on Ru-zeolite/NiMo HDN catalysts. *Appl. Catal. A Gen.* **1996**, *146*, 317–321. [CrossRef]

37. Betancourt, P.; Rives, A.; Hubaut, R.; Scott, C.E.; Goldwasser, J. A study of the ruthenium–alumina system. *Appl. Catal. A Gen.* **1998**, *170*, 307–314. [CrossRef]

38. Gorbanev, Y.Y.; Kegnæs, S.; Riisager, A. Effect of support in heterogeneous ruthenium catalysts used for the selective aerobic oxidation of HMF in water. *Top. Catal.* **2011**, *54*, 1318–1324. [CrossRef]

39. Laursen, A.B.; Gorbanev, Y.Y.; Cavalca, F.; Malacrida, P.; Kleiman-Shwarsctein, A.; Kegnæs, S.; Riisager, A.; Chorkendorff, I.; Dahl, S. Highly dispersed ruthenium oxide as an aerobic catalyst for acetic acid synthesis. *Appl. Catal. A* **2012**, *433–434*, 243–250. [CrossRef]
40. Cruz, J.; Baglio, V. Preparation and Characterization of RuO_2 Catalysts for Oxygen Evolution in a Solid Polymer Electrolyte. *Int. J. Mol. Sci.* **2011**, *6*, 6607–6619.
41. Hwang, S.; Lee, C.-H.; Ahn, I.-S. Product identification of guaiacol oxidation catalyzed by manganese peroxidase. *J. Ind. Chem. Eng.* **2008**, *14*, 487–492. [CrossRef]
42. Gorbanev, Y.; Kegnæs, S.; Hanning, C.W.; Hansen, T.W.; Riisager, A. Acetic Acid Formation by Selective Aerobic Oxidation of Aqueous Ethanol over Heterogeneous Ruthenium Catalysts. *ACS Catal.* **2012**, *2*, 604–612. [CrossRef]
43. Gorbanev, Y.Y.; Kegnæs, S.; Riisager, A. Selective aerobic oxidation of 5-hydroxymethylfurfural in water over solid ruthenium hydroxide catalysts with magnesium-based supports. *Catal. Lett.* **2011**, *141*, 1752–1760. [CrossRef]
44. Song, Q.; Wang, F.; Xu, J. Hydrogenolysis of lignosulfonate into phenols over heterogeneous nickel catalysts. *Chem. Commun.* **2012**, *48*, 7019–7021. [CrossRef] [PubMed]

Article

Sulfonated Hydrothermal Carbons from Cellulose and Glucose as Catalysts for Glycerol Ketalization

Pablo Fernández [1], José M. Fraile [2,*], Enrique García-Bordejé [3] and Elísabet Pires [1,2,*]

[1] Dpto. Química Orgánica, Facultad de Ciencias, Universidad de Zaragoza, E-50009 Zaragoza, Spain; pablozgz@hotmail.com

[2] Instituto de Síntesis Química y Catálisis Homogénea (ISQCH), Facultad de Ciencias, C.S.I.C. - Universidad de Zaragoza, E-50009 Zaragoza, Spain

[3] Instituto de Carboquímica (ICB-CSIC), Miguel Luesma Castán 4, E-50018 Zaragoza, Spain; jegarcia@icb.csic.es

* Correspondence: jmfraile@unizar.es (J.M.F.); epires@unizar.es (E.P.); Tel.: +34-976-553-514 (J.M.F.); +34-976-553-501 (E.P.)

Received: 1 July 2019; Accepted: 23 September 2019; Published: 25 September 2019

Abstract: Solketal is one of the most used glycerol-derived solvents. Its production via heterogeneous catalysis is crucial for avoiding important product losses typically found in the aqueous work-up in homogeneous catalysis. In this work, we present a study of the catalytic synthesis of solketal using sulfonated hydrothermal carbons (SHTC). They were prepared from glucose and cellulose resulting in different textural properties depending on the hydrothermal treatment conditions. The sulfonated hydrothermal carbons were also coated on a graphite microfiber felt (SHTC@GF). Thus, up to nine different solids were tested, and their activity was compared with commercial acidic resins. The solids presented very different catalytic activity, which did not correlate with their physical-chemical properties indicating that other aspects likely influence the transport of reactants and products to the catalytic surface. Additionally, the SHTC prepared from cellulose showed better reusability in batch reaction tests. This work also presents the first results for the production of solketal in a flow reactor, which opens the way to the use of SHTC@GF for this kind of reactions.

Keywords: sulfonated hydrothermal carbon; solketal; sulfonic solids; ketalization; continuous flow

---

1. Introduction

Lignocellulosic biomass is a renewable and accessible feedstock for the production of commodity chemicals and fuels [1]. The main component of raw biomass is cellulose, a biopolymer made from glucose monomers. The hydrothermal treatment of cellulose was first developed to hydrolyze cellulose into liquid fuels or platform chemical molecules such as furfurals [2–5]. The treatment under pressure of cellulose in water, at temperatures in the 150–400 °C range, gives rise to a mixture of soluble organic substances and a carbon-rich solid product, whose relative amount depends on the conditions. Only until recently, attention has been paid to the solids resulting from hydrothermal treatment [6–8]. It is widely reported that the hydrothermal treatment of saccharides such as glucose produces carbon microspheres of uniform sizes under very mild process conditions [9–12]. Compared to glucose, cellulose is a more convenient feedstock because it is more abundant and inexpensive and studies can be found in the literature on the hydrothermal carbonization of cellulose [13–16]. Sulfonated hydrothermal carbons derived from renewable raw materials are envisaged as sustainable catalysts and are often applied in biomass transformations [17]. However, the hydrothermal carbons prepared from cellulose have been rarely used as catalyst precursors.

Together with the use of renewable raw materials, the use of green solvents is a trending topic in biorefineries. Among the different alternatives, the use of glycerol itself and its derivatives as renewable solvents has attracted great attention in the last decade [18] and, in particular, glycerol carbonate [19] and solketal [20] can be considered the most applied glycerol derived solvents up to now. In order to envisage the possibility of producing large amounts of solketal (Scheme 1), a sustainable synthetic methodology must be developed and, thus, many works have been published so far related to the study of different catalytic systems for the reaction of glycerol with acetone [21]. One of such examples is the use of homogeneous acid catalysts, which implies an aqueous work-up of the reaction. Due to the large transfer of the ketal to the aqueous phase, the global isolated ketal yield is seriously lowered, up to 70% [22]. Thus, many efforts have been devoted to the search of efficient heterogeneous catalysts for this reaction. Among the different catalytic systems, heterogeneous Brønsted acids, such as the sulfonic resin Amberlyst 36 [23], sulfonic-functionalized SBA-15 [24] or mesoporous zeolites [25], and heterogeneous Lewis acids, such as Zr- and Hf-TUD-1 and Sn-MCM-41 [26], have been described as effective catalysts for solketal production. In the case of zeolites, the surface area and the presence of mesoporosity seems to also play a crucial role for obtaining high glycerol conversions (80%) and total solketal selectivity. In a recent work, glycidol has been described as starting material for solketal production and several heterogenous catalysts were tried such as Nafion®NR50, supported metal triflates, K10 montmorillonite, and Amberlyst 15 [27]. In this case, Nafion®NR50 is described as the best catalyst, although high acetone/glycidol ratios and longer reaction times were needed in order to obtain high glycidol conversions.

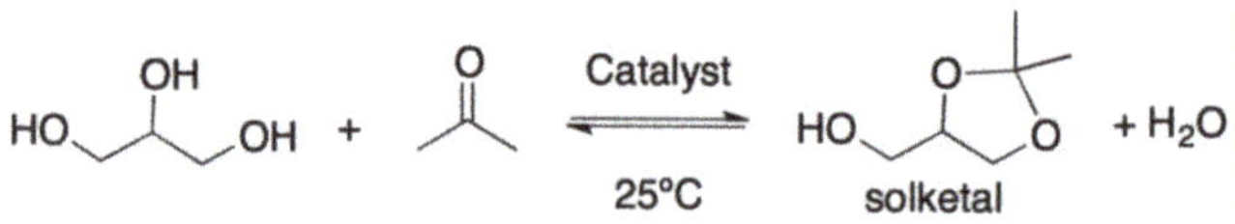

Scheme 1. Overall reaction for the production of solketal.

It is also worth mentioning that in all the cases, reaction temperatures in between 343–353 K are used in order to achieve good solketal yields.

An interesting point when developing sustainable processes is the possibility of using heterogeneous catalysts that are also derived from renewable raw materials. As mentioned above, this is the case of carbons. Thus, sulfonated activated carbons obtained from olive stones [28] and hydrothermal carbons prepared from glycerine as biodiesel waste [29] have been proposed as catalysts in the reaction of glycerol and acetone. In both cases, the reaction proceeded smoothly at room temperature using 3 wt% catalyst. Conversions of glycerol of 80% and solketal selectivity over 95% were described.

Continuing the efforts we have done in our laboratory to develop low-cost and versatile biomass derived catalysts, in this work, sulfonated hydrothermal carbons were selected because of their renewable origin, mild preparation conditions and good catalytic performance shown in other acid-catalyzed reactions. More specifically, we present here for the first time the preparation and characterization of sulfonated hydrothermal carbons from cellulose, both bulk and deposited on graphite felt, as well as a comparative study of their activity with a previously described catalyst derived from glucose and commercial sulfonic resins, as acid catalysts in the synthesis of solketal, their reusability and the preliminary results in a flow reactor.

2. Results and Discussion

In previous works, we studied the synthesis of hydrothermal carbons from both glucose and cellulose [30–32]. Thus, the hydrothermal treatment of glucose leads to a carbon material (HTC) in the form of microspheres [30], with rather high density of oxygenated functional groups (oxygen/carbon molar ratio = 0.3) that confers to the solid a highly hydrophilic character. The surface area, determined by nitrogen adsorption, was very low (<10 m^2/g) as well as the pore volume (0.014 cm^3/g). However, the use of CO_2 as adsorbate indicated the presence of a larger surface area and pore volume

(>140 m^2/g and 0.06 cm^3/g, respectively), which is interpreted as an indication of the presence of ultramicropores (<0.7 nm) [31]. In agreement with the relatively high oxygen content, the different types of spectroscopic techniques, such as XPS (X-ray photoelectron spectroscopy) and CP-MAS-NMR (Cross Polarization-Magic Angle Spinning-NMR), indicated the presence of carbonyl and carboxylic groups, as well as furanic and benzofuranic (and probably phenolic) aromatic groups, together with a significant amount of aliphatic chains, associated with terminal or bridge groups between the aromatic rings [30]. The presence of such carboxylic groups confers a weak acidity (3.4 mmol/g, determined by back titration) to the HTC.

In the studies of hydrothermal carbonization of microcrystalline cellulose [32], different reaction conditions of temperature and time were tried, in the absence or in the presence of HCl at different concentrations to promote the partial hydrolysis of cellulose. The samples are named as Cel-temperature-HCl concentration (when used)-time, for example, Cel-195-20 h indicates an HTC prepared from cellulose at 195 °C for 20 h in the absence of HCl and Cel-215-2 M-40 h indicates a HTC prepared from cellulose at 215 °C, with 2 M HCl for 40 h. Analogously, the HTC from glucose is named as Glu-195-20 h, as in that case no HCl was used in the hydrothermal process. Although the hydrothermal carbons from cellulose showed similar general features to that prepared from glucose, the hydrothermal conditions (temperature, time, acid) significantly modified the morphology and textural properties of the HTC. Less developed microspheres, with oxygen contents ranging from 12.8 to 27.9% were obtained, together with surface areas measured with CO$_2$ from 104 to 386 m^2/g. In fact, the surface area was used here to establish a sort of harshness scale of the hydrothermal conditions that we called "hydrothermal index" (H.I.) with an arbitrary scale of 0–20 [32]. The acidity of the cellulose derived HTCs, corresponding to carboxylic groups, determined by the difference between total acidity and the number of SO$_3$H groups, also varied with the hydrothermal conditions, from 0.85 to 2.31 mmol/g, but it was always lower than the acidity of the HTC from glucose (3.42 mmol/g).

2.1. Synthesis and Characterization of Sulfonated Hydrothermal Carbon Catalysts from Cellullose

The HTC samples obtained from glucose and microcrystalline cellulose were sulfonated under the standard conditions, concentrated H$_2$SO$_4$ at 150 °C for 15 h (see Materials and Methods). The sulfonated solid samples are named using the HTC precursor nomenclature followed by an S. That is, Cel-195-20 h-S indicates a sulfonated hydrothermal carbon (SHTC) prepared from cellulose at 195°C in the absence of HCl for 20 h and subsequently sulfonated with H$_2$SO$_4$. Results of elemental analysis, textural properties and acidity of SHTC are collected in Table 1 together with the ones of the non-sulfonated samples for the sake of comparison.

As we previously reported [30], the sulfonation of Glu-195–20 h (sample 1 in Table 1) with concentrated sulfuric acid at 150 °C for >4 h preserves the morphology of microspheres, as well as the textural properties, with only an increase in CO$_2$ surface area and porosity (sample 1 vs. 1-S in Table 1). However, the sulfonation produces significant changes in the chemical composition. The sulfonated hydrothermal carbon (Glu-195-20 h-S, sample 1-S) is a more oxygenated carbon material, with an oxygen/carbon molar ratio around 0.5 and a sulfur content of 0.60–0.77 mmol/g, evenly distributed along the particles [31]. The number of total acid sites was larger than the amount of sulfonic sites, indicating that, besides sulfonation, the treatment with sulfuric acid produces additional reactions (mainly oxidations) on the HTC.

The sulfonated cellulose samples prepared in this work also suffered composition and textural changes upon sulfonation that will be now commented. As an example, Figure 1 gathers SEM (Scanning Electron Microscopy) images, CO$_2$ adsorption isotherm plots and pore volume distributions for two representative solids, Cel-195-2 M-20 h (sample 3, Table 1) and Cel-195-2 M-20 h-S (sample 3-S, Table 1). The plots for all the SHTC samples are gathered in the ESI. As it can be seen, changes in morphology are observed by SEM analysis, although some microspheres were still present upon sulfonation.

Table 1. Composition, textural properties and acidity of HTC and SHTC samples.

Id.	Sample	H.I.	wt%				S_A (m^2/g) [a]	$V_{\mu P}$ (cm^3/g) [b]	SO$_3$H [c]	T.A. [d]	COOH [g]
			C	H	O	S			(mmol/g)		
1	Glu-195-20 h [e]	0.0	66.3	4.4	29.9	–	143	0.057	–	3.42	3.42
1-S	Glu-195-20 h-S [e]	-	55.2	2.3	40.1	2.5	224	0.090	0.77	5.43	4.66
2	Cel-195-20 h [f]	2.4	68.3	4.5	27.1	–	201	0.083	–	2.31	2.31
2-S	Cel-195-20 h-S	-	53.3	2.8	40.5	3.4	272	0.099	1.07	4.45	3.38
2-S′	Cel-195-20 h-S-used	-	52.8	3.0	41.9	2.3			0.74	n.d.	–
3	Cel-195-2 M-20 h [f]	2.7	67.6	4.5	25.6	–	208	0.085	–	1.94	1.94
3-S	Cel-195-2 M-20 h-S	-	56.4	3.1	38.1	2.4	284	0.104	0.74	5.20	4.46
3-S′	Cel-195-2 M-20 h-S-used	-	58.0	3.2	37.0	1.8			0.58	n.d	–
4	Cel-215-20 h [f]	3.2	70.0	4.5	25.3	–	220	0.092	–	2.09	2.09
4-S	Cel-215-20 h-S	-	52.6	3.5	40.9	3.0	226	0.083	0.96	3.34	2.38
5	Cel-195-5 M-20 h [f]	5.4	73.5	5.1	21.4	–	274	0.118	–	0.85	0.85
5-S	Cel-195-5 M-20 h-S	-	58.2	3.3	36.0	2.5	303	0.111	0.77	4.08	3.31
5-S′	Cel-195-5 M-20 h-S-used	-	58.9	3.3	35.6	2.2			0.69	n.d	–
6	Cel-215-2 M-20 h [f]	9.5	74.5	5.1	20.4	–	374	0.168	–	n.d	–
6-S	Cel-215-2 M-20 h-S	-	52,3	2,5	42,8	2.4	279	0.103	0.73	3.42	2.69
6-S′	Cel-215-2 M-20 h-S-used	-	57.5	3.3	37.1	2.1			0.67	n.d	–
7	Cel-195-2 M-40 h [f]	10.0	69.6	4.7	25.7	–	386	0.166	–	n.d	–
7-S	Cel-195-2 M-40 h-S	-	57.2	3.6	35.4	3.8	283	0.105	1.18	4.27	3.09
8	Cel-195-5M-40h [f]	10.6	75.3	5.6	19.0	–	104	0.045	–	1.28	1.28
8-S	Cel-195-5 M-40 h-S	-	57.7	4.1	34.0	4.2	335	0.125	1.33	2.47	1.14
9	Cel-215-2 M-40 h [f]	17.7	77.2	5.7	17.1	–	170	0.072	–	1.37	1.37
9-S	Cel-215-2 M-40 h-S	-	59.3	4.0	32.4	4.3	347	0.133	1.35	2.90	1.55
9-S′	Cel-215-2 M-40 h-S-used	-	58.0	2.6	37.5	1.9			0.60		–

[a] S_A = Surface area. [b] $V_{\mu P}$ = Micropore volume. [c] Calculated from sulfur content. [d] T.A. = total acidity. Determined by back titration. [e] Data from reference [30]. [f] Data from reference [32]. [g] determined by the difference between total acidity and the number of sulfonic sites.

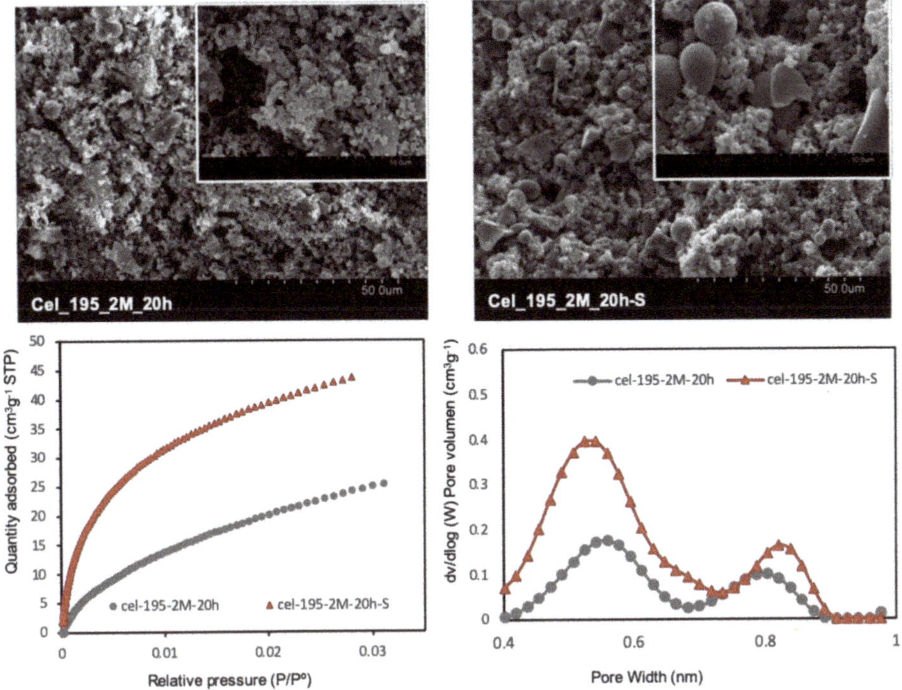

Figure 1. Comparison of textural properties of Cel-195-2 M-20 h (sample 3) and Cel-195-2 M-20 h-S (sample 3-S).

CO_2 isotherms adsorption plots and pore width distribution are typical of microporous solids with narrow pores with significant adsorption at low relative pressures [33,34]. The pore size distribution is bimodal with peaks centered at around 0.55 nm and 0.80 nm, that is in the range of ultramicropores. The bimodal shape of the pore size distributions is characteristic for many adsorbents possessing a small amount of micropores and results from the similarity of the local adsorption isotherm in the range of the pore widths for which the gap between peaks (related to the primary and secondary micropore filling mechanism) exists [35].

In spite of the variability of the surface area in the HTC samples, the sulfonation of these solids provided samples with surface areas in a narrower range (224–347 $m^2\,g^{-1}$) and with a linear relationship with the hydrothermal index values of the original HTC (Figure 2a), which contrasts with the volcano representation observed for the HTC samples [32]. These facts evidenced that the treatment with sulfuric acid is able to complete the hydrothermal process when this is performed under mild conditions (low H.I.), whereas sulfonation induces a partial degradation of the carbon framework in the case of the carbon samples prepared under harsh conditions (high H.I.), leading to a more open structure with higher surface area and porosity (Table 1). The micropore volume follows a similar linear trend (ESI). As linear trends are observed between H.I. and surface area or micropore volume for SHTC, H.I. based on preparation conditions of original HTC might be of usefulness to predict textural properties of SHTC.

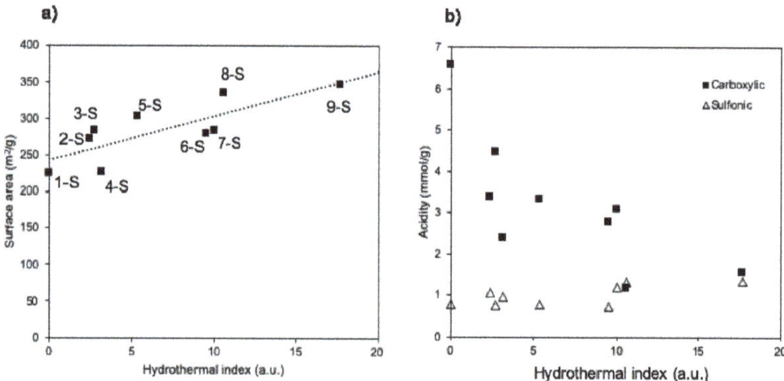

Figure 2. Relationship between hydrothermal index (H.I.) from original HTC and properties of SHTC: (**a**) Surface area; (**b**) sulfonic and carboxylic acidity (determined by difference between total acidity an sulfonic acidity).

In terms of solid composition, and as expected, sulfonation increases the oxygen content (Table 1), which is also confirmed by XPS. In the C1s spectrum, the contributions of C–O (286.2 eV), C=O (287.3 eV) and COOH (289.0 eV) bonds [36] increase with respect to that of C without bonds to oxygen (284.6 eV) (Figure 3a), whereas in the O1s spectrum the contribution of C=O (531.6 eV) increases with respect to the other oxygenated groups (Figure 3b), confirming in this way the partial oxidation of HTC upon sulfonation.

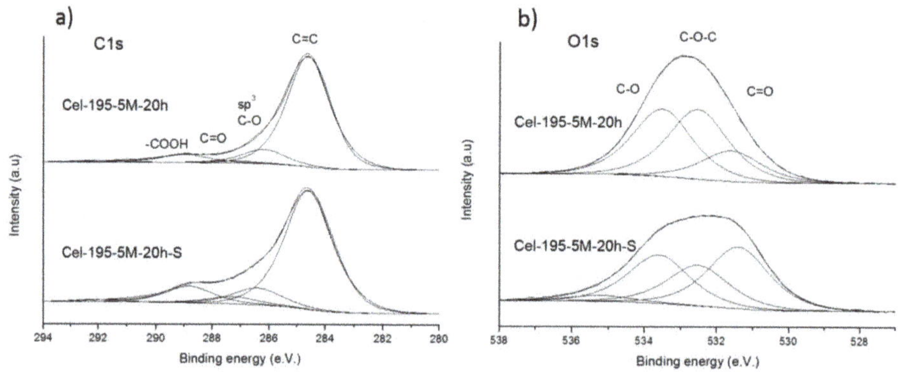

Figure 3. XPS spectra of Cel-195-5 M-20 h and Cel-195-5 M-20 h-S (samples 5 and 5-S, respectively): (**a**) C1s; (**b**) O1s.

The carbon/oxygen composition of the SHTC based on cellulose is linear with the hydrothermal index (H.I., Figure 4a), and the lines for both carbon and oxygen are nearly parallel to those of the non-sulfonated HTC samples (Figure 4a). Thus, harsher hydrothermal conditions produce higher deoxygenation of the generated HTC, a trend that is reproduced in the corresponding SHTC. However, the effect is not so linear in the case of sulfur content, which shows a larger variability (Figure 4b). The explanation for this behavior is not straightforward, as the sulfonation of the aromatic groups takes place at the same time as other side reactions and strongly depend on the chemical nature of the solids, which present different features as it will be shown in NMR study (Figure 5).

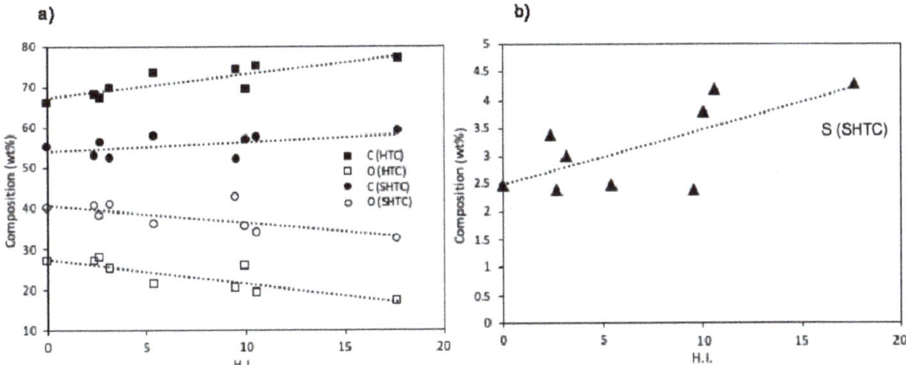

Figure 4. Relationship between hydrothermal index (H.I.) of HTC and composition of HTCs and SHTCs: (**a**) Carbon and oxygen content (wt%); (**b**) sulfur content (wt%).

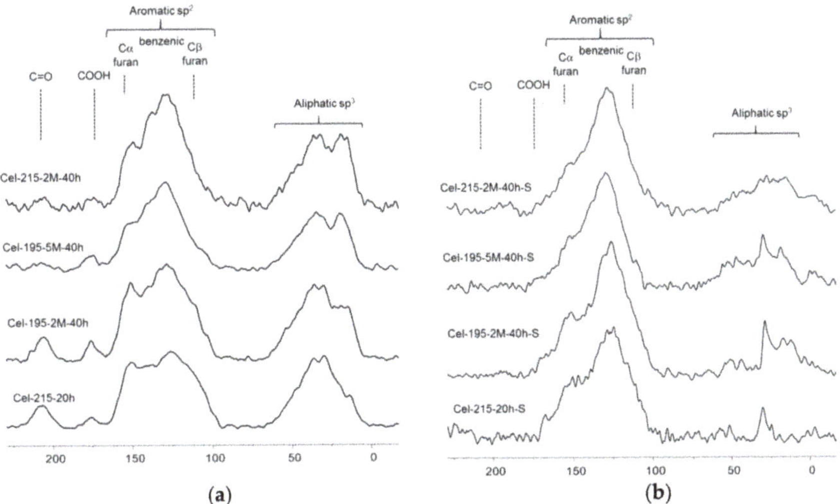

Figure 5. ^{13}C-CP-MAS-RMN spectra of: (**a**) HTC (samples 4, 7, 8, 9); (**b**) SHTC (samples 4-S, 7-S, 8-S, 9-S).

This sulfur content (Table 1 and Figure 4b) indicates sulfonic acidities in the range of 0.75–1.35 mmol/g, with a certain dispersion but slight increasing trend with H.I., which contrasts with the case of carboxylic acidity (Figure 2b), calculated by the difference between total acidity (determined by titration) and number of sulfonic groups (calculated from sulfur analysis). The materials prepared under harsher conditions (higher hydrothermal index) present significantly lower acidity, in agreement with the lower oxygen content (Figure 4a), which would indicate the formation of higher aromatic and condensed materials.

As previously described [32], ^{13}C-CP-MAS-NMR spectra of HTC from cellulose show three main types of carbon atoms: carbonyl groups (C=O and COOH), aromatic sp^2 carbons and aliphatic sp^3 carbons (Figure 5a). Upon sulfonation, the spectra of SHTC samples (Figure 5b) show in all the cases a drastic reduction in the contribution of aliphatic groups, together with a decrease in the contribution of carbonyl groups, and a higher graphitization degree, shown by the lower contribution of furanic aromatic carbons. These results seem to be contradictory with the increase in oxygen content determined by elemental analysis and XPS. However, the spectra of HTC and SHTC are not directly comparable, as they were registered using the cross-polarization (CP) technique, which enhances the signal of the carbon atoms close to hydrogen atoms. The sulfonation also produces a decrease in the hydrogen content of the hydrothermal carbons, a consequence of the larger condensation and graphitization processes, lowering in this way the intensity of the ^{13}C-CP signals.

The analysis of the nature of the acid sites on the solids has been carried out by ^{31}P-MAS-NMR using triethyl phosphine oxide (TEPO) as probe molecule (Figure 6). The spectra show a very broad signal in the range of 50–95 ppm due to the contribution of different acidic sites. The signal deconvolution evidences the presence of arylsulfonic sites at 82–85 ppm [37,38], although in a much lower contribution than expected. In fact, the contribution of carboxylic sites (signal at 60–64 ppm [37]) is also important. However, a signal at 70–73 ppm, in the range of alkylsulfonic sites [37,38], appears as the major contribution in some of the spectra (Figure 6). In fact, this signal had been already detected by other authors, but it has not been interpreted [39], as it is difficult to envisage the formation of such kind of sites in a sulfonation process with sulfuric acid, that should take place on aromatic groups (electrophilic aromatic substitution) leading to arylsulfonic sites. Given that the adsorbed TEPO/SO$_3$H molar ratio is only 0.8, and taking into account the big difference in pK$_a$ between sulfonic and carboxylic acids, the important signal at 60–64 ppm seems to indicate the existence of diffusion limitations to get access to part of the sulfonic sites, at least in the adsorption conditions (r.t., methanol as solvent), that may condition the catalytic activity of the SHTCs. This fact is due to the flexibility of these materials, which strongly depends on the polarity of the media, as it has been previously shown for Glu-195-20 h [31].

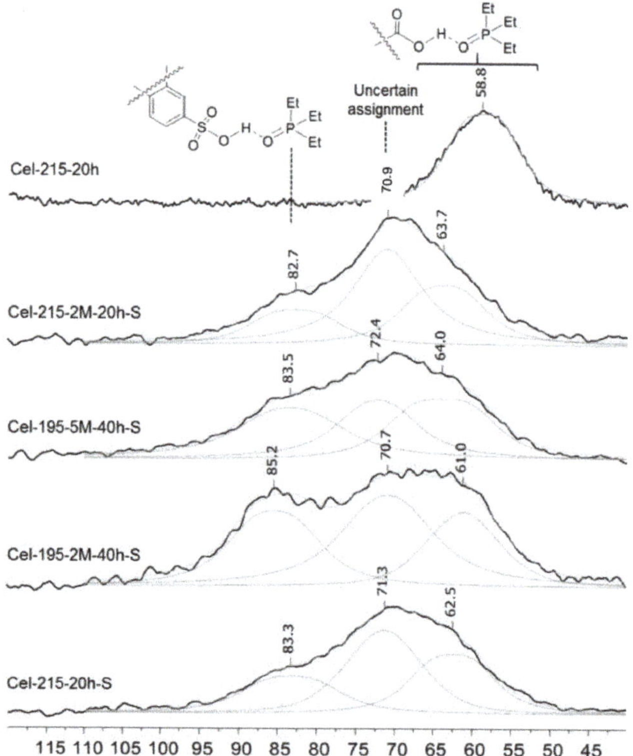

Figure 6. ^{31}P- MAS-RMN spectra of TEPO adsorbed on one HTC sample (Cel-215-20 h, sample 4) and SHTC (samples 6-S, 8-S, 7-S, and 4-S from the top to the bottom).

2.2. SHTC on Graphite Felt: Preparation and Characterization

In a previous work [12], we described the preparation of SHTC-covered graphite felt (SHTC@GF) from glucose and its characterization by different methods. SEM images showed that the felt microfibers were homogeneously coated by a 300–350 nm HTC layer, which was stable to sulfonation. The SHTC loading was determined from the weight loss in TPO-MS (Temperature Programme Oxidation- Mass Spectroscopy) experiments

Although the SHTC@GF samples could also be used in batch reactors, the main purpose for these samples in this work to demonstrate the reactions in continuous flow reactors. Thus, the felt mats of 5 mm thickness were cut into disks of 16 mm of diameter (Figure 7) to allow the tightly fitting inside the reactor (Figure 7). In this way, the length of the bed can be increased by numbering up several felt disks in a pile. To convert the graphite felts into the structured catalyst SHTC@GF, the microfibers were coated by a HTC layer and then sulfonated as described in the experimental section.

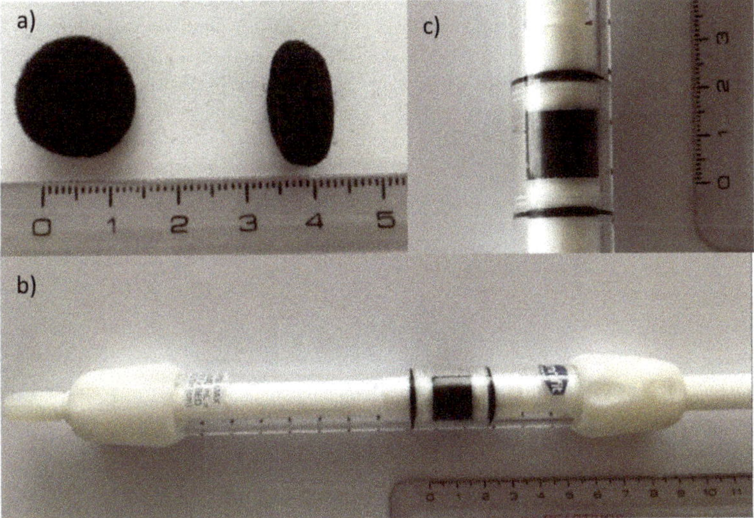

Figure 7. Disks of SHTC@GF (**a**) reactor with three tightly piled disks (**b**) and enlarged image of the piled disks (**c**).

The percentage of hydrothermal carbon coating was estimated by TPO-MS. The weight loss in the range of 300–500 °C was 19.4 wt% which corresponds to HTC, whereas the graphite felt starts to burn out above 610 °C. The sulfur content of SHTC@GF was 0.3 wt%, corresponding to a sulfur loading on the SHTC coating of 0.48 mmol g^{-1}, lower than the content in unsupported SHTC, 0.77 mmol g^{-1} (Table 1 sample 1-S). This suggest that the sulfonation process is more effective in the powdered sample than in the HTC@GF.

The graphite felt was also covered with Cel-215-2 M-20 h (sample 6), the SHTC from cellulose that led to the best catalytic results in batch reactions (see Section 2.3). In this case, the carbon coating was estimated to be 21.0% by TPO-MS. The sulfur content was 0.29 wt%, corresponding to a sulfur loading on the SHTC coating of 0.43 mmol g^{-1} similar to the functionalization of SHTC@GF and again lower than the one of the powdered sample, 0.73 mmol g^{-1} (Table 1, sample 6-S), thus confirming a more difficult sulfonation of hydrothermal carbons coated on felts.

2.3. Catalytic Performance of SHTCs in the Synthesis of Solketal: Batch Reactions

The SHTCs were tested as catalysts in the synthesis of solketal (Scheme 1) at 25 °C using an acetone: glycerol molar ratio of 7:1 and 1 wt% of the catalyst with respect to glycerol. Solketal yields were determined by GC. As functionalization of the SHTCs was different, the initial glycerol/SO$_3$H ratio varied from 823 to 1522. Thus, their catalytic activity is compared using initial TOF (Turn Over Frequency) values (h^{-1}) (evolution of solketal yields with time is gathered in supplementary information). A broad dispersion of the results was obtained, from 193 h^{-1} for Cel-195-2 M-40 h-S (sample 7-S in Table 1, H.I. 10.0) up to 2194 h^{-1} for Cel-215-2 M-20 h-S ((sample 6-S in Table 1, H.I. 9.5) and with an activity result (571 h^{-1}) for Glu-195-20 h-S ((sample. 1-S, H.I. 0.0). From the results obtained it is difficult to find a clear relationship between the catalytic activity and any of the parameters obtained by the different characterization techniques (acid content, surface area, pore volume, acid density, acidity both total and sulfonic, and hydrothermal index). The plot of TOF vs. sulfonic content is represented in Figure 8. It seems that there is a trend of decrease in activity for increasing sulfonic content, with the catalysts prepared from cellulose with HCl for 20 h as the most active ones. Interestingly, the other two catalysts with better performance than Glu-195-20 h-S are also those prepared from cellulose for 20 h of hydrothermal synthesis. As pointed by the ^{31}P-NMR experiments of TEPO adsorption, part of the sulfonic sites seems not to be accessible for this probe molecule, depending on the nature of the SHTC. The situation is even more complicated in the case of the reaction, as acetone and glycerol are only partially miscible and the adsorption of both reactants might be conditioned also by the hydrophilicity/hydrophobicity character of the catalyst. This fact would introduce an unknown factor to the catalytic activity, and probably the hydrothermal synthesis for longer times is detrimental in this respect.

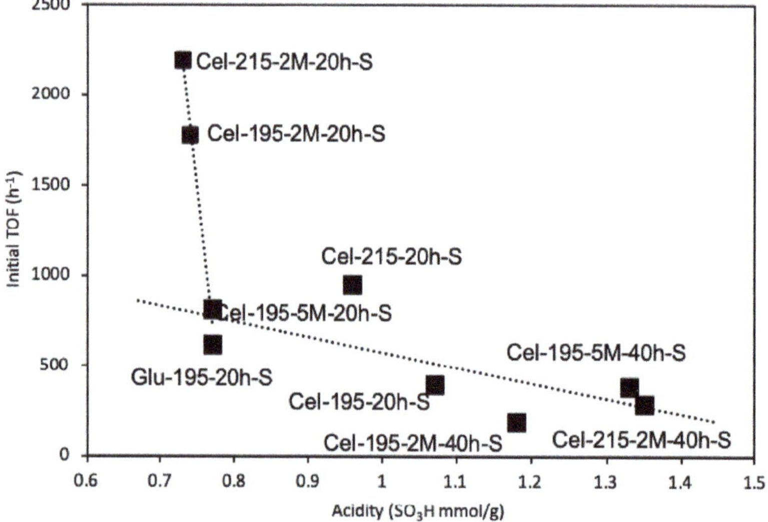

Figure 8. Initial TOF in solketal synthesis vs. sulfonic acidity plot with fresh SHTCs as catalysts (reaction conditions: acetone: glycerol molar ratio 7:1, catalyst 1% *w/w* with respect to glycerol, 25 °C).

In all the cases, yields in the range of 80–86% were obtained after 2–4 h, depending on the nature of the SHTC. It is noteworthy that these yields are comparable to previously described in the literature with different heterogeneous catalysts [25,29], but in our case reactions were carried out at r.t. while reactions with activated carbons and zeolites are described at 80 °C and 70 °C, respectively.

In order to compare the activity of SHTC with commercial sulfonic solids, two arylsulfonic resins (Dowex 50Wx2 and Amberlyst A15), Nafion-silica SAC-13 with perfluoroalkylsulfonic sites, and Deloxan ASP with alkylsulfonic sites were tested in the solketal synthesis using the same reaction conditions as with SHTCs (see experimental section). These commercial catalysts showed similar or significantly lower activity per sulfonic site, an effect that had been already observed in esterification reactions [37,40]. The activity of aryl sulfonic resins seems to depend on the cross-linking degree, with values similar to those of the SHTCs for the resin with low cross-linking degree (653 h^{-1} for Dowex 50W with 2% cross-linking) and very poor catalytic activity with a macroreticular resin (35 h^{-1} with Amberlyst 15) This effect evidences the importance of the swelling of the resin in order to facilitate the accessibility to the catalytic sites, thus Dowex 50Wx2 with a lower crosslinking degree has a bigger swelling capability that allowed a better diffusion of the reactants and a better accessibility to the active sites. Perfluoroalkyl sulfonic sites do not show better activity in spite of their stronger acidity and the lack of diffusion problems in Nafion-silica SAC-13 (408 h^{-1}). Alkyl sulfonic sites also display lower activity (148 h^{-1} with Deloxan ASP) in agreement with their weaker acidity. Weaker acid sites, such as carboxylic acids present in an acrylic resin (Dowex CCR2), were not active at all. (plots of productivity values, calculated as mmol of solketal produced per mmol of sulfonic sites, vs. time for these reactions are gathered in the ESI).

A deactivation mechanism of Glu-195-20 h-S based on the esterification of the surface acid sites, promoted by their close proximity, was evidenced by ^{13}C-CP-MAS-NMR when this catalyst was used in the esterification of fatty acids in methanol at high temperature [30,33]. However, the mild conditions of temperature in the solketal synthesis (25 °C) seemed to be favorable for an efficient recovery and reuse of the SHTCs. Surprisingly, Glu-195-20 h-S was strongly deactivated as the initial TOF decreased from 618 to 0 in the second run (Figures 8 and 9). On the contrary, the SHTCs from cellulose were recoverable. TOF values for reused catalysts (calculated with the functionalization data from the non-used catalyst) are shown in Figure 9. As it can be seen, a narrower range of TOF values was observed than in the case of the first run (fresh catalysts), with values between 548 and 800 h^{-1}, and only one exception, Cel-195-5 M-20 h-S (sample. 5-S in Table 1, H.I. 5.4) with TOF of 1048 h^{-1}. Similar values of TOF are obtained when using the functionalization values of the used catalysts for its calculation (see Figure S18 in ESI). This seems to indicate that, in spite of the very large amount of water used for washing during the preparation of SHTC, the best fresh catalysts may still retain some weakly adsorbed sulfuric acid, which is removed during their first use in the solketal synthesis. This is evidenced by the loss of sulfur observed in used SHTCs samples (Table 1) On the contrary, the improvement upon recycling of the worst fresh catalysts might be explained by some kind of pore clogging produced during preparation, which might be removed in the first reaction, and hence all the catalysts show a similar performance in the second run. Plots of productivity values (mol of solketal produced per mol of sulfonic sites) vs. time upon reuse for all the catalysts are provided in the supplementary information. From the results herein described, no relationship between TOF of the reused catalysts and the loss of sulfur could be stablished, indicating that the loss of active sites by leaching was not the only deactivation mechanism. Other possible reasons for the changes in TOF values for fresh and reused catalysts might be the adsorption of glycerol on the highly hydrophilic catalyst, the chemical reaction of the acidic sites [30,37] or the pore blocking due to by-products adsorption; however, due to the low temperature reaction conditions (r.t), the most probable deactivation mechanism is more likely to be the adsorption of glycerol or by-products on the surface of the catalyst.

Finally, it is noteworthy that the SHTC, from cellulose, performed reasonably well even at room temperature in the synthesis of solketal, and the TOFs herein described (from 193 to 2194 h^{-1}) are relatively high and comparable to the ones of commercial sulfonic resin Dowex 50Wx2 described in this work (653 h^{-1}).

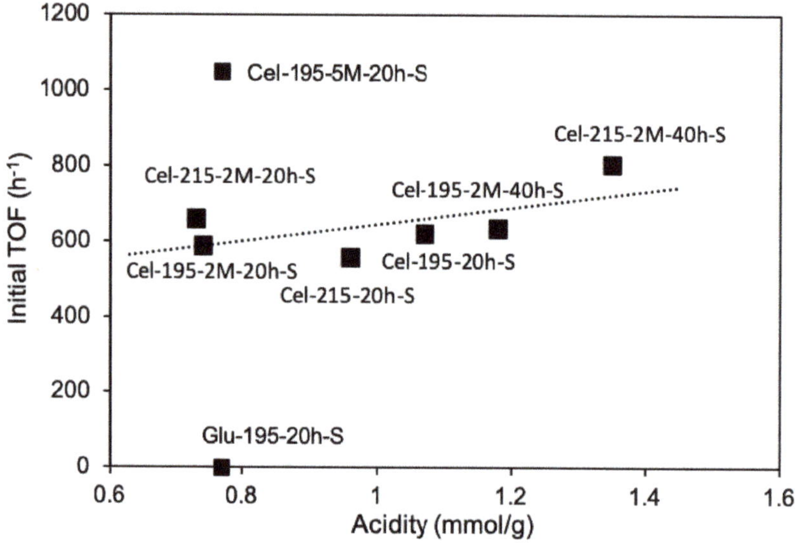

Figure 9. Initial TOF in solketal synthesis vs. sulfonic acidity plot with second-used SHTC as catalyst (reaction conditions: acetone: glycerol molar ratio 7:1, catalyst 1% *w/w* with respect to glycerol, 25 °C).

2.4. Catalytic Performance of SHTC in the Synthesis of Solketal: Continuous Flow Reactions

Once all the SHTC solids were tested and proved their activity in the reaction of glycerol with acetone, two of the catalysts were selected in order to carry out continuous flow reactions for the synthesis of solketal: the best SHTC coming from cellulose, Cel-215-2 M-20 h-S (sample 6-S in Table 1, H.I. 9.5) and Glu-195-20 h-S (sample 1-S in Table 1), in order to compare two carbons from different sources. Thus, graphite felts covered with Glu-195-20 h-S and Cel-215-2 M-20 h-S were tested in the continuous acetalization of glycerol and acetone at 25 °C using the reactor shown in Figure 7 and the flow system described in the Materials and Methods section.

In the developing of a continuous production of solketal, parameters such as acetone/glycerol molar ratio and flow rate were optimized using the Glu-195-20 h-S@Graphite Felt. As described in the experimental section, SHTC@GF disks were stacked inside the reactor, through which glycerol and acetone passed and reacted. We carried out two sets of experiments at 25 °C with two different acetone/glycerol molar ratios, 9:1 and 4:1, while varying the weight hourly space velocity (WHSV) of glycerol in between 336 and 1680 g glycerol h^{-1} g^{-1} of catalyst (Figure 10).

The results show that glycerol conversion declined as the WHSV increased, but passing through a maximum at an intermediate value which depended on the acetone/glycerol molar ratio. For example, the solketal productivity (mmol of solketal produced per gram of hydrothermal carbon per hour) exhibited a maximum of 2048 mmol g^{-1} h^{-1} for an intermediate WHSV of 1175 g of glycerol $h^{-1}g^{-1}$ of catalyst with an acetone/glycerol ratio of 9, whereas for an acetone/glycerol molar ratio of 4 the maximum productivity was 671 mmol g^{-1} h^{-1} for an intermediate WHSV of 627 g of glycerol h^{-1} g^{-1} of catalyst. Considering these results, a ratio of acetone/glycerol 9:1 and a WHSV of 1175 h^{-1} were selected to carry out a preliminary study of solketal productivity vs. time.

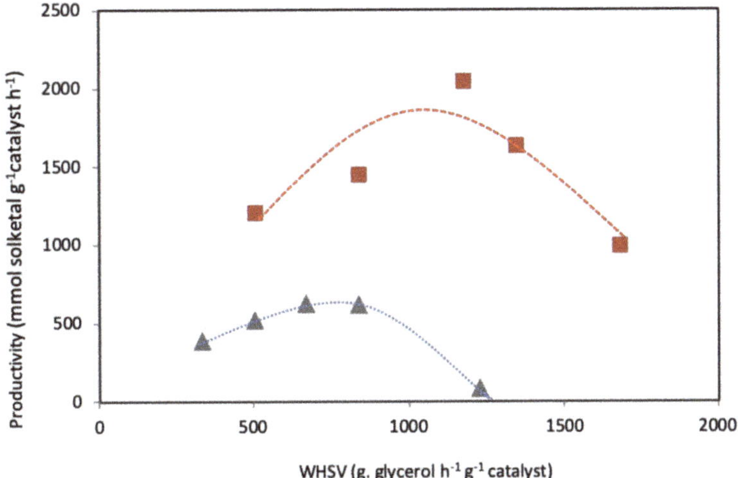

Figure 10. Solketal productivity as a function of WHSV for constant acetone/glycerol ratios of 9:1 (■) and 4:1 (▲). (Reaction conditions: 45 mg of Glu-195-20 h-S deposited on GF, 25 °C, residence time 1 min).

Figure 11 shows that solketal productivity (defined as moles of product per g of catalyst and per hour) for the Glu-195-20 h-S based catalyst decreased gradually with time on stream, with almost complete deactivation after only 1 h. The temperature programmed oxidation of the used catalyst (Glu-195-20 h-S@GF) showed a 65% weight loss at 170–200 °C, which can be unambiguously ascribed to the accumulation of glycerol on the structured catalyst since the flash point of glycerol is 176 °C. Therefore, the accumulation of unconverted glycerol on the catalyst surface, appears to be due to the high hydrophilicity of the SHTC and the high viscosity of glycerol, resulting in the blocking of the catalytic sites and deactivation of the catalyst under flow conditions, which can also explain the deactivation of the catalysts in the case of batch reactions. In fact, when the deactivated catalyst (Glu-195-20 h-S@GF) was washed with hot water at 60 °C for 30 min at a flow of 3 mL/min, in order to remove glycerol in the solid, the activity was partially recovered, thus confirming the hypothesis of the presence of glycerol blocking the catalytic sites and the need of a regeneration step under these reaction conditions.

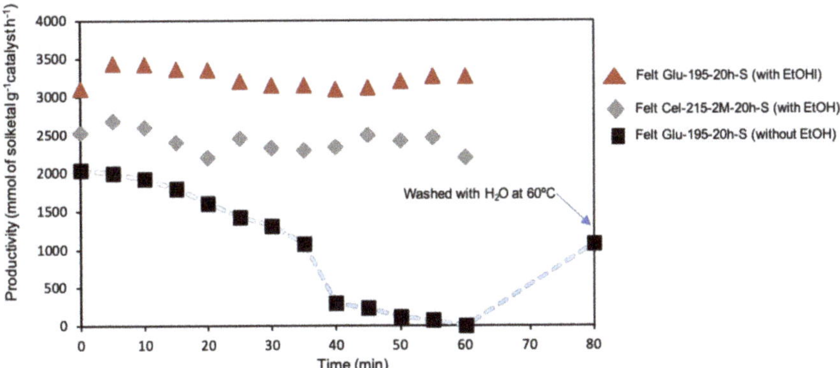

Figure 11. Solketal productivity in continuous flow at 25 °C using 10 mmol/min of glycerol (WHSV 1175 h^{-1}) and acetone/glycerol molar ratio of 9.

In order to overcome deactivation by glycerol adsorption and the need of catalyst regeneration, EtOH (20% *v/v*) was fed together with the mixture of acetone and glycerol. In this way, reactants dilution and mixing in one single liquid phase is favored (since acetone and glycerol are only partially miscible) which resulted in a relatively constant solketal productivity in between 3100 and 3500 mmol g^{-1} of catalyst h^{-1} maintained for 1 h (Figure 11), indicating that this strategy also prevented the accumulation of unconverted glycerol on the catalyst, at least at short times on stream. This productivity is much higher than the initial one using SHTC@GF in a batch process in the same conditions (ca. 1770 mmol g^{-1} of catalyst h^{-1} in 15 min residence time). These productivities are far higher than previous ones described in the literature [23], where a productivity of 20.65 (mmol of solketal g^{-1} of catalyst h^{-1}) was achieved using Amberlyst 36 as catalyst with a WHSV of 2 g of glycerol g^{-1} of catalyst h^{-1}, a molar ratio of acetone/glycerol 4:1 and at 25 °C.

Finally, felts covered with Cel-215-2 M-20 h-S were also tested in the flow reaction using optimized conditions. In this case, productivities similar to those obtained with Glu-195-20 h-S@GF were achieved, resulting in steady performance for 1 h time on stream (Figure 11). Contrary to what has been observed in batch reactor, higher productivities were obtained when using Glu-195-20 h-S@GF. In this case, deposition of the carbon over the felt can slightly modify textural properties and accessibility to sulfonic sites, thus modifying the activity of the catalyst to some extent.

Overall, these results demonstrate the utility of using this type of supported sulfonated hydrothermal carbons in order to implement continuous flow systems for the production of biomass derived chemicals, such as solketal.

3. Materials and Methods

Sulfuric acid (>95%, Fischer, Madrid, España), acetone (99.8%, Fischer), absolute ethanol (Labkem-Labbox, Villasar de Dalt, Barcelona, España), glycerol (99.5%, Sigma-Aldrich, Madrid, España), microcrystalline cellulose (Merck, Kenilworth, NJ, USA) and 1-methyl-naphtalene (96%, Alfa Aesar, Ward Hill, MA, USA) were used as received without further purification. The catalysts Dowex 50Wx2, Amberlyst 15, Nafion-silica SAC-13 and Dowex CCR2 were purchased from Sigma-Aldrich, and Deloxan ASP was purchased form Degussa (Frankfurt, Germany).

3.1. Catalysts Synthesis and Characterization

Glucose sulfonated hydrothermal carbon was prepared in a two-step procedure: hydrothermal synthesis at 195 °C and sulfonation with concentrated sulfuric acid at 150 °C [30]. The carbon preparation from cellulose follows a similar procedure, in this case HCl can be used to favor the pre-hydrolysis of the raw material [32]. For simplicity, all the solids are noted in the main text indicating their origin (glucose by Glu, or cellulose by Cel) and the preparation conditions such as temperature (195 °C or 215 °C), concentration of HCl acid (2 M or 5 M) and time of hydrothermal treatment (20 h or 40 h).

In a typical procedure, 2 g of microcrystalline cellulose in 10 mL of pure distilled water (alternatively, 10 mL of either 2 M or 5 M HCl solution were used) were introduced in a 40 mL Teflon-lined autoclave. The autoclave was closed and introduced in an oven at the desired temperature (195 °C or 215 °C) and it was maintained inside the oven during the desired time (20 h or 40 h). The obtained carbon was filtered off and thoroughly washed with water and acetone. Subsequently, the carbons were treated with concentrated (>96%) sulfuric acid (20 mL H_2SO_4/g solid) in a round-bottom flask furnished with a reflux condenser under argon atmosphere for 15 h at 150 °C. The sulfonated samples were then washed thoroughly with hot distilled water and acetone and dried in an oven overnight in static air at 105 °C.

A 5 mm-thick graphite felt (GF) sheet was cut into 16 mm diameter circular pieces, which were washed with acetone and treated with nitric acid (65%) in a flask furnished with a reflux condenser with gentle stirring at 80 °C for 16 h. The hydrothermal carbon covered graphite felt (HTC@GF) was prepared from GF and glucose or cellulose by hydrothermal synthesis into an autoclave vessel, followed by sonication in water for 10 min and final washing with ethanol to eliminate the hydrothermal carbon loosely bound to the graphite felt. The sulfonation step was carried out under the same conditions used for the HTC, which led to the formation of SHTC@GF. Additional experimental details have been described elsewhere [12].

The obtained solids were characterized by the following techniques. Elemental analysis (C,H,S), were carried out in an Thermofisher Flash 1112 elemental analyzer (Waltham, MA, USA) and CO_2 adsorption was carried out for surface area and pore volume distribution determination. CO_2 adsorption (Dubinin-Radushkevich) was determined at 0 °C using a Micromeritics ASAP 2020 apparatus (Norcross, GA, USA) after outgassing for 4 h at 150 °C. For the estimation of the surface area, the Dubinin–Astakhov equation was used.

Scanning electron microscopy (SEM) was carried out with a SEM EDX Hitachi S-3400 N microscope (Tokyo, Japan) with variable pressure up to 270 Pa and with an EDX Röntec XFlash of Si(Li) analyzer (Berlin, Germany). The samples were sputtered with gold previously to measurements and the images were obtained from the secondary electron signal. The mean particle size was determined by measuring 100 particles from images at different locations of the sample.

XPS spectra were recorded with an ESCAPlus Omnicrom system (Taunusstein, Germany) equipped with an Al K radiation source to excite the sample. Calibration of the instrument was done with Ag 3d5/2 line at 368.27 eV. All measurements were performed under UHV, better than 10–10 Torr. Internal referencing of spectrometer energies was made using the dominating C 1s peak of the support at 284.6 eV. The program used to do curve fitting of the spectra was CasaXPS using baseline Shirley method.

Back titration with NaOH was used for the determination of the total amount of acid sites. The solid sample (30 mg) was added to 25 mL of 0.01 M NaOH solution and allowed to equilibrate under stirring for 1 h. Thereafter, it was titrated with 0.05 M potassium hydrogen phthalate solution using a Crison pH Burette 24. The use of potassium hydrogen phthalate (pKa 5.4) avoids the reaction of this acid with the neutralized acid sites on the solid, thus precluding the underestimation of the acid sites, besides it avoids the filtering of the solid before titration that might be sources of errors in the site determination.

Total acid density was defined as the number of total acid sites per m^2 and calculated from titration values and surface area data.

Sulfonic sites density was defined as number of sulfonic sites per m^2 and calculated from sulfur content and surface area data.

Solids were also studied by ^{13}C-CP-MAS-NMR (cross polarization–magic angle spinning–nuclear magnetic resonance). NMR spectra were recorded in a Bruker Avance III WB400 spectrometer (Billerica, MA, USA) with 4 mm zirconia rotors spun at magic angle in N_2 at 10 kHz. ^{1}He-^{13}C CP (cross-polarization) spectra (up to 10,000 scans) were measured using a ^{1}H π/2 pulse length of 2.45 µs, with a contact time of 2 µs, and spinal-64 proton decoupling sequence with a pulse length of 4.6 µs.

Triethylphosphine oxide (TEPO) adsorption and ^{31}P-MAS-NMR analysis were also carried out. In a typical procedure, 25 mg of SHTC was suspended in a TEPO methanol solution. The mixture was stirred for 2 h, the solvent was removed by vacuum distillation and the solid TEPO-SHTC sample was analyzed by NMR.

3.2. Acetalization Reactions of Glycerol with Acetone Catalyzed by SHTC in a Batch Reactor

In a typical catalytic test, 4.5 g (0.05 mol) of highly purified glycerol, 25 mL of acetone (0.34 mol, 1 wt% (45 mg) of the catalyst and 0.675 g of 1-methylnaphtalene (GC internal standard), were weighed in a 50 mL glass round bottom flask. The mixture was stirred at 800 rpm at room temperature. The reaction was monitored by gas chromatography by taking samples were taken at different times, which were diluted in methanol and micro-filtrated prior to injection and analyzed in a Agilent 7890 GC with a FID detector and a Zebron inferno column.

In all cases, the liquid reaction mixture was originally biphasic and became monophasic during the reaction.

After the reaction, the catalysts were filtered off, washed with methanol and acetone and dried at 105 °C overnight before reuse.

In the case of the comparison study of the activity of SHTC with commercial sulfonic solids the amount of catalysts was adjusted depending on the functionalization of the resins Dowex 50Wx2 (4.7 mmol SO_3H g^{-1}), Amberlyst A15 (4.76 mmol SO_3H g^{-1}), Nafion-silica SAC-13 (0.16 mmol SO_3H g^{-1}) and Deloxan ASP (0.8 mmol SO_3H g^{-1}). Thus in a typical experiment, 0.15 mol of highly purified glycerol, 2 g of 1-methylnaphtalene (GC internal standard), 77 ml of acetone and the amount of catalyst corresponding to a 15% mol ratio of sulfonic sites/mol of glycerol were weighed in a 100 mL glass round bottom flask. The mixture was stirred at 800 rpm at room temperature. The reaction was monitored by gas chromatography by taking samples were taken at different times, which were diluted in methanol and micro-filtrated prior to injection and analyzed in an Agilent 7890 GC with a FID detector and a Zebron inferno column.

In all the cases, TOF was calculated as mol of glycerol reacted per hour and per mol of sulfonic sites, as carboxylic sites were not active in this reaction as it is demonstrated by the lack of activity of carboxylic resin Dowex CCR2 and non-sulfonated HTC (Glu-195-20 h), whereas productivity was calculated as mol of solketal per mol of sulfonic sites.

3.3. Acetalization Reaction of Glycerol with Acetone in a Continuous Flow Reactor

The continuous acetalization of glycerol with acetone was carried out in a flow system, as that schematized in Figure 12. The reactor was an Omnifit®chromatography column 15 mm in internal diameter and a variable length of up to 100 mm where three SHTC felts were stacked (ca. 75 mg of SHTC). The tubing was 1.6 mm O.D. made of PTFE and connected with luer type connections. The reaction was performed at room temperature and atmospheric pressure. Glycerol and acetone were fed with a "Gemini 88" infusion pump through two independently controlled glass syringes with a total flow from 2 to 17 ml/min. For the case of a single homogeneous liquid phase, glycerol and acetone were previously mixed and 20% *v/v* of absolute EtOH was added. A total flow of 5 ml/min was used. The reaction samples were collected each minute in sealed vials, diluted in methanol and analyzed by GC using 1-methylnaphthalene as standard.

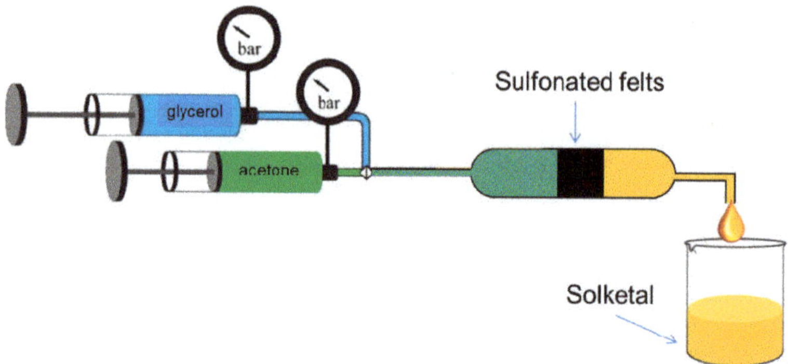

Figure 12. Experimental set-up for continuous production of solketal in a flow reactor.

The reaction products were analyzed by gas chromatography (GC) using an Agilent 7890 GC with a FID detector and a Zebron inferno column.

4. Conclusions

Active carbon-based acid catalysts were prepared by hydrothermal treatment of glucose and cellulose and their subsequent sulfonation. It was found that the starting material (glucose or cellulose) and the synthesis parameters strongly influenced the nature of the carbon catalyst. All the sulfonated carbons showed a high activity in the synthesis of solketal at room temperature in a batch reactor, this activity was comparable to the one by Dowex 50Wx2 resin and higher than the ones obtained with Amberlyst 15, Deloxan or composites such as Nafion-silica-SAC-13. Differences in initial TOFs observed for the sulfonated carbons were difficult to relate to any structural parameter (acid content, surface area, pore volume, acid density, acidity both total and sulfonic, and hydrothermal index). Nevertheless, the solids obtained at longer hydrothermal treatment times were generally less active, probably due to a poor access of the reactants to catalytic sites. The use of felts covered with sulfonated carbons derived from both glucose and cellulose allowed the design of a flow system for continuous solketal production. The use of ethanol in the feeding mixture was crucial to avoid catalyst deactivation and to maintain solketal productivity at least at short times on stream (one hour).

Supplementary Materials: The following are available online at http://www.mdpi.com/2073-4344/9/10/804/s1, Hydrothermal index definition, SEM images of HTC and SHTC, Plots of CO_2 adsorption isotherms and pore distribution of HTC and SHTC, Relationship between H.I. and micropore volume, Productivity vs time plots for the synthesis of Solketal catalyzed by sulfonated hydrothermal carbons from cellulose prepared at 195 °C, Productivity vs time plots for the synthesis of Solketal catalyzed by sulfonated hydrothermal carbons from cellulose prepared at 215 °C, Productivity vs time plots for the synthesis of Solketal catalyzed by commercial resins and Glu-195-20 h-S, Initial TOF in solketal synthesis vs sulfonic acidity of reused catalyst plot with second-used SHTC as catalyst.

Author Contributions: E.G.-B. was responsible for hydrothermal carbon synthesis, J.M.F. carried out catalyst characterizations and E.P. and P.F. performed catalytic tests. J.M.F., E.G.-B. and E.P. are responsible for conceptualization and discussion of the results and contributed equally to the writing and design of the manuscript. J.M.F. and E.G.-B. secured funding for the research work.

Funding: Financial support from Ministerio de Ciencia, Innovación y Universidades (project RTI2018-093431-B-I00 and ENE2016-79282-C5-1-R) and the Gobierno de Aragón (Group E37_17R) co-funded by FEDER 2014-2020 "Construyendo Europa desde Aragón" is acknowledged.

Conflicts of Interest: The authors declare no conflict of interest.

References

1. Abate, S.; Lanzafame, P.; Perathoner, S.; Centi, G. New Sustainable model of biorefineries: Biofactories and challenges of integrating bio- and solar refineries. *ChemSusChem* **2015**, *8*, 2854–2866. [CrossRef] [PubMed]

2. Bobleter, O.; Niesner, R.; Röhr, M. The hydrothermal degradation of cellulosic matter to sugars and their fermentative conversion to protein. *J. Appl. Polym.* **1976**, *20*, 2083–2093. [CrossRef]

3. Bonn, G.; Concin, R.; Bobleter, O. Hydrothermolysis—A new process for the utilization of biomass. *Wood Sci. Technol.* **1983**, *17*, 195–202. [CrossRef]

4. Karagöz, S.; Bhaskar, T.; Muto, A.; Sakata, Y.; Oshiki, T.; Kishimoto, T. Low-temperature catalytic hydrothermal treatment of wood biomass: Analysis of liquid products. *Chem. Eng. J.* **2005**, *108*, 127–137. [CrossRef]

5. Sakaki, T.; Shibata, M.; Miki, T.; Hirosue, H.; Hayashi, N. Reaction model of cellulose decomposition in near-critical water and fermentation of products. *Bioresour. Technol.* **1996**, *58*, 197–202. [CrossRef]

6. Titirici, M.-M.; Antonietti, M. Chemistry and materials options of sustainable carbon materials made by hydrothermal carbonization. *Chem. Soc. Rev.* **2010**, *39*, 103–116. [CrossRef]

7. Titirici, M.-M.; Thomas, A.; Antonietti, M. Back in the black: Hydrothermal carbonization of plant material as an efficient chemical process to treat the CO_2 problem? *New J. Chem.* **2007**, *31*, 787–789. [CrossRef]

8. Hu, B.; Wang, K.; Wu, L.; Yu, S.-H.; Antonietti, M.; Titirici, M.-M. Engineering carbon materials from the hydrothermal carbonization process of biomass. *Adv. Mater.* **2010**, *22*, 813–828. [CrossRef]

9. Titirici, M.-M.; Antonietti, M.; Baccile, N. Hydrothermal carbon from biomass: A comparison of the local structure from poly- to monosaccharides and pentoses/hexoses. *Green Chem.* **2008**, *10*, 1204–1212. [CrossRef]

10. Demir-Cakan, R.; Baccile, N.; Antonietti, M.; Titirici, M.-M. Carboxylate-rich carbonaceous materials via one-step hydrothermal carbonization of glucose in the presence of acrylic acid. *Chem. Mater.* **2009**, *21*, 484–490. [CrossRef]

11. Sevilla, M.; Fuertes, A.B. Chemical and structural properties of carbonaceous products obtained by hydrothermal carbonization of saccharides. *Chem. Eur. J.* **2009**, *15*, 4195–4203. [CrossRef] [PubMed]

12. Roldán, L.; Santos, I.; Armenise, S.; Fraile, J.M.; García-Bordejé, E. The formation of a hydrothermal carbon coating on graphite microfiber felts for using as structured acid catalyst. *Carbon* **2012**, *50*, 1363–1372. [CrossRef]

13. Alatalo, S.-M.; Pileidis, F.; Mäkilä, E.; Sevilla, M.; Repo, E.; Salonen, J.; Sillanpää, M.; Titirici, M.-M. Versatile cellulose-based carbon aerogel for the removal of both cationic and anionic metal contaminants from water. *ACS Appl. Mater. Interfaces* **2015**, *7*, 25875–25883. [CrossRef] [PubMed]

14. Falco, C.; Baccile, N.; Titirici, M.-M. Morphological and structural differences between glucose, cellulose and lignocellulosic biomass derived hydrothermal carbons. *Green Chem.* **2011**, *13*, 3273–3281. [CrossRef]

15. Sevilla, M.; Fuertes, A.B. The production of carbon materials by hydrothermal carbonization of cellulose. *Carbon* **2009**, *47*, 2281–2289. [CrossRef]

16. Tekin, K.; Pileidis, F.D.; Akalin, M.K.; Karagöz, S. Cellulose-derived carbon spheres produced under supercritical ethanol conditions. *Clean Technol. Environ. Policy* **2016**, *18*, 331–338. [CrossRef]

17. Zhong, R.; Sels, B.F. Sulfonated mesoporous carbon and silica-carbon nanocomposites for biomass conversion. *Appl. Catal. B* **2018**, *236*, 518–545. [CrossRef]

18. García, J.I.; García-Marín, H.; Pires, E. Glycerol based solvents: Synthesis, properties and applications. *Green Chem.* **2014**, *16*, 1007–1033. [CrossRef]

19. Sonnati, M.O.; Amigoni, S.; Taffin de Givenchy, E.P.; Darmanin, T.; Choulet, O.; Guittard, F. Glycerol carbonate as a versatile building block for tomorrow: Synthesis, reactivity, properties and applications. *Green Chem.* **2013**, *15*, 283–306. [CrossRef]

20. García, H.; García, J.I.; Fraile, J.M.; Mayoral, J.A. Solketal: Green and catalytic synthesis and its classification as a solvent. *Chim. Oggi* **2008**, *26*, 10–12.

21. Nanda, M.R.; Zhang, Y.; Yuan, Z.; Qin, W.; Ghaziaskar, H.S.; Xu, C. Catalytic conversion of glycerol for sustainable production of solketal as a fuel additive: A review. *Renew. Sustain. Energy Rev.* **2016**, *56*, 1022–1031. [CrossRef]

22. De Torres, M.; Jiménez-Osés, G.; Mayoral, J.A.; Pires, E.; De los Santos, M. Glycerol ketals: Synthesis and profits in biodiesel blends. *Fuel* **2012**, *94*, 614–616. [CrossRef]

23. Nanda, M.R.; Yuan, Z.; Qin, W.; Ghaziaskar, H.S.; Poirier, M.A.; Xu, C. Catalytic conversion of glycerol to oxygenated fuel additive in a continuous flow reactor: Process optimization. *Fuel* **2014**, *128*, 113–119. [CrossRef]

24. Vicente, G.; Melero, J.A.; Morales, G.; Paniagua, M.; Martín, E. Acetalisation of bio-glycerol with acetone to produce solketal over sulfonic mesostructured silicas. *Green Chem.* **2010**, *12*, 899–907. [CrossRef]

25. Kowalska-Kus, J.; Held, A.; Frankwoski, M.; Nowinska, K. Solketal formation from glycerol and acetone over hierarchical zeolites of different structure as catalysts. *J. Mol. Catal. A* **2017**, *426*, 205–212. [CrossRef]

26. Li, L.; Koránayi, I.; Sels, B.F.; Pescarmona, P. Highly-efficient conversion of glycerol to solketal over heterogeneous Lewis acid catalysts. *Green Chem.* **2012**, *14*, 1611–1619. [CrossRef]

27. Ricciardí, M.; Falivene, L.; Tabanelli, T.; Proto, A.; Cucciniello, R.; Cavani, F. Bio-Glycidol Conversion to Solketal over Acid Heterogeneous Catalysts: Synthesis and Theoretical Approach. *Catalysts* **2018**, *8*, 391. [CrossRef]

28. Rodrigues, R.; Gonçalves, M.; Mandelli, D.; Pescarmona, P.P.; Carvalho, W.A. Solvent-free conversion of glycerol to solketal catalysed by activated carbons functionalised with acid groups. *Catal. Sci. Technol.* **2014**, *4*, 2293–2301. [CrossRef]

29. Gonçalves, M.; Rodrigues, R.; Galhardo, T.S.; Carvalho, W.A. Highly selective acetalization of glycerol with acetone to solketal over acidic carbon-based catalysts from biodiesel waste. *Fuel* **2016**, *181*, 46–54. [CrossRef]

30. Fraile, J.M.; García-Bordejé, E.; Roldán, L. Deactivation of sulfonated hydrothermal carbons in the presence of alcohols: Evidences for sulfonic esters formation. *J. Catal.* **2012**, *289*, 73–79. [CrossRef]

31. Fraile, J.M.; García-Bordejé, E.; Pires, E.; Roldán, L. New insights into the strength and accessibility of acid sites of sulfonated hydrothermal carbon. *Carbon* **2014**, *77*, 1157–1167. [CrossRef]

32. Fraile, J.M.; García-Bordejé, E.; Pires, E. Parametric study of the hydrothermal carbonization of cellulose and effect of acidic conditions. *Carbon* **2017**, *123*, 421–432. [CrossRef]

33. Lozano-Castelló, D.; Cazorla-Amorós, D.; Linares-Solano, A. Usefulness of CO_2 adsorption at 273 K for the characterization of porous carbons. *Carbon* **2004**, *42*, 1233–1242. [CrossRef]

34. Urbonaite, S.; Juárez-Galán, J.M.; Leis, J.; Rodríguez-Reinoso, F.; Svensson, G. Porosity development along the synthesis of carbons from metal carbides. *Microporous Mesoporous Mater.* **2008**, *113*, 14–21. [CrossRef]

35. Gauden, P.A.; Terzyk, A.P.; Jaroniec, M.; Kowalczyk, P. Bimodal pore size distributions for carbons:Experimental results and computational studies. *J. Colloid Interface Sci.* **2007**, *310*, 205–216. [CrossRef] [PubMed]

36. Okpalugo, T.I.T.; Papakonstantinou, P.; Murphy, H.; McLaughlin, J.; Brown, N.M.D. High resolution XPS characterization of chemical functionalised MWCNTs and SWCNTs. *Carbon* **2005**, *43*, 153–161. [CrossRef]

37. Fraile, J.M.; García-Bordejé, E.; Pires, E.; Roldán, L. Catalytic performance and deactivation of sulfonated hydrothermal carbon in the esterification of fatty acids: Comparison with sulfonic solids of different nature. *J. Catal.* **2015**, *324*, 107–118. [CrossRef]

38. Margolese, D.; Melero, J.A.; Christiansen, S.C.; Chmelka, B.F.; Stucky, G.D. Direct syntheses of ordered SBA-15 mesoporous silica containing sulfonic acid groups. *Chem. Mater.* **2000**, *12*, 2448–2459. [CrossRef]

39. Russo, P.A.; Antunes, M.M.; Neves, P.; Wiper, P.V.; Fazio, E.; Neri, F.; Barreca, F.; Mafra, L.; Pillinger, M.; Pinna, N.; et al. Solid acids with SO_3H groups and tunable surface properties: Versatile catalysts for biomass conversion. *J. Mater. Chem. A* **2014**, *2*, 11813–11824. [CrossRef]

40. De la Calle, C.; Fraile, J.M.; García-Bordejé, E.; Pires, E.; Roldán, L. Biobased catalyst in biorefinery processes: Sulphonated hydrothermal carbon for glycerol esterification. *Catal. Sci. Technol.* **2015**, *5*, 2897–2903. [CrossRef]

Article

C-O Bond Hydrogenolysis of Aqueous Mixtures of Sugar Polyols and Sugars over ReO$_x$-Rh/ZrO$_2$ Catalyst: Application to an Hemicelluloses Extracted Liquor

Modibo Mounguengui-Diallo [1], Achraf Sadier [1], Eddi Noly [1], Denilson Da Silva Perez [2], Catherine Pinel [1], Noémie Perret [1,*] and Michèle Besson [1,*]

[1] Univ Lyon, Universite Claude Bernard, CNRS, IRCELYON, UMR5256, 2 Avenue Albert Einstein, F-69626 Villeurbanne, France
[2] Institut FCBA, InTechFibres, Domaine Universitaire, CS 90251, F-38044 Grenoble, France
* Correspondence: noemie.perret@ircelyon.univ-lyon1.fr (N.P.); michele.besson@ircelyon.univ-lyon1.fr (M.B.); Tel.: +33-472-445-446 (N.P.); +33-472-445-356 (M.B.)

Received: 18 July 2019; Accepted: 29 August 2019; Published: 31 August 2019

Abstract: The recovery and upgrade of hemicelluloses, a family of heteropolysaccharides in wood, is a key step to making lignocellulosic biomass conversion a cost-effective sustainable process in biorefinery. The comparative selective catalytic C-O bond hydrogenolysis of C5-C6 polyols, sugars, and their mixtures for the production of valuable C6 and C5 deoxygenated products was studied at 200 °C under 80 bar H$_2$ over ReO$_x$-Rh/ZrO$_2$ catalysts. The sugars were rapidly converted to the polyols or converted into their hydrogenolysis products. Regardless of the reactants, C-O bond cleavage occurred significantly via multiple consecutive deoxygenation steps and led to the formation of linear deoxygenated C6 or C5 polyols. The distribution of products depended on the nature of the substrate and C-C bond scission was more important from monosaccharides. In addition, we demonstrated effective hydrogenolysis of a hemicellulose-extracted liquor from delignified maritime pine containing monosaccharides and low MW oligomers. Compared with the sugar-derived polyols, the mono- and oligosaccharides in the liquor were more rapidly converted to hexanediols or pentanediols. C-O bond scission was significant, giving a yield of desired deoxygenated products as high as 65%, higher than in the reaction of the synthetic mixture of glucose/xylose of the same C6/C5 sugar ratio (yield of 30%).

Keywords: hydrogenolysis; polyols; monosaccharides; hemicelluloses extracted liquor; ReO$_x$-Rh/ZrO$_2$ catalysts

---

1. Introduction

Until recently, chemicals have mostly been derived from fossil resources. The utilization of non-edible, abundant and renewable lignocellulosic biomass from agriculture and forestry has emerged as an interesting alternative for the chemical industry and polymer production in particular [1–6]. Lignocellulose consists of carbohydrate polymers (cellulose and hemicelluloses) embedded in a lignin matrix [4,7]. Cellulose (30–50%) is a water insoluble linear homopolysaccharide of β (1 → 4) linked glucose units with strong intra- and intermolecular hydrogen bonds and with a degree of polymerisation as high as 15,000 units. Different to cellulose, hemicelluloses (20–35%) are branched polymers made of structures of hexoses (D-galactose, D-mannose, D-glucose) cand pentoses (D-xylose and L-arabinose) and some sugar acid subunits, with a low degree of polymerization of 80–200 units. The composition of hemicelluloses varies depending on the source of biomass: in general, softwood hemicelluloses are mainly composed of O-acetyl-galacto-glucomannans and some arabino-4-O-methyl-glucuronoxylan [8–10], whereas in hardwoods, *O*-acetyl-4-*O*-methylglucuronoxylans are the most abundant [11,12]. In addition, lignin is an aromatic polymer with a complex composition of phenylpropane units linked by ether bonds. In order to utilize the woody biomass efficiently, all three main components should be used in a biorefinery to maximize economic returns [13].

Although cellulose has been much studied for the conversion to fuels and chemicals, there has been less effort in the conversion of hemicelluloses, which are the second most abundant polysaccharides. During the conventional Kraft process in the pulp and paper industry, the cellulose fraction is the core product to be used in papermaking, while most hemicelluloses are currently degraded and accumulated in the black liquor along with lignin, and are burnt to produce energy. However, if the hemicelluloses are pre-extracted or hydrolyzed to their structural C5 and C6 sugars—which are building blocks for numerous biomaterials or chemicals—with a minimum effect on subsequent pulping step, this makes a better valorization of this under-utilized fraction in paper mills feasible [14,15]. The potential of valorization is up to the millions of tons of wood treated by the Kraft process.

The relatively low degree of polymerization and the branched side groups allow a large fraction of hemicelluloses to dissolve in aqueous solution. Various effective methods have been proposed for hemicelluloses extraction from wood chips prior to cooking, including the use of acids, water, and alkaline solutions [16–27]. Many applications for the extracted hemicelluloses have been developed [28]. Moderate treatment provides polymeric hemicelluloses with sufficient molecular mass to be recovered by precipitation and valorized after chemical modification [29–31] as hydrocolloids [32,33], surfactants [34–36], coating materials [37], emulsion stabilizers [38] or as packaging materials [39], among others. In some more severe extraction processes where autohydrolysis takes place, the backbone of hemicelluloses is hydrolysed into oligomers with a low degree of polymerization, hexoses, pentoses, and uronic acids, which may be subsequently upgraded in further steps for various applications.

Though the pH may influence the composition of liquor after extraction (sugars, distribution of oligomers), extraction using only hot water (120–240 °C) has been shown to be an attractive environmentally friendly method [21,40]. Some auto-hydrolysis occurs due to release of acetic acid from the deacetylation of hemicelluloses and uronic acids. The glycosidic bonds between the sugar units of hemicelluloses are cleaved by a hydrolysis reaction to the partially hydrolyzed dimers, trimers and other oligomers or the sugar monomers, depending on the operating conditions and on hemicelluloses source [41]. Once separated, they can be further processed into chemicals. They may be a source for furans and their derivatives for ethanol [14,28,42–44], but also for the production of suitable monomers by catalytic hydrogenation [45–47] or oxidation [48] of the extracted sugars.

Like most of the biomass derivatives, the compounds in the extracted hemicellulosic stream are characterized by a much higher oxygen content compared with fossil-derived feedstocks. Usually, the target products, such as monomers for polymer synthesis, have a comparatively lower oxygen content. Triols and diols are useful as chemical intermediates for the production of polymers, agrochemicals and surfactants. For example, α,ω-diols, especially C5 and C6 α,ω-diols, can be used as plasticizers and as co-monomers in polyesters and polyurethanes manufacture [49–51]. Therefore, methodologies for upgrading of the hemicelluloses stream to monomers necessitate a controlled removal of oxygen—containing functional groups from sugars. One route is catalytic-selective hydrolysis/hydrogenolysis of soluble hemicelluloses/polysaccharides/monomers in the presence of supported metallic catalysts. Hydrogenolysis of C-O bonds has been applied to alcohols and polyols to produce more or less deoxygenated polyols. The transfer of hydrogenolysis has been applied to polyols over heterogeneous catalysts using H-donors (alcohols, formic acid) [52,53]. Under hydrogen pressure, although monometallic catalysts such as Cu, Ni, and Ru have been reported for dehydroxylation [54–57], a better control is obtained using catalysts that contain a noble metal (Pt, Pd, Rh, Ir) modified by an oxophilic metal (Re, Mo, W) [58]. These bimetallic catalysts have been found to be highly active and selective for —C-O- cleavage reactions of various biomass-derived substrates, including glycerol, 2-(hydroxymethyl)tetrahydropyran or tetrahydrofurfuryl alcohol (the full saturation product of furfural), and others [59,60]. Partial hydrodeoxygenation of glycerol to 1,2- or 1,3-propanediol was investigated intensively and reviewed [61–65]. It is much more difficult to selectively remove OH groups from substrates with four or more OH groups such as erythritol, xylitol, or sorbitol [66–72]. During this process, several products are very often obtained by unwanted reactions (hydrogenation, isomerization, dehydration, C-C bond cleavage by retro-aldol condensation, decarbonylation and decarboxylation).

The objective of this work was to evaluate the feasibility of catalytic hydrogenolysis of an hemicellulose extract from delignified maritime pine wood chips that contained simple sugars such as glucose or xylose or partially hydrolyzed dimers, trimers and other oligomers, to prepare more or less deoxygenated linear C6 and C5 polyols. We used ReO_x-Rh/ZrO_2 catalysts, which were particularly efficient in our previous work on the hydrogenolysis of erythritol, xylitol and sorbitol [68,69]. Before performing the catalytic hydrogenolysis of the hemicelluloses liquor, we studied the reaction of aqueous solutions of sugar polyols and sugars of different compositions.

2. Results and Discussions

2.1. Characterization of Catalysts

Three batches of ReO_x-Rh/ZrO_2 catalysts were prepared with Re/Rh molar ratios of 1.5 (8.8%Re–3.3%Rh), 1.6 (9.4%Re–3.2%Rh), and 1.8 (10.2%Re–3.0%Rh). The BET surface areas were 109 m^2 g^{-1}, slightly lower than that of the support of 129 m^2 g^{-1}. The X-ray diffraction patterns of the support and 9.4%Re–3.2%Rh/ZrO_2 are presented in Figure S1 as a representative example. Except the peaks characteristic of ZrO_2, there were no other peaks that could be assigned to Rh or Re species, suggesting a very good dispersion of Rh and Re. As observed previously, TEM and CO chemisorption confirmed a low mean particle size of ca. 3 nm [70].

2.2. Hydrogenolysis of Polyols or Sugars Separately

The reaction conditions (200 °C under 80 bar of H_2) were selected from our previous study on hydrogenolysis of sugar polyols (erythritol, xylitol, sorbitol) which allowed us to obtain the highest selectivity to deoxy C4 to C6 polyols over the ReO_x-modified Rh catalysts [70,71]. Since the extraction of hemicelluloses liberates sugars and their oligomers, we first compared the results for the hydrogenolysis of the C5 or C6 sugars with those of the corresponding C5 or C6 polyols.

2.2.1. Comparison of Sorbitol and Glucose Hydrogenolysis

As shown previously [70], many reactions take place during hydrogenolysis (Scheme 1, Scheme S1), which produces different families of products: (1) isomers of sorbitol by epimerization reaction (mannitol, iditol, and dulcitol), (2) more or less deoxygenated linear C6 compounds by multiple C-O bond cleavage reactions; (3) cyclic C6 compounds by dehydration reaction, mainly sorbitans and isosorbide and (4) products of C-C bond cleavage (C1-C5), including C5 products (such as 1,2-pentanediol, 1,5-pentanediol, 1-pentanol and tetrahydrofurfuryl alcohol among others), C4 products (such as 1,2-butanediol and 1-butanol), C3 products (such as glycerol, 1,2-propanediol and 1-propanol), as well as ethylene glycol and ethanol, and finally methanol. Specifically, C6 deoxygenated products consist of different sub-families: hexanepentaols (one C-O bond cleavage), hexanetetraols (two C-O bond cleavages), hexanetriols (3 C-O cleavages), hexanediols (4 C-O cleavages) and hexanols (5 C-O cleavages).

Hydrogenolysis reactions of sorbitol and glucose (6 g of substrate in 120 mL water, i.e., 0.25 M and 0.29 M, respectively) were compared over Rh-Re/ZrO_2 (Re/Rh molar ratio = 1.5 or 1.6) at 200 °C under 80 bar H_2 (Figure 1). Figure 1a,b shows the evolution of the concentrations of substrate (sorbitol or glucose) and families of products as a function of the time of reaction (Scheme S1). Figure 1a′,b′ shows the evolution of the sub-families of linear deoxyhexitols.

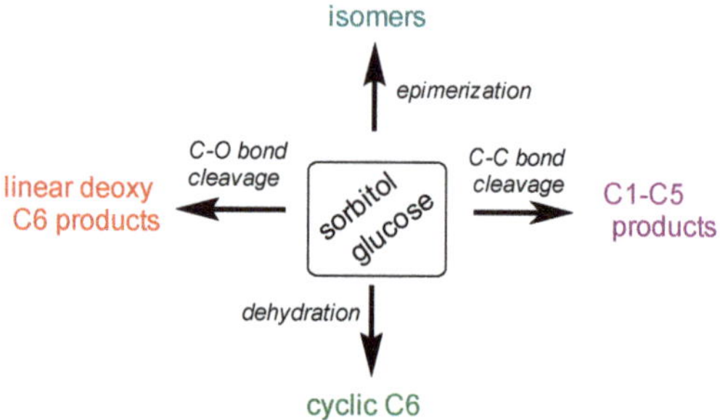

Scheme 1. The reactions taking place during the hydrogenolysis of sorbitol or glucose.

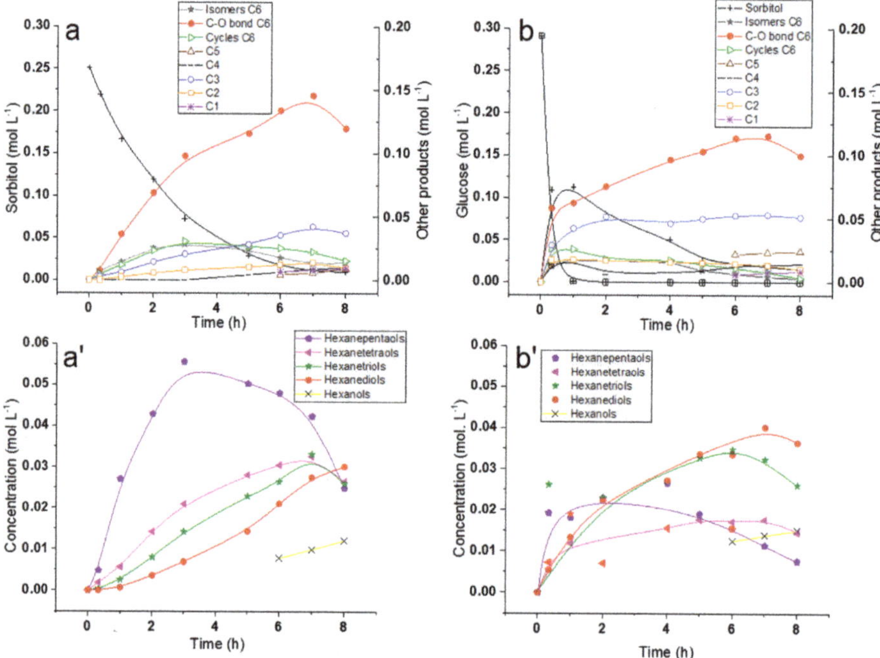

Figure 1. The evolution of the concentration of (**a**,**b**) the different families of products and (**a′**,**b′**) the sub-families of linear deoxygenated C6 compounds during hydrogenolysis of (**a**,**a′**) sorbitol (0.25 M), and (**b**,**b′**) glucose (0.29 M), over ReO_x-Rh/ZrO_2 catalysts (Re/Rh molar ratio = 1.5 for sorbitol, 1.6 for glucose) at 200 °C under 80 bar H_2, $n_{sorbitol}/n_{Rh}$ = 188, $n_{glucose}/n_{Rh}$ = 218.

Figure 1a, focused on sorbitol hydrogenolysis, indicates that the substrate was almost fully converted after an 8-h reaction. The main products formed were a mixture of linear deoxygenated C6 compounds, showing that selective C-O bond cleavage is achievable with appreciable selectivity. The concentration was increasing substantially over time to attain ca. 0.15 M (60% yield) at 7 h, in line with previous observations [57,71]. The other reaction products were formed in lower amounts. Isomerisation of sorbitol and transformation to cyclic compounds was significant from the beginning of the reaction, however, the concentrations of the isomers and cycles decreased after 3 h of reaction (maximum concentration of each ca. 0.025 M). Like sorbitol, the isomers were easily hydrogenolysed, whereas the cyclic compounds were more inert to further hydrogenolysis. C-C bond cleavage reactions also occurred and competed with C-O hydrogenolysis; the concentration of the C5, C4, C3, and C2 products increased slowly during the 8 h-reaction and accounted for very low concentrations. Finally, the C3 polyols were the main C-C bond cleavage products, with a maximum concentration of 0.04 M after 7 h (ca. 16% yield) (Figure 1a). These C3 polyols are mainly formed by retro-aldol reaction of C6 polyols [55,72].

In comparison to sorbitol (0.25 M, $n_{sorbitol}/n_{Rh}$ = 188), glucose (0.29 M, $n_{sorbitol}/n_{Rh}$ = 218) was very rapidly fully converted within only 1 h (Figure 1b). Sorbitol was the main product formed by the hydrogenation of glucose, along with all the compounds observed above in the hydrogenolysis of sorbitol. The presence of all the other products of the reaction besides sorbitol (deoxygenated C6 products, C6 cyclic compounds, and C1C5 products) from the beginning of the hydrogenolysis reaction is indicative of the direct hydrogenolysis of glucose as well as the fast hydrogenation of glucose to sorbitol. Sorbitol concentration was maximum at total conversion of glucose at 1 h (0.07 M, ca. 24% yield), then it decreased up to almost full conversion at 8 h. The reaction rates of formation of the different products were significantly higher during the hydrogenolysis of glucose than during the reaction of sorbitol. This was especially observable for C3 products that were formed much more rapidly with a plateau of 0.055 M (ca. 19%yield) after 2 h. The formation of these C3 compounds was probably due to the retro-aldol reaction of glucose, which led to the significant formation of glycerol [73,74]. At the end of the reaction, the distribution of the families of products was not too different when starting either with glucose or sorbitol. Deoxygenated C6 compouds (Figure 1b) represented the majority of the compounds after 2 h of reaction, with a final concentration (0.12 M, i.e., 41% yield) lower than that formed from sorbitol (0.15 M, i.e., 60% yield). Kinetic models based on the complex pathway proposed in Scheme S1 should be developed in order to confirm the mechanism.

The distribution of the sub-families of linear deoxy C6 products during sorbitol and glucose hydrogenolysis is compared in Figure 1a′,b′. During sorbitol reaction, hexanepentaols, -tetraols, -triols, -diols and hexanols were observed. Hexanepentaols were formed initially at a significant rate to reach a maximum of ~0.055 M at 3 h (22% yield). The successive formation of hexanetetraols, -triols, -diols and hexanols confirms the previously observed multiple sequential deoxygenation pathway [61,70]. Finally, there was a mixture of products of C-O bond cleavages (ca. 0.03 M, i.e., 12% yield of each after 8 h), while the yield of hexanols after 6 h remained below 4%.

During the reaction of glucose under similar conditions (Figure 1b′), hexanepentaols, -tetraols and -triols were more rapidly converted into hexanediols, which finally were the major products from the reaction time of 6 h; their concentration was 0.04 M (14% yield) at the maximum. Hexanols were present in slightly higher concentrations than during sorbitol conversion (ca. 0.012 M, <4% yield).

2.2.2. Comparison of Hydrogenolysis of Xylitol and Xylose

Like sorbitol and glucose, the hydrogenolysis of xylitol and xylose was compared under the same operating conditions (200 °C, 80 bar H_2, 6 g substrate over 500 mg of ReO_x-Rh/ZrO_2). The families of products from the main reactions (epimerization to arabitol and adonitol), linear C5 deoxygenated polyols, C-C bond cleavage to C1-C4 polyols, and dehydration (yielding anhydroxylitol and tetrahydrofurfuryl alcohol) are shown in Scheme S2.

Figure 2 compares the evolution of the concentration of C5 reactant (xylitol or xylose) and all the products as a function of reaction time. The same trends as in the comparison of sorbitol and glucose hydrogenolysis were observed.

Xylitol (0.42 M) was gradually converted up to full conversion at 8 h (Figure 2a). Under these conditions, and as described previously [71], xylitol hydrogenolysis afforded mainly the desired deoxy C5 compounds, with a maximum concentration of ~0.20 M (48% yield) at 6 h. After exhibiting a maximum yield of 7%, the concentration of the C5 isomers decreased. Relatively stable 5-membered cyclic compounds were formed. As for sorbitol, C3 compounds still represent the main products of C-C bond breakage (0.06 M i.e., 14% yield at the end of the reaction).

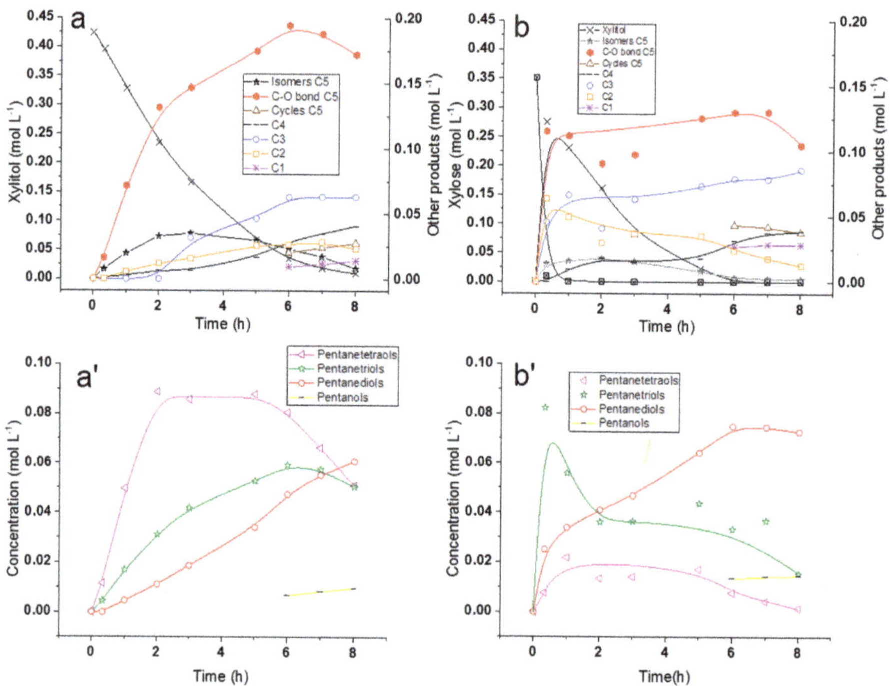

Figure 2. Evolution of (**a,b**) products families and (**a′,b′**) C-O bond C6 sub-families concentrations during hydrogenolysis of (**a,a′**) xylitol (0.42 M), and (**b,b′**) xylose (0.35 M), over ReO$_x$-Rh/ZrO$_2$ catalysts (Re/Rh molar ratio = 1.5 for xylitol, 1.8 for xylose) at 200 °C under 80 bar H$_2$, $n_{xylitol}/n_{Rh}$ = 316, n_{xylose}/n_{Rh} = 264.

The transformation of xylose (0.35 M) was significantly faster than for xylitol and the monosaccharide was nearly totally converted within only 20 min (Figure 2b). Xylitol was formed and was in turn fully consumed within 6 h. All the products detected in xylitol hydrogenolysis were also analysed during sugar hydrogenolysis, i.e., the C5 isomers that were further converted, cyclic C5 compounds, and different C1-C4 polyols. The deoxy C5 products were formed with a maximum yield of 37%. One may note that compared to xylitol hydrogenolysis (Figure 2a), the concentrations of C2 and C3 products (initially, and during all the reaction) were clearly higher (Figure 2b). The products with shorter chains of carbon were formed to the detriment of the C5 deoxygenated products. As mentioned above, appropriate kinetic models based on the suggested reaction network in Scheme S2, which agree well with the experimental concentration-time profiles starting from xylitol or xylose, should be developed.

The distribution of the C5 deoxypentanols during hydrogenolysis of xylitol followed the successive mechanism of deoxygenation reported previously [71]: pentanetetraols > pentanetriols > pentanediols > pentanols (Figure 2a′). A higher amount of diols was obtained from xylitol compared with the reaction of sorbitol, simply due to the lower number of hydroxyl groups to be broken in the 5-carbon polyol; the yield of pentanediols was 17% after 8 h in xylitol hydrogenolysis.

A higher yield of 22% pentanediols was observed from xylose at 8 h (Figure 2b′), since pentanetetraols and pentanetriols were deoxygenated deeper.

2.2.3. Discussion on Hydrogenolysis of Polyols vs. Hydrogenolysis of Sugars

For an easier comparison of the results of hydrogenolysis of the sugars vs polyols, the carbon selectivities to the different families of products for the different substrates (sorbitol, xylitol, glucose, and xylose), (Figure 3), were compared at about 80% conversion of the polyols (sorbitol, xylitol) or at about 20% sorbitol or xylitol remaining to be converted (glucose, xylose).

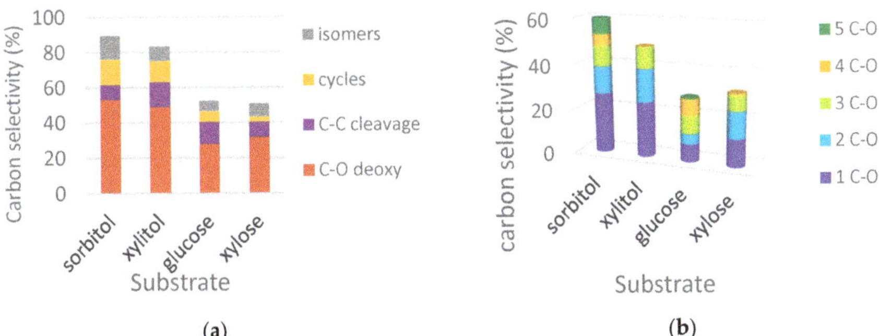

Figure 3. Hydrogenolysis of polyols and sugars over ReO$_x$-Rh/ZrO$_2$ at ca. 20% sorbitol or xylitol remaining to be converted: (**a**) carbon selectivity to families of products and (**b**) selectivity to C-O bond sub-families, 200 °C under 80 bar H$_2$.

First, it was noted that the total carbon selectivity in the liquid phase was higher for the polyols than for the sugars. This was confirmed by the comparison of the carbon balance calculated from chromatographic analysis and the measurement of total organic carbon (TOC) in the liquid phase, which are in relatively good agreement within the experimental errors (not shown). This also confirms the relatively adequate analysis of all the products in the liquid phase. This indicates an important loss of carbon material transferred to the gas phase, which was the result of all C-C bond scissions occurring during the reaction of the sugars, probably by decarbonylation and retro-aldol reactions [57].

The polyols and sugars can undergo both C-C and C-O bond cleavage processes. Figure 3a clearly shows that the products formed by successive C-O bond cleavages were the main products, regardless of the substrate. The selectivity to deoxygenated polyols was ca. 28% and ca. 32% from glucose and xylose, respectively, it was higher from sorbitol and xylitol (ca. 55% and ca. 49%, respectively). The selectivity to C-C bond cleavage products was in the range 8–15%. It appeared to be higher in the hydrogenolysis of C5 substrates than in the reaction of C6 substrates. Cyclic compounds are more easily formed from C6 substrates.

Figure 3b, representing the carbon selectivity to the different sub-families of C-O bond cleavage products, shows a higher degree of dehydroxylation of the sugars compared to the sugar polyols.

2.3. Hydrogenolysis of Mixtures of Polyols and Sugars

The composition of hemicelluloses composition may vary depending on their origin. After extraction, the hemicellulose streams contain C5 and C6 sugar monomers and oligomers of various compositions. In order to analyze the reaction medium of hydrogenolysis of hemicellulose extracts, this study was continued by investigating the hydrogenolysis of mixtures of polyols (sorbitol and xylitol) and then of sugars (glucose and xylose) of various composition at 200 °C under 80 bar H$_2$.

2.3.1. Hydrogenolysis of Sorbitol/Xylitol Mixtures

Different compositions of polyols between pure xylitol and pure sorbitol were examined. As an example, the results for the hydrogenolysis of an equimolar mixture of sorbitol/xylitol (1/1, 0.29 M) is shown in Figure 4a,a′.

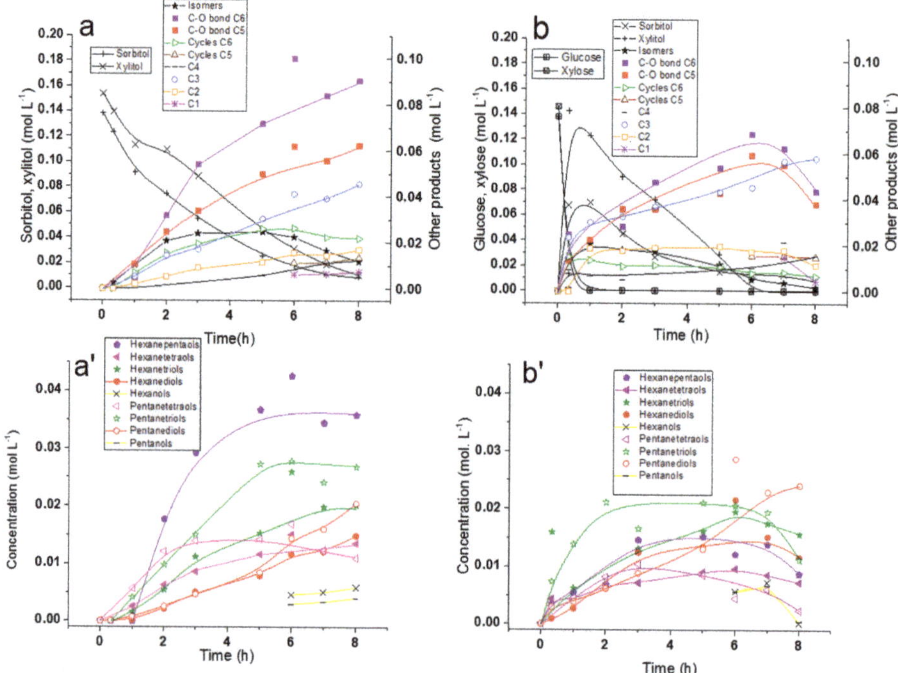

Figure 4. Evolution of concentrations of: (**a**,**b**) families of products and (**a′**,**b′**) C–O bond cleavage C6 and C5 sub-families during hydrogenolysis of (**a**,**a′**) mixture of sorbitol (0.14 M) and xylitol (0.15 M) and (**b**,**b′**) of mixture of glucose (0.14 M) and xylose (0.15 M) over ReO_x-Rh/ZrO_2 (Re/Rh = 1.6) at 200 °C under 80 bar H_2, $n_{substrate}/n_{Rh}$ = 218.

The concentration profiles as a function of time show that both sugar polyols were converted simultaneously and totally within 8 h (Figure 4a). All the reaction products corresponding to those obtained during hydrogenolysis of sorbitol and xylitol alone could be clearly detected after 1 h of reaction. No new products were detected, which eliminates the possibility of inter-reaction between both substrates or their products. The concentrations of the families of products for both substrates in the mixture evolved similarly to those during the hydrogenolysis of sorbitol and xylitol alone. In fact, the sum of the concentrations of C5 and C6 isomers reached a maximum of ca. 0.02 M at 3 h (7% yield, Figure 4a), the cyclic compounds followed the same trends as in separate experiments, C3 polyols were still the main C–C bond cleavage products (0.045 M, ca. 15%) and the concentration of the C–O bond cleavage products attained 0.16 M (55% yield, Figure 4a) vs. 60% and 48% from sorbitol and xylitol, respectively. The C6 deoxy products were produced in slightly higher concentration than the C5 deoxy products.

More or less all deoxygenated polyols were formed by successive C–O bond cleavage (Figure 4a′). Among these deoxy compounds, hexanepentaols (one C–O cleavage) and pentanetriols (2 C–O bond cleavage) were the major products of the sorbitol/xylitol mixture. Pentanetetraols that were formed initially were rapidly further deoxygenated. The final concentrations of the hexanediols and pentanediols were 0.015 M and 0.02 M (11% and 13% yield, respectively) after 8 h. Only the low concentration of hexanols and pentanols (0.005 M) were measured (yield < 2% each).

Figure 5a compares the distribution of the different families of products as the sorbitol/xylitol molar ratio was varied from pure sorbitol to pure xylitol at around 80% conversion of sorbitol or xylitol (initial substrate or formed during reaction). The selectivity to C–O cleavage products, C–C bond cleavage cyclic products, or isomers represent all products issued both from the C6 and C5 polyols.

The 1/1 sorbitol/xylitol mixture corresponds to Figure 4, 1/0 corresponds to sorbitol separately (Figure 2), and 0/1 to xylitol (Figure 3).

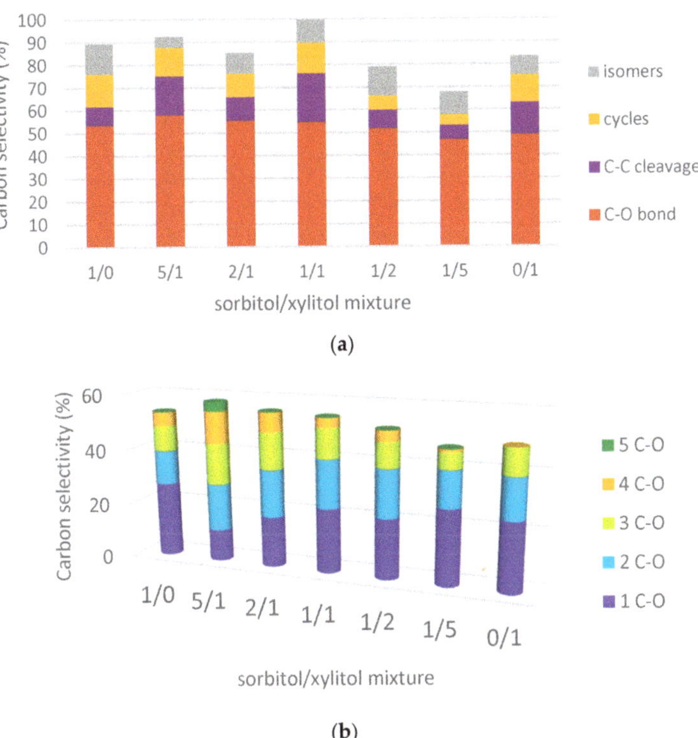

Figure 5. Evolution of carbon selectivity (**a**) to families of products and (**b**) to C-O bond sub-families versus sorbitol/xylitol mixtures during hydrogenolysis of polyols over ReO_x-Rh/ZrO_2 at ca. 80% conversion, 200 °C under 80 bar H_2, substrate/nRh ~ 200–300.

With increasing amounts of xylitol in the mixture, the carbon balance in the aqueous phase decreased, in line with the data discussed above, comparing pure sorbitol and xylitol. The products issued from different degrees of C-O bond cleavage always represented the main products. The selectivity to these desired compounds was in the range 47–58%. It was not possible to identify any directional trend from sorbitol to xylitol with increasing amounts of xylitol, due to a possible margin of error of ±10% in the determination. Selectivity to C-C bond breakage was in the range 6–17% at 80% conversion, while the selectivity to C5 and C6 cycles varied between 6% and 15%. Less cycles were formed from xylitol-rich solutions than from sorbitol-rich ones. Isomers were formed with selectivity in the range of 7% to 13%.

Figure 5b details the carbon selectivity to linear deoxygenated products after 1 to 5 C-O bond cleavages of C5 or C6 polyols, at ca. 80% conversion. Globally, a sequential deoxygenation was confirmed for all compositions, and the selectivity to the different deoxygenation degrees was in the order:

1 C-O bond (12–27%) > 2 C-O bonds (12–17%) > 3 C-Os bond (6–15%) > 4 C-O bonds (0–12%) > 5 C-O bonds (0–3%).

2.3.2. Hydrogenolysis of Mixtures of Glucose/Xylose

In fact, extracts from hemicelluloses contain mixtures of sugars whose hydrogenolysis behavior were essential for this study. The hydrogenolysis treatment of an equimolar mixture of glucose and xylose under the same operating conditions as for polyols is shown in Figure 4b. The conversion of both sugars was total within only 1 h, in line with their high reactivity when reacted separately (Figures 2b and 3b). Xylose and glucose were essentially transformed to xylitol or sorbitol, and the usually observed hydrogenolysis products. The maximum concentration of the corresponding polyols was very rapidly attained after 20 min at 0.08 M (57% yield) and 0.04 M (27% yield), respectively. The polyols were then progressively converted; xylitol disappeared at a higher rate than sorbitol and transformation was total after 6 h vs. >8 h. The evolution of the concentrations of the different families of products in the mixture of sugars was very similar to that of sugars alone (Figure 3b). The maximum total concentration of the deoxy C5 and C6 compounds was 0.13 M (yield ca. 45%) after 6 h.

As expected from the comparison of polyols and sugars, (Figure 4b), deoxygenation after 7 h of hydrogenolysis of the glucose/xylose 1/1 mixture was more complete than from the mixture of sorbitol/xylitol 1/1 because of the very high reactivity of the sugars. At the end of the reaction, pentanediols were dominant (0.025 M), followed by pentanetriols (0.02 M), hexanetriols (0.018 M), hexanediols (0.016 M), hexanepentaols (0.014 M), hexanetetraols (0.009 M) and finally by pentanetetraols, hexanols and pentanols (~0.006 M each), which almost follows the distribution of the sub families of C-O bond cleavage products during hydrogenolysis of glucose and xylose alone (see in Section 3.1, Figure 4b′).

As for the polyol mixtures, the C6/C5 composition of the sugars was varied between pure glucose (1/0) and pure xylose (0/1). The results summarized in Figure 6a confirm the higher loss of material in the liquid phase for the mixture of monosaccharides than for polyols. Selectivity to C5 plus C6 deoxygenated products was in the range of 17% to 32%.

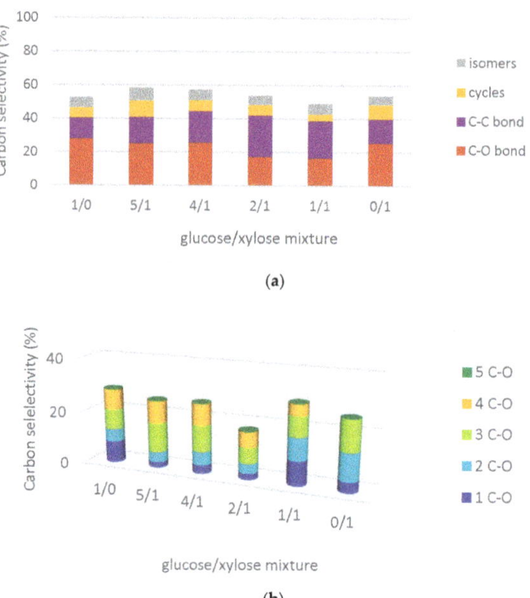

Figure 6. Evolution of carbon selectivity: (**a**) to the different families of products and (**b**) to the sub-families of C-O bond breakage products during hydrogenolysis of sugar mixtures over ReO$_x$-Rh/ZrO$_2$, 200 °C under 80 bar H$_2$, $n_{substrate}/n_{Rh}$ ~ 200–300, 20% of the sugars remaining as polyols.

All deoxygenated compounds (1 to 4 –C-O- bond scissions) were formed (Figure 4b).

2.4. Hydrogenolysis of Sugar Isomers

Table 1 compares the hydrogenolysis of glucose, mannose, galactose (ca. 0.25 M) on one hand, and xylose and arabinose (ca. 0.35 M) on the other hand (Scheme S3) over a ReO_x-Rh/ZrO_2 catalyst (with same Re/Rh ratio of 1.8). The results associated with glucose and xylose correspond to Figure 6 (1/0 and 0/1, respectively).

Table 1. Hydrogenolysis of C6 and C5 monosaccharides. Data at 20% of the corresponding polyol remaining.

Carbon Selectivity to Products (%)	Glucose	Mannose	Galactose	Xylose	Arabinose
C-O bond scission	26.3	34.3	47.4	31.9	32.9
C-C products	9.1	23.5	10.1	15.2	18.9
Cycles	6.8	11.3	5.3	2.9	1.4
Isomers	7.8	1.2	6.3	7.6	3.5
1 C-O breaking	10	13.4	17.1	12.3	12.3
2 C-O breakings	5.0	3.7	11.7	12.1	12.6
3 C-O breakings	5.5	8.5	10.1	7.0	7.5
4 C-O breakings	5.8	8.7	8.5	0.5	0.5
5 C-O breakings	-	-	-	-	-

The results were compared at total conversion of the sugar (that occurs in less than 1 h) and ca. 20% of the corresponding polyol remaining (sorbitol, mannitol, or dulcitol; xylitol or arabitol). The conversion of the different C6 and C5 sugar isomers was little affected and they were converted in less than 1 h. The composition of products at around 20% remaining polyol is given in Table 1.

The global selectivity to the deoxy C6 products depending on the nature of the isomer was as follows: glucose (23.3%) < mannose (different from glucose by the orientation of OH-group at carbon in position 2; 34.3%) < galactose (different in orientation of OH-group at carbon in position 4; 47.4%). Hydrogenolysis of mannose was characterized by a higher level of C-C bond breaking, leading to a selectivity of 23.5% to C-C products; these were mainly C3 and C2 compounds (not shown). On the other hand, xylose and arabinose behaved very similarly and the composition to the different families were comparable. Among the C-O bond scission products, polyols with one or two dehydroxylation processes were formed as major products at the end of the reaction.

These results may be analysed in parallel with the mechanistic study by Kühne et al. on C-O hydrogenolysis of polyols over Raney® Cu [57]. The steric orientation of OH-groups in sorbitol and mannitol did not affect the activity and global selectivity to C-O deoxy C6 products. However, the arrangement of these OH-groups strongly modified the hydrogenolysis pathway, as shown in the study of the reactivity of various polyol compounds.

The results obtained during hydrogenolysis of mixtures of polyols and sugars and of different component sugars of hemicelluloses should help us to better analyze and understand the behavior during hydrogenolysis of the extracted hemicellulose liquor presented in the next section.

2.5. Hydrogenolysis of Hemicelluloses Liquor

Prior to hemicelluloses extraction, a mild and selective delignification treatment using sodium chlorite in buffer aqueous medium was carried out. Hemicelluloses extraction took place on fully delignified samples by applying water extraction at 160 °C during 30 min. Breakdown of hemicelluloses, dissolution of the corresponding oligomers, and degradation of oligomers to monomeric saccharides occur during hot water extraction. The extracted liquid was filtrated and the hemicelluloses high weight fraction precipitated by addition of ethanol. The final recovered liquid phase containing low MW oligosaccharides and sugars (see experimental part) was treated over active carbon and subjected to catalytic hydrogenolysis after the elimination of ethanol, under the same reaction conditions (over ReO_x-Rh/ZrO_2 at 200 °C under 80 bar H_2).

Figure 7 shows the concentrations of the sugar oligomers and monomers detected by HPLC during hydrogenolysis of this hemicelluloses liquor. Galactose and mannose could not be separated properly on the Ca^{2+} column used for the sugars analysis, therefore, they were analysed together. The analysis revealed the presence of oligosacharides such as verbascose (C30, stachyose with an additional unit of α-D-galactopyranose), stachyose (C24, two α-D-galactose units, one α-D-glucose unit, and one β-D-fructose unit sequentially linked), raffinose (C18, trisaccharide composed of galactose, glucose, and fructose), cellobiose (C12) and xylobiose (C10). It is important to mention that the peaks of raffinose, stachyose and verbascose taken as standards for HPLC analysis may also be attributed to maltopentaose (C30), maltotetraose (C24) and maltotriose (C18), respectively. With the same DP, they could not be separated and analysed in a satisfactory manner, as also previously observed [75].

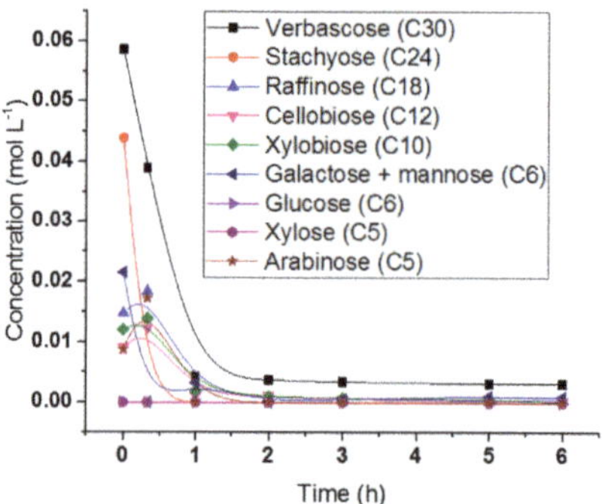

Figure 7. The evolution of oligomers and sugars during the hydrogenolysis of hemicellulose over ReO_x-Rh/ZrO_2 at 200 °C under 80 bar H_2.

Verbascose and stachyose were the main products detected before the reaction, which is not surprising since the low weight molecular fraction of hemicellulose after extraction with hot water contains small oligomers. Raffinose, lactose and xylobiose were also observed with a maximum concentration at 0.33 h of reaction, due to partial conversion of verbascose and stachyose to these secondary products. The concentration profile of xylobiose indicates that there were probably C5 oligomers with DP > 10, that may have been transformed into xylobiose. They were presumably not detected by the HPLC system used or they were not soluble. In fact, C5 hemicelluloses, such as xylan, may be insoluble in water and they very often require alkaline medium to be extracted and stabilized at room temperature [16]; moreover, a low pH facilitates the formation of humins [69]. Galactose, mannose and arabinose were also detected faintly as free-sugars in the hemicelluloses extract. All the products were almost fully transformed within 1 h, except verbascose, whose concentration remained at a low stable value of 0.003 mol L^{-1}.

Secondly, in order to determine the composition of constituent sugars, hydrolysis of hemicelluloses was performed in HCl at 90 °C for 24 h (see experimental part). These hemicelluloses are composed of mannose and galactose (0.17 mol L^{-1}), xylose (0.053 mol L^{-1}), arabinose (0.051 mol L^{-1}) and glucose (0.036 mol L^{-1}). This composition compared well with those reported by Gonzalez-Munoz et al. [9,10], and Rivas et al. [76] for maritime pine. A few oligomers with low DP were still present as stachyose (0.025 mol L^{-1}), raffinose (0.015 mol L^{-1}), and cellobiose (0.011 mol L^{-1}), indicating incomplete hydrolysis. As expected, oligomers are the primary products of hemicellulose hydrolysis, which are further degraded to sugars.

Finally, the hydrogenolysis reaction was performed on the hemicelluloses extract. The results are shown in Figure 8. The total initial concentration (ca. 0.35 M) of free sugars and small oligomers were based on the concentrations measured after hydrolysis of the hemicelluloses extract with HCl at 90 °C. The C6/C5 s molar ratio of sugars was ca. 2/1. Therefore, the reaction was compared to the reaction of an equivalent 2/1 synthetic mixture of glucose and xylose. Interestingly, the hemicelluloses extract liquor gave almost the same products and demonstrated a comparable evolution of the family products as the synthetic mixture of sugars. Levulinic acid, hydroxymethylfurfural (HMF), and furfural were also observed, but in very low amounts (not shown).

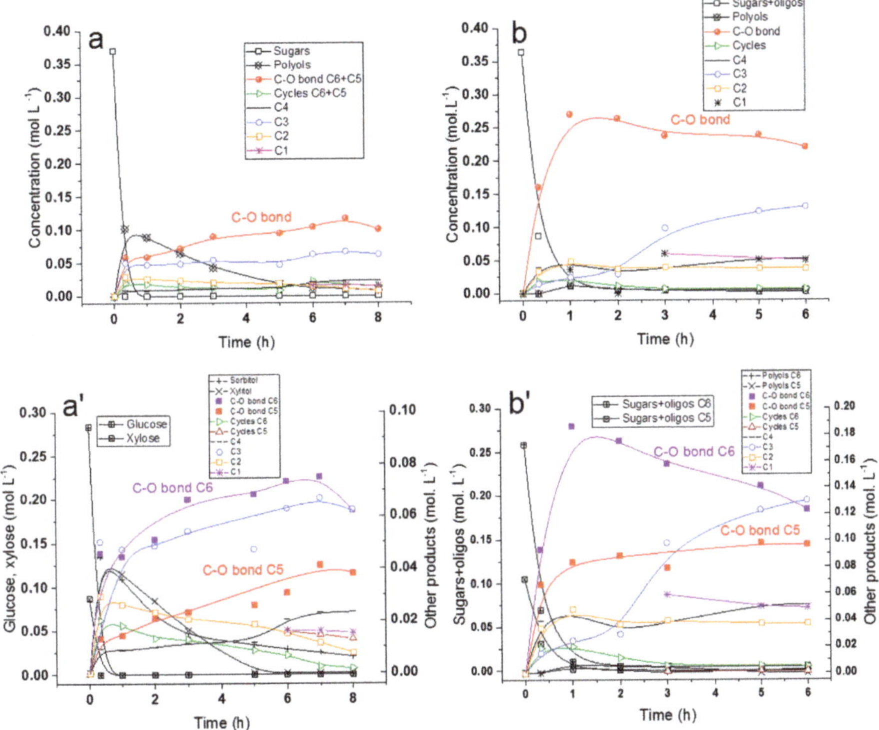

Figure 8. Evolution of families of products concentrations during hydrogenolysis of (**a**,**a′**) glucose/xylose mixture (2/1) and of (**b**,**b′**) hemicelluloses liquor over ReO_x-Rh/ZrO_2 at 200 °C under 80 bar H_2.

However, the reaction rate and the products distribution were quite different (Figure 8a,b). First, one may notice a very low formation of gaseous products from hemicelluloses liquor. Moreover, a rapid and more important formation of products was issued from C-O bond scission reactions was attained (0.26 mol L^{-1} after 1 h). Taking into account the initial concentration of free sugars and oligomers determined after acid hydrolysis of the extract, the selectivity to C-O bond cleavage products was around 65% vs. 30% for the synthetic solution. Overall, the distribution of the sub-families of C-O bond C6 products and C-O bond C5 products was coherent with the distribution of these products from the glucose/xylose (2/1) sugars mixture (Figure 8a′,b′). Nevertheless, concentrations of deoxy C6, deoxy C5, C4, C3, C2 concentrations were doubled after 6 h for hemicellulose extract compared to those in the glucose/xylose 2/1 mixture. This may be attributed to the higher concentration of mannose and galactose in the hemicelluloses liquor.

Furthermore, the repartition of the differently deoxygenated C6 and C5 products is shown in Figure 9, both for the hemicelluloses liquor and for the synthetic mixture.

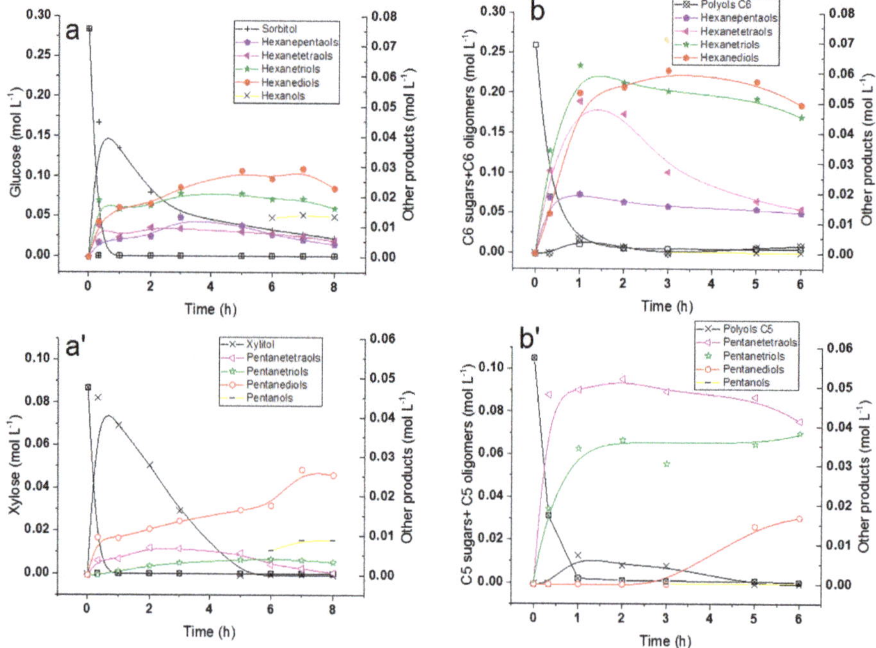

Figure 9. Temporal profiles of concentrations of sub-families during hydrogenolysis of (**a,a′**) glucose/xylose mixture (2/1) and of (**b,b′**) liquor of pine maritime over Rh-Re/ZrO$_2$ at 200 °C under 80 bar H$_2$.

The rapid transformation of the C6 polyols in the hemicelluloses stream clearly yielded successively hexanepentaols and hexanetetraols at the beginning of the reaction, followed by hexanetriols and hexanediols during the reaction (Figure 9b). Hexanols were not observed. The C5 polyols were also very rapidly converted into pentanetetraols (Figure 9b′), which, in this case, was hardly transformed into pentanetriols and pentanediols in comparison with the model mixture (Figure 9a′). Nor were pentanols observed, confirming that gaseous products were very poorly formed during hydrogenolysis of the hemicelluloses liquor.

Finally, the relatively good accuracy of analysis of all categories of products was confirmed by TOC measurement (Figure 10). There was a good fit between TOC calculated from chromatography and TOC measured. Both indicate the loss of material in the aqueous phase during reaction (ca. 10 gC L^{-1}). It is important to note that a small fraction of ethanol could not be completely removed after extraction and was still present in the reaction medium.

The reason for the large difference between the hydrogenolysis of the synthetic solution of mixture of sugars and the real hemicelluloses sample is not clear. Although the hemicelluloses were extracted from delignified wood and further purified by treatment with active carbon, which is a good adsorbent for removing unsaturated products, during the hydrothermal treatment, furfural and 5-hydroxymethylfurfural may be formed from the dehydration of pentoses and hexoses, respectively. In addition, uronic acids (glucuronic, galacturonic) may be released during hydrolysis. These by-products may be strongly adsorbed on the catalyst surface and play the role of selective poisoning of the C-C bond breaking. Further research is needed to explain such different behaviours and the possible effect of these residual by-products.

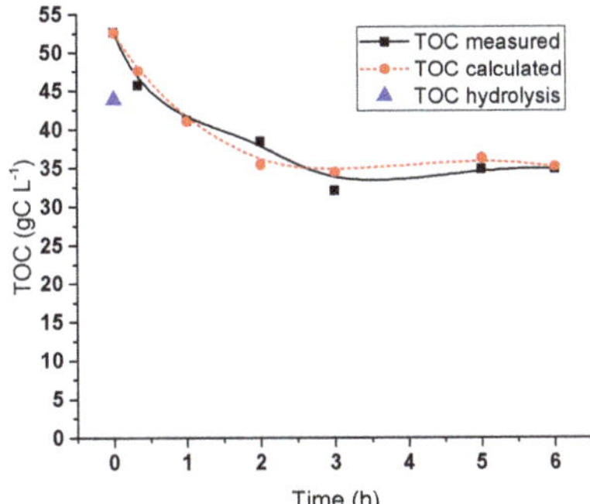

Figure 10. Hydrogenolysis of hemicelluloses liquor: comparison TOC measured, TOC calculated from chromatography, and TOC measured after hydrolysis with HCl.

3. Experimental

3.1. Catalyst Preparation

The ReO_x-Rh/ZrO_2 catalysts were prepared by successive impregnation as previously described [70]. ZrO_2 was supplied by MEL Chemicals (XZO 632/18, 129 m^2 g^{-1}, mesoporous, average pore diameter 9 nm). In a typical preparation, ZrO_2 was first wet-impregnated with an aqueous solution of $RhCl_3$ at room temperature for 7 h. After the evaporation of water, the resulting solid was dried at 100 °C overnight and calcinated under air at 500 °C for 3 h. The second impregnation was conducted with an aqueous solution of ammonium perrhenate (NH_4ReO_4), followed by a second calcination, a reduction step under H_2 at 450 °C, and passivation under 1% v/v O_2/N_2.

3.2. Maritime Pine Extraction and Hydrolysis

3.2.1. Extraction of Hemicelluloses

The chemical composition of maritime pine wood used in this work (on a dry weight basis) was 41.4% glucan, 26.1% lignin, 13.1% mannan, 7.3% xylan, 4.6% extractibles, 2.5% galactan, and 1.3% arabinan (balance 96.3%).

Wood chips (1.2 kg) were ground and sieved to size particle below 1 mm. Wood powder was first delignified using sodium chlorite in an acetate buffer medium (liquor/wood weight ratio = 6) in a 5 L glass reactor, at 60 °C, under mechanical stirring and nitrogen gas flow [77]. Chlorite sodium was progressively added to the medium until complete delignification (around one week of extraction for each batch). The delignification was followed by the measurement of Kappa number (carried our using the ISO 302-2015 standard method) of the treated wooden materiel. At the end of the delignification, the wood particles were totally converted into fibres, which were thoroughly washed with demineralized water until neutral pH.

The hot water extraction of hemicelluloses was performed after elimination of lignin by a sodium chlorite treatment. Typically, the delignified wood chips (300 g, dry basis, particles below 1 mm) and deionized water (1800 mL) at a water/wood ratio of 6 were filled into a 5 L batch stainless-steel oil-heated autoclaves. Temperature was raised from 25 °C to 120 °C in 30 min, followed by a plateau at 160 °C for 30 min. Subsequently, the liquor was separated from the solid material by filtration over a 4-L Büchner porosity 2. The pH of the collected filtrate was acidic (due to the release of small amounts of acetic or uronic acids upon extraction) and its value was then raised to 6 with the addition of NaOH.

The extracted stream was then further fractionated. The supernatant solution was concentrated by evaporation, and subsequent addition of ethanol (1/3 water–2/3 ethanol) caused the precipitation of high molecular weight hemicelluloses polymers. The remaining aqueous extract, containing monosaccharides and low molecular weight oligosaccharides [78], was finally purified with active carbon in order to remove all unsatured by-products. This liquor phase was used for the hydrogenolysis reactions after the removal of EtOH using a rotary evaporator at 40 °C under vacuum (175 mbar).

3.2.2. Analysis of Carbohydrates in Hemicellulose Extract

The low MW hemicelluloses extracted liquor was analysed by high performance liquid chomatography (HPLC, see below) for monosaccharide contents (glucose, xylose, mannose, galactose and arabinose). The carbohydrate composition was determined after a second acid post-hydrolysis of the liquor to depolymerize all remaining oligomers to monosaccharides using 37% HCl aqueous solution (pH 1) for 24 h at 90 °C in an autoclave, according to the protocol described in [79,80].

Table 2 shows the chemical compositions of the unhydrolyzed and post-hydrolyzed hemicelluloses liquor.

Table 2. Amounts of monosaccharides in the extracted purified liquor before and after acid post-hydrolysis.

	Glucose (mg L^{-1})	Xylose (mg L^{-1})	Mannose (mg L^{-1})	Galactose (mg L^{-1})	Arabinose (mg L^{-1})
Non-hydrolyzed liquor	35.3	235.7	113.6	95.0	818.8
Post-hydrolysis liquor	1229.9	2195.2	5575.5	437.4	1088.8

The sugars in the non-hydrolyzed extraction liquor were essentially present as oligomers. Arabinose was extracted early as monosaccharide, consistent with results of Grénman et al. [81] in their study of aqueous extraction of hemicelluloses from spruce chips. These authors demonstrated that this sugar experienced further reaction and degradation during extraction.

The post-hydrolysis fraction consisted of a mixture of 7.24 g/L of hexoses and 3.28 g/L of pentoses, with a hexoses/pentoses ratio of 2.2. Mannose accounted for about 50% in the extracted purified liquor. This observation is in accordance with the fact that the main hemicelluloses of softwood are galactoglucomannans.

3.3. Catalyst Evaluation Experiments and Products Analysis

The model substrates D-Xylitol (99%), D-sorbitol (99%), D-xylose (99%), and D-mannose (99%) were purchased from Sigma-Aldrich (St. Louis, MO, USA), D-glucose (99%), D-galactose (98%), and L-arabinose from Alfa Aesar. All reagents were used without further purification.

All the hydrogenolysis reactions of polyols, sugars, their mixtures, and the hemicelluloses liquor were performed in a batch Hastelloy Parr 4560 reactor (300 mL). For each run, the autoclave was loaded with 120 mL of aqueous solution and 0.5 g of catalyst. For the polyols (xylitol, sorbitol) and sugars (xylose, arabinose, glucose, mannose and galactose), 6 g of each product were previously dissolved in 120 mL of water. An amount of 120 mL of hemicelluloses liquor was loaded after elimination of ethanol. After sealing and flushing three times with 10 bar Ar to remove residual air, the reactor was pressurized with H$_2$ and heated to 200 °C under stirring (1000 rpm). The pressure increased during heating; hence, the pressure given of 80 bar reflects the pressure under the reaction conditions. Liquid samples were regularly collected during the reaction.

A detailed description of the analytical procedure of the reaction medium during hydrogenolysis of polyols can be found in our previous study [71]. Hexanepentaols, hexanetetraols, hexanetriols, hexanediols, pentanetetraols, pentanetriols, isosorbide, butanediols (mainly 1,2-butanediol), glycerol (GLY), 1,2 propanediol (1,2-PrDO), ethylene glycol (EG) were analyzed using a Shimadzu LC 20 A HPLC equiped with a Rezex ROA-Organic Acid H$^+$ column (crosslinked sulfonated styrene divinylbenzene copolymer) with a 0.005 N H$_2$SO$_4$ mobile phase at a flow rate of 0.5 mL min^{-1}) kept at 40 °C and connected to a RID detector.

The C1-C6 mono alcohols and tetrahydrofurfuryl alcohol (THFA), 1,2-pentanediol (1,2 PDO), 1,5-pentanediol (1,5 PDO), tetrahydrofurane (THF) present in the aqueous reaction medium were determined on a Shimadzu GC-2010 analyzer equiped with a Phenomenex FFAP 30 m × 0.25 μm column (the temperature was increased from 40 °C to 100 °C at 5 °C min^{-1}, and then to 250 °C at 8 °C min^{-1}.

C6 and C5 sugars and sugar polyols (glucose, xylose, galactose, mannose, arabinose, sorbitol, xylitol, dulcitol, mannitol, arabitol, iditol) were analyzed on a Shimadzu LC 20 A HPLC using Rezex RCM-Monosaccharide Ca^{2+} column heated at 65 °C with water as eluent and with refractive index detector (RID-10A) [67]. In fact, this column allows analysis of sugars, polyols, and can additionally provide analysis of oligosaccharides with low DP < 6. Therefore, in addition to the monosugars, available soluble oligosaccharides were used as analytical reference standards to attempt to identify the structures or homologs released during extraction, such as cellobiose, the raffinose family of oligosaccharides comprising raffinose (trisaccharide), stachyose (-galactosido-1,6-raffinose) and verbascose (α-galactosido-1,6-stachyose), but also sorbitan, 2-MeTHF, HMF, furfural, THFA, tetrahydropyran-2-methanol. Figure S2 presents the HPLC chromatogram of the hydrolysed extract, which shows that not only sugar monomers were formed, but also a high percentage of the hemicelluloses as dimers, trimers, and larger oligomers.

The calibration of products was carried out by injecting known concentrations of commercial standards. For a same family of product, the same response coefficient as the one of the commercially available product was considered.

The substrate conversion was calculated from Equation (1):

$$\text{Conversion (\%)} = \frac{[Substrate]_0 - [Substrate]_t}{[Substrate]_0} \times 100 \tag{1}$$

where $[Substrate]_0$ is the initial concentration of s, and $[substrate]_t$ the concentration at time t.

The yield of a given product was calculated according to Equation (2):

$$\text{Product yield (\%)} = \frac{[Product]_t}{[Substrate]_0} \times 100 \tag{2}$$

The carbon selectivity S_t^i to a desired product is based on Equation (3):

$$S_t^i(\%) = \frac{[Product]_t^i \times N_C^{product\ i}}{([Substrate]_0 - [Substrate]_t) \times n_C^{substrate}} \times 100 \tag{3}$$

where $[Product]_t^i$ is the concentration of product i formed at time t, $N_C^{substrate}$ and $N_C^{product\ i}$ is the number of carbon atoms in the substrate and product I, respectively.

Total Organic Carbon (TOC) was measured using a TOC-V$_{CSH}$ Shimadzu analyzer. The values were compared with the TOC calculated from the analysis of the products in liquid phase by HPLC and GC by the following Equation (4):

$$\text{Calculated TOC}_t\left(g\ L^{-1}\right) = M_C\left[\left(\sum_i^n N_{carbon}^i \times C_t^i\right)HPLC + \left(\sum_i^n N_{carbon}^i \times C_t^i\right)GC\right] \tag{4}$$

where M_C is the molar mass of carbon (12 g mol^{-1}), N_{carbon}^i is the number of carbon atoms in compound i, and C_t^i is the concentration of compound i at time t (in mol L^{-1}). The comparison between TOC calculated and TOC measured provides an indication as to the carbon balance in liquid phase.

4. Summary

The pre-extraction of hemicelluloses from lignocellulosic materials before pulping processes generates sugars as feedstocks for hemicelluloses-based chemicals. In this work, the comparative hydrogenolysis of C5-C6 separate sugar polyols and sugars was studied over ReO$_x$-Rh/ZrO$_2$ catalysts before evaluating mixtures and a liquid stream of hemicelluloses-based sugars.

A complete conversion of sugars was achieved very rapidly: the predominant reaction was hydrogenation to the corresponding sugar alcohol to yield the corresponding polyols by hydrogenation, but direct hydrogenolysis of the monosaccharides was observed from the beginning of the reaction, yielding all the products formed by hydrogenolysis of the sugar-derived polyols. Due to higher C-C cleavage and mainly related to the retro-aldol and decarbonylation reactions, the maximum global selectivity to deoxy linear polyols was lower from sugars (28% and 32%, from glucose and xylose) than from polyols (35% and 49% from sorbitol and xylitol). However, sugars underwent deeper C-O bond hydrogenolysis reaction, which favored the rapid formation of diols.

Hydrogenolysis of mixtures of polyols and sugars allowed the further analysis of the complete product spectra. The comparative hydrogenolysis of epimers of glucose and xylose also suggested that orientation of the OH-groups may affect the selectivity to the deoxy-products.

In the end, hot water extraction of hemicelluloses from a delignified maritime pine wood was performed. Conversion of the low molecular weight mono- and oligosaccharides in the recovered extract was studied by hydrogenolysis and the results were compared to a synthetic aqueous solution of glucose/xylose mixture of same C6/C5 composition of 2/1. The global selectivity to linear deoxygenated polyols was found to be higher than in the reaction of the synthetic solution. The yield of the deoxy polyols from the extracted hemicelluloses liquor was as high as 65% whereas it was only 30% from the 2/1 glucose/xylose mixture. This selectivity difference is mainly linked to the high concentration of mannose and galactose leading to improved selectivity compared to glucose. The presence of by-products from the extraction process, such as furfural or HMF formed from the dehydration of sugars, may play a role of selective poisoning adsorbate limiting the reduction of the undesired C-C bond breaking side reactions. These runs show the possibility of the production of value-added C6 and C5 tetraols, triols, and diols by catalytic hydrogenolysis of sugars and their oligomers present in the hemicellulose extracts. The next step should be to study the catalyst stability. Further research is also needed to develop kinetic models that describe adequately the experimental concentration of liquid products–time data and the complex networks of hydrogenolysis of either the polyols or the sugars. This kinetic analysis is essential for maximizing the yield of desired chemical products.

Supplementary Materials: The following are available online at http://www.mdpi.com/2073-4344/9/9/740/s1, Figure S1: XRD pattern of support ZrO_2 and of catalyst 9.4%Re–3.2%Rh/ZrO_2, Figure S2: HPLC chromatogram of pine maritime hemicellulose after chlorite treatment, hot water extraction, and acid hydrolysis with HCl at 90 °C during 24 h using Rezex RCM-Monosaccharide Ca^{2+} HPLC column, Scheme S1: Reaction pathway during hydrogenolysis of sorbitol and glucose. Adapted from [71], Scheme S2: Reaction pathway during hydrogenolysis of xylitol and xylose, Scheme S3: C6 and C5 sugars in the hemicelluloses liquor Adated from [71].

Author Contributions: Conceptualization: D.D.S.P., M.B., and C.P.; Design of experiments: M.B. and N.P.; Carrying the experiments and analyzing the data: A.S., M.M.-D., and E.N.; Writing original draft: M.M.-D.; Writing, review and editing: M.B., N.P.; Supervision: C.P., N.P. and M.B.

Funding: This research was funded by the French Agence Nationale de la Recherche (ANR) within the French-Luxembourg program under Award Number ANR-15-CE07-0017-02.

Acknowledgments: The authors would like to express their grateful appreciation to the scientific platforms of IRCELYON and FCBA.

Conflicts of Interest: The authors declare no conflict of interest.

References

1. Corma, A.; Iborra, S.; Velty, A. Chemical routes for the transformation of biomass into chemicals. *Chem. Rev.* **2007**, *107*, 2411–2502. [CrossRef] [PubMed]

2. FitzPatrick, M.; Champagne, P.; Cunningham, M.F.; Whitney, R.A. A biorefinery prospective: Treatment of lignocellulosic materials for the production of value-added products. *Bioresour. Technol.* **2010**, *101*, 8915–8922. [CrossRef] [PubMed]

3. Vennestrøm, P.N.R.; Osmundsen, C.M.; Christensen, C.H.; Taarning, E. Beyond petrochemicals: The renewable chemicals industry. *Angew. Chem. Int. Ed.* **2011**, *50*, 10502–10509. [CrossRef] [PubMed]

4. Belgacem, M.N.; Gandini, A. *Monomers, Polymers and Composites from Renewable Resources*; Elsevier: Amsterdam, The Netherlands, 2008; ISBN 978-0-08-045316-3.

5. Corma, A.; Huber, G.W.; Sauvanaud, L.; O'Connor, P. Biomass to chemicals: Catalytic conversion of glycerol/water mixtures into acrolein, reaction network. *J. Catal.* **2008**, *257*, 163–171. [CrossRef]

6. Lanfazame, P.; Barbera, K.; Papanikolaou, G.; Perathoner, S.; Centi, G.; Migliori, M.; Catizzone, E.; Giordano, G. Comparison of H+ and NH4+ forms of zeolites as acid catalysts for HMF etherification. *Catal. Today* **2018**, *304*, 97–102. [CrossRef]

7. Sjöström, E. *Wood Chemistry-Fundamentals and Applications*, 2nd ed.; Academic Press: New York, NY, USA, 1993; ISBN 978-0-08-092589-9.

8. Willför, S.; Sundberg, A.; Hemmings, J.; Holmbom, B. Polysaccharides in some industrially important softwood species. *Wood Sci. Technol.* **2005**, *39*, 245–258. [CrossRef]

9. González-Muñoz, M.J.; Rivas, S.; Santos, V.; Parajó, J.C. Fractionation of extracted hemicellulosic saccharides from Pinus pinaster wood by multistep membrane processing. *J. Membr. Sci.* **2013**, *428*, 281–289. [CrossRef]

10. González-Muñoz, M.J.; Rivas, S.; Santos, V.; Parajó, J.C. Aqueous processing of Pinus pinaster wood: Kinetics of polysaccharide breakdown. *Chem. Eng. J.* **2013**, *231*, 380–387. [CrossRef]

11. Willför, S.; Sundberg, A.; Hemmings, J.; Holmbom, B. Polysaccharides in some industrially important hardwood species. *Wood Sci. Technol.* **2005**, *39*, 601–617. [CrossRef]

12. Fradinho, D.M.; Pascola Neto, C.; Evtuguin, D.; Jorge, F.C.; Irle, M.A.; Gil, M.H.; Pedrosa de Jesus, J. Chemical characterisation of bark and of alkaline bark extracts from maritime pine grown in Portugal. *Ind. Crops Prod.* **2002**, *16*, 23–32. [CrossRef]

13. Liu, S.; Lu, H.; Hu, R.; Shupe, A.; Lin, L.; Liang, B. A sustainable woody biomass biorefinery. *Biotechnol. Adv.* **2012**, *30*, 785–810. [CrossRef] [PubMed]

14. Gürbüz, E.I.; Gallo, J.M.R.; Alonso, D.M.; Wettstein, S.G.; Lim, W.Y.; Dumesic, J.A. Conversion of Hemicellulose into Furfural Using Solid Acid Catalysts in γ-Valerolactone. *Angew. Chem. Int. Ed.* **2013**, *52*, 1270–1274. [CrossRef] [PubMed]

15. Yang, B.; Whyman, C.E. Pretreatment: The key to unlocking low-cost cellulosic ethanol. *Biofuel Bioprod. Bioref.* **2008**, *2*, 26–40. [CrossRef]

16. Timell, T.E. Recent progress in the chemistry of wood hemicelluloses. *Wood Sci. Technol.* **1967**, *1*, 45–70. [CrossRef]

17. Song, T.; Pranovich, A.; Sumerskiy, I.; Holmbom, B. Extraction of galactoglucomannan from spruce wood with pressurised hot water. *Holzforschung* **2008**, *62*, 659–666. [CrossRef]

18. Song, T.; Pranovich, A.; Holmbom, B. Characterisation of Norway spruce hemicelluloses extracted by pressurised hot-water extraction (ASE) in the presence of sodium bicarbonate. *Holzforschung* **2011**, *65*, 35–42. [CrossRef]

19. Song, T.; Pranovich, A.; Holmbom, B. Effects of pH control with phthalate buffers on hot-water extraction of hemicelluloses from spruce wood. *Bioresour. Technol.* **2011**, *102*, 10518–10523. [CrossRef] [PubMed]

20. Li, Z.; Qin, M.; Xu, C.; Chen, X. Hot water extraction of hemicelluloses from aspen wood chips of different sizes. *BioResources* **2013**, *8*, 5690–5700. [CrossRef]

21. Bravo, C.; Garcés, D.; Faba, L.; Sastre, H.; Ordóñez, S. Selective arabinose extraction from Pinus sp. sawdust by two-step soft acid hydrolysis. *Ind. Crops Prod.* **2017**, *104*, 229–236. [CrossRef]

22. Cebreiros, F.; Guigou, M.D.; Cabrera, M.N. Integrated forest biorefineries: Recovery of acetic acid as a by-product from eucalyptus wood hemicellulosic hydrolysates by solvent extraction. *Ind. Crops Prod.* **2017**, *109*, 101–108. [CrossRef]

23. Martin-Sampedro, R.; Eugenio, M.E.; Moreno, J.A.; Revilla, E.; Villar, J.C. Integration of a kraft pulping mill into a forest biorefinery: Pre-extraction of hemicellulose by steam explosion versus steam treatment. *Bioresour. Technol.* **2014**, *153*, 236–244. [CrossRef] [PubMed]

24. Fišerová, M.; Opálená, E.; Illa, A. Hot water and oxalic acid pre-extraction of beech wood integrated with kraft pulping. *Wood Res.* **2013**, *58*, 381–394.

25. Vena, P.F.; García-Aparicio, M.P.; Brienzo, M.; Görgens, J.F.; Rypstra, T. Effect of alkaline hemicellulose extraction on kraft pulp fibers from Eucalyptus Grandis. *J. Wood Chem. Technol.* **2013**, *33*, 157–173. [CrossRef]

26. Sim, K.; Youn, H.J.; Cho, H.; Shin, H.; Lee, H.L. Improvements in pulp properties by alkali prextraction and subsequent kraft pulping with controlling H-factor and alkali charge. *BioResources* **2012**, *7*, 5864–5878. [CrossRef]

27. Duarte, G.V.; Ramarao, B.V.; Amidon, T.E.; Ferreira, P.T. Effect of hot water extraction on hardwood kraft pulp fibers (Acer saccharum, sugar maple). *Ind. Eng. Chem. Res.* **2011**, *50*, 9949–9959. [CrossRef]

28. Naidu, D.S.; Hlangothi, S.P.; John, M.J. Bio-based products from xylan: A review. *Carbohydr. Polym.* **2018**, *179*, 28–41. [CrossRef] [PubMed]

29. Belmokaddem, F.-Z.; Pinel, C.; Huber, P.; Petit-Conil, M.; da Silva Perez, D. Green Synthesis of xylan hemicellulose esters. *Carbohydr. Res.* **2011**, *346*, 2896–2904. [CrossRef]

30. Bigand, V.; Pinel, C.; da Silva Perez, D.; Rataboul, F.; Huber, P.; Petit-Conil, M. Cationisation of galactomannan and xylan hemicelluloses. *Carbohydr. Polym.* **2011**, *85*, 138–149. [CrossRef]

31. Bigand, V.; Pinel, C.; da Silva Perez, D.; Rataboul, F.; Petit-Conil, M.; Huber, P. Influence of liquid or solid phase preparation of cationic hemicelluloses on physical properties of paper. *BioResources* **2013**, *8*, 2118–2134. [CrossRef]

32. Sedlmeyer, F.B. Xylan as by-product of biorefineries: Characteristics and potential use for food applications. *Food Hydrocoll.* **2011**, *25*, 1891–1898. [CrossRef]

33. Willför, S.; Sundberg, K.; Tenkanen, M.; Holmbom, B. Spruce-derived mannans—A potential raw material for hydrocolloids and novel advanced natural materials. *Carbohydr. Polym.* **2008**, *72*, 197–210. [CrossRef]

34. Dax, D.; Eklund, P.; Hemming, J.; Sarfraz, J.; Backman, P.; Xu, C.; Willför, S. Amphiphilic spruce galactoglucamannan derivatives based on naturally-occurring fatty acids. *BioResources* **2013**, *8*, 3771–3790. [CrossRef]

35. Marinkovic, S.; Estrine, B. Direct conversion of wheat bran hemicelluloses into n-decyl-pentosides. *Green Chem.* **2010**, *12*, 1929–1932. [CrossRef]

36. Sanglard, M.; Chirat, C.; Jarman, B.; Lachenal, D. Biorefinery in a pulp mill: Simultaneous production of cellulosic fibers from Eucalyptus globulus by soda-anthraquinone cooking and surface-active agents. *Holzforschung* **2013**, *67*, 481–488. [CrossRef]

37. Hansen, N.M.L.; Plackett, D. Sustainable films and coatings from hemicelluloses: A review. *Biomacromolecules* **2008**, *9*, 1493–1505. [CrossRef] [PubMed]

38. Xu, C.; Eckerman, C.; Smeds, A.; Reunanen, M.; Eklund, P.C.; Sjöholm, R.; Willför, S. Carboxymethylated spruce galactoglucomannas: Preparation, characterization, dispersion stability, water-in-oil emulsion stability, and sorption on cellulose surface. *Nord. Pulp. Pap. Res. J.* **2011**, *26*, 167–178. [CrossRef]

39. Mikkonen, K.S.; Tenkanen, M. Sustainable food-packaging materials based on future biorefinery products: Xylans and mannans. *Trends Food Sci. Technol.* **2012**, *28*, 90–102. [CrossRef]

40. Palm, M.; Zacchi, G. Extraction of hemicellulosic oligosaccharides from spruce using microwve oven or steam tretment. *Biomacromolecules* **2003**, *4*, 617–623. [CrossRef]

41. Chheda, J.N.; Huber, G.W.; Dumesic, J.A. Liquid-phase catalytic processing of biomass-derived oxygenated hydrocarbons to fuels and chemicals. *Angew. Chem. Int. Ed.* **2007**, *46*, 7164–7183. [CrossRef]

42. Morais, A.R.C.; Matuchaki, M.D.D.J.; Andreaus, J.; Bogel-Lukasik, R. A green and efficient approach to selective conversion of xylose and biomass hemicellulose into furfural in aqueous media using high-pressure CO_2 as a sustainable catalyst. *Green Chem.* **2016**, *18*, 2985–2994. [CrossRef]

43. Wei, L.; Shrestha, A.; Tu, M.; Adhikari, S. Effects of surfactant on biochemical and hydrothermal conversion of softwood hemicellulose to ethanol and furan derivatives. *Process Biochem.* **2011**, *46*, 1785–1792. [CrossRef]

44. Boucher, J.; Chirat, C.; Lachenal, D. Extraction of hemicelluloses from wood in a pulp biorefinery, and subsequent fermentation into ethanol. *Energy Convers. Manag.* **2014**, *88*, 1120–1126. [CrossRef]

45. Yi, G.; Zhang, Y. One-Pot Selective Conversion of hemicellulose (Xylan) to xylitol under mild Conditions. *ChemSusChem* **2012**, *5*, 1383–1387. [CrossRef] [PubMed]

46. Murzin, D.Y.; Kusema, B.; Murzina, E.V.; Aho, A.; Tokarev, A.; Boymirzaev, A.S.; Wärna, J.; Dapsens, P.Y.; Mondelli, C.; Perez-Ramirez, J.; et al. Hemicellulose arabinogalactan hydrolytic hydrogenation over Ru-modified H-USY zeolites. *J. Catal.* **2015**, *330*, 93–105. [CrossRef]

47. Tathod, A.P.; Dhepe, P.L. Towards efficient synthesis of sugar alcohols from mono-and poly-saccharides: Role of metals, supports & promoters. *Green Chem.* **2014**, *16*, 4944–4954. [CrossRef]

48. Derrien, E.; Ahmar, M.; Martin-Sisteron, E.; Raffin, G.; Queneau, Y.; Marion, P.; Beyerle, M.; Pinel, C.; Besson, M. Oxidation of aldoses contained in softwood hemicellulose acid hydrolysates into aldaric acids under alkaline or non-controlled pH conditions. *Ind. Eng. Chem. Res.* **2018**, *57*, 4543–4552. [CrossRef]

49. Sanborn, A.; Binder, T. Synthesis of R-Glucosides, Sugar Alcohols, Reduced Sugar Alcohols, and Furan Derivatives of Reduced Sugar Alcohols. U.S. Patent WO2015/156806, 9 July 2015.

50. Cao, Y.; Niu, W.; Guo, J.; Xian, M.; Liu, H. Biotechnological production of 1,2,4-butanetriol: An efficient process to synthesize energetic material precursor from renewable biomass. *Sci. Rep.* **2015**, *5*, 18149–18159. [CrossRef]

51. Mueller, H.; Mesch, W.; Broellos, K. Preparation of 1,2,4-Butanetriol. U.S. Patent US4973769, 27 November 1990.

52. Hasegawa, R.; Hayashi, K. Polyester Containing Impure 1,2-Butanediol. U.S. Patent US 4596886, 24 June 1986.

53. Espro, C.; Gumina, B.; Szumelda, T.; Paone, E.; Mauriello, F. Catalytic transfer hydrogenolysis as an effective tool for the reductive upgrading of cellulose, hemicellulose, lignin, and their derived molecules. *Catalysts* **2018**, *8*, 313. [CrossRef]

54. Furikado, I.; Miyazawa, T.; Koso, S.; Shimao, A.; Kunimori, K.; Tomishige, K. Catalytic performance of Rh/SiO$_2$ in glycerol reaction under hydrogen. *Green Chem.* **2007**, *9*, 582–588. [CrossRef]

55. Ruppert, A.M.; Weinberg, K.; Palkovits, R. Hydrogenolysis goes bio: From carbohydrates and sugar alcohols to platform chemicals. *Angew. Chem. Int. Ed.* **2012**, *51*, 2564–2601. [CrossRef]

56. Blanc, B.; Bourrel, A.; Gallezot, P.; Haas, T.; Taylor, P. Starch-derived polyols for polymer technologies: Preparation by hydrogenolysis on metal catalysts. *Green Chem.* **2000**, *2*, 89–91. [CrossRef]

57. Kühne, B.; Vogel, H.; Meusinger, R.; Kunz, S.; Kunz, M. Mechanistic study on –C–O–and –C–C– hydrogenolysis over Cu catalysts: Identification of reaction pathways and key intermediates. *Catal. Sci. Technol.* **2018**, *8*, 755–767. [CrossRef]

58. Tomishige, K.; Nakagawa, Y.; Tamura, M. Selective hydrogenolysis and hydrogenation using metal catalysts directly modified with metal oxide species. *Green Chem.* **2017**, *19*, 2876–2924. [CrossRef]

59. Chia, M.; Pagan-Torres, H.J.; Hibbits, D.; Davis, R.J.; Dumesic, J.A. Selective hydrogenolysis of polyols and cyclic ethers over bifunctional surface sites on rhodium-rhenium catalysts. *J. Am. Chem. Soc.* **2011**, *133*, 12675–12689. [CrossRef] [PubMed]

60. Chen, K.; Mori, K.; Watanabe, H.; Tomishige, K. C-O bond hydrogenolysis of cyclic ethers with OH groups over rhenium-modified supported iridium catalysts. *J. Catal.* **2012**, *294*, 171–183. [CrossRef]

61. Behr, A.; Eilting, J.; Irawadi, K.; Lescinski, J.; Lindner, F. Improved utilisation of renewable resources: New important derivatives of glycerol. *Green Chem.* **2008**, *10*, 13–20. [CrossRef]

62. Zhou, C.-H.; Beltrami, J.N.; Fan, Y.-X.; Lu, G.Q. Chemoselective catalytic conversion of glycerol as a biorenewable source to valuable commodity chemicals. *Chem. Soc. Rev.* **2008**, *37*, 527–549. [CrossRef] [PubMed]

63. Ten Dam, J.; Hanefeld, U. Renewable chemicals: Dehydroxylation of glycerol and polyols. *ChemSusChem* **2011**, *4*, 1017–1034. [CrossRef]

64. Nakagawa, Y.; Tomishige, K. Heterogeneous catalysis of the glycerol hydrogenolysis. *Catal. Sci. Technol.* **2011**, *1*, 179–190. [CrossRef]

65. Nakagawa, Y.; Tamura, M.; Tomishige, K. Catalytic materials for the hydrogenolysis of glycerol to 1,3-propanediol. *J. Mater. Chem. A* **2014**, *2*, 6688–6702. [CrossRef]

66. Besson, M.; Gallezot, P.; Pinel, C. Conversion of biomass into chemicals over metal catalysts. *Chem. Rev.* **2014**, *114*, 1827–1870. [CrossRef] [PubMed]

67. Amada, Y.; Watanabe, H.; Hirai, Y.; Kajikawa, Y.; Nakagawa, Y.; Tomishige, K. Production of butanediols by the hydrogenolysis of erythritol. *ChemSusChem* **2012**, *5*, 1991–1999. [CrossRef] [PubMed]

68. Chen, K.; Tamura, M.; Yuan, Z.; Nakagawa, Y.; Tomishige, K. One-pot conversion of sugar and sugar polyols to n-alkanes without C-C dissociation over the Ir-ReO$_x$/SiO$_2$ catalyst combined with H-ZSM65. *ChemSusChem* **2013**, *6*, 613–621. [CrossRef] [PubMed]

69. Liu, S.; Okuyama, Y.; Tamura, M.; Nakagawa, Y.; Imai, A.; Tomishige, K. Selective transformation of hemicellulose (xylan) into n-pentane, pentanols or xylitol over a rhenium-modified iridium catalyst combined with acids. *Green Chem.* **2016**, *16*, 165–175. [CrossRef]

70. Said, A.; Da Silva Perez, D.; Perret, N.; Pinel, C.; Besson, M. Selective C−O Hydrogenolysis of erythritol over Supported Rh-ReOx Catalysts in the Aqueous Phase. *ChemCatChem* **2017**, *9*, 2768–2783. [CrossRef]

71. Sadier, A.; Perret, N.; Da Silva Perez, D.; Besson, M.; Pinel, C. Effect of carbon chain length on catalytic C-O bond cleavage of polyols over Rh-ReO$_x$/ZrO$_2$ in aqueous phase. *Appl. Catal. A Gen.* **2019**. [CrossRef]

72. Rivière, M.; Perret, N.; Cabiac, A.; Delcroix, D.; Pinel, C.; Besson, M. Xylitol hydrogenolysis over ruthenium-based catalysts: Effect of alkaline promoters and basic oxide-modified catalysts. *ChemCatChem* **2017**, *9*, 2145–2159. [CrossRef]

73. Kanie, K.; Akiyama, K.; Iwamoto, M. Reaction pathways of glucose and fructose on Pt nanoparticles in subcritial water under a hydroge a hydrogen atmosphere. *Catal. Today* **2011**, *178*, 58–63. [CrossRef]

74. Lazaridis, P.A.; Karakoulia, S.A.; Teodorescu, C.; Apostol, N.; Macovei, D.; Panteli, A.; Delimitis, A.; Coman, S.M.; Parvulecu, V.I.; Triantafyllidis, K.S. High hexitols selectivity in cellulose hydrolytic hydrogenation over platinum (Pt) vs. ruthenium (Ru) catalysts supported on micro/mesoporous carbon. *Appl. Catal. B Environ.* **2017**, *214*, 1–14. [CrossRef]

75. Cerning-Béroard, J.; Filiatre-Verel, A. Characterization and distribution of soluble and insoluble carbohydrates in lupin seeds. *Z. Lebensm. Unters. Forchung* **1980**, *171*, 281–285. [CrossRef]

76. Rivas, S.; Raspolli-Galletti, A.M.; Antonetti, C.; Santos, V.; Parajo, J.C. Sustainable conversion of Pinus pinaster wood into biofuel precursors: A biorefinery approach. *Fuel* **2016**, *164*, 51–58. [CrossRef]

77. Ahlgren, P.A.; Goring, D.A.I. Removal of wood components during chlorite delignification of black spruce. *Can. J. Chem.* **1971**, *49*, 1272–1275. [CrossRef]

78. Buranov, A.U.; Mazza, G. Extraction and characterization of hemicelluloses from flax shives by different methods. *Carbohydr. Polym.* **2010**, *79*, 17–25. [CrossRef]

79. Sifontes Herrera, V.A.; Saleem, F.; Kusema, B.; Eränen, K.; Salmi, T. Hydrogenation of L-arabinose and D-galactose mixtures over a heterogeneous Ru/C catalyst. *Top. Catal.* **2012**, *55*, 550–555. [CrossRef]

80. Kusema, B.T.; Xu, C.; Mäki-Arvela, P.; Willför, S.; Holmbom, B.; Salmi, T.; Murzin, D.Y. Kinetics of acid hydrolysis of arabinogalactans. *Int. J. Chem. Reactor Eng.* **2010**, *8*, 1–16. [CrossRef]

81. Grénman, H.; Eränen, K.; Krogell, J.; Willför, S.; Salmi, T.; Murzin, D.Y. Kinetics of aqueous extraction of hemicelluloses from spruce in an intensified reactor system. *Ind. Eng. Chem. Res.* **2011**, *50*, 3818–3828. [CrossRef]

Article

Waste Seashells as a Highly Active Catalyst for Cyclopentanone Self-Aldol Condensation

Xueru Sheng [1,†], Qianqian Xu [1,†], Xing Wang [1], Na Li [1], Haiyuan Jia [2], Haiqiang Shi [1], Meihong Niu [1], Jian Zhang [1,*] and Qingwei Ping [1,*]

[1] College of Light Industry and Chemical Engineering, Dalian Polytechnic University, Dalian 116034, China
[2] Shandong Provincial Key Laboratory of Molecular Engineering, School of Chemistry and Pharmaceutical Engineering, Qilu University of Technology—Shandong Academy of Sciences, No. 3501, Daxue Road, Jinan 250353, China
* Correspondence: zhangjian@dlpu.edu.cn (J.Z.); pingqw@dlpu.edu.cn (Q.P.); Tel.: +86-411-86323327-603 (J.Z.); +86-411-8632-3327-601 (Q.P.)
† These authors contributed equally to this work.

Received: 29 July 2019; Accepted: 31 July 2019; Published: 1 August 2019

Abstract: For the first time, waste-seashell-derived CaO catalysts were used as high-performance solid base catalysts for cyclopentanone self-condensation, which is an important reaction in bio-jet fuel or perfume precursor synthesis. Among the investigated seashell-derived catalysts, *Scapharca Broughtonii*-derived CaO catalyst (S-shell-750) exhibited the highest dimer yield (92.1%), which was comparable with commercial CaO (88.2%). The activity sequence of different catalysts was consistent with the CaO purity sequence and contact angle sequence. X-ray diffraction (XRD) results showed that $CaCO_3$ in waste shell were completely converted to CaO after calcination at 750 °C or above for 4 h. CO_2 temperature-programmed desorption (CO_2-TPD) results indicate that both the amount and strength of base sites increase significantly when the calcination temperature climbs to 750 °C. Therefore, we can attribute the excellent performance of S-shell-750/850/950 catalysts to the higher CaO content, relatively low hydrophilicity, and stronger acidity and basicity of this catalyst. This study developed a new route for waste shell utilization in bio-derived ketone aldol condensation.

Keywords: biomass; waste seashell; aldol condensation; heterogeneous catalyst

---

1. Introduction

With the depleting supplies of fossil fuel and increasing environmental problems, the catalytic conversion of biomass to fuel and chemicals has been gaining great attention [1–3]. Compared with other biomass, lignocellulose, which is derived from agricultural waste and forest residues, is much cheaper and more abundant. Therefore, the synthesis of fuel and chemicals with lignocellulose-derived platform molecules has been a research hotspot [4–8].

Cyclopentanone is a promising lignocellulose-derived platform molecule in the conversion of biomass to fuels and chemicals. It can be produced via aqueous-phase selective hydrogenation of furfural derived from hemicellulose [9,10]. Cyclopentanone can undergo a self-aldol condensation pathway, and as-obtained dimer can be used as either high-density fuel [11,12] or perfume precursors [13]. Generally, this reaction was catalyzed by solid base or acid catalysts, for example: commercial CaO, hydrotalcites [11], MOF-encapsulating phosphotungstic acid [14] or MgO-based catalysts [15] (as shown in Scheme 1). However, all of these catalysts need to be prepared by multiple steps or purchased additionally. From the point of view of green chemistry and economic cost, it is still expected to develop bio-based and cost-effective catalysts with high activity for cyclopentanone condensation.

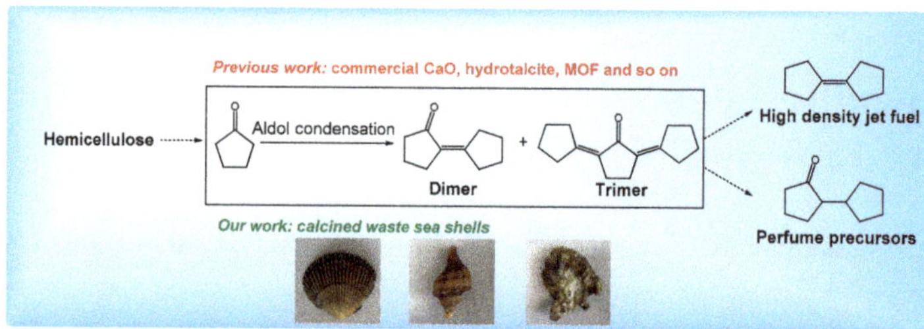

Scheme 1. Reaction pathway of cyclopentanone aldol-condensation.

Waste seashells are one of the major food residues in China, especially in the southeastern coastal areas. The annual production of seafood in China was 35 million tons in 2017. Among these, the production of mollusks was over 12.7 million tons, which accounts for over 36% [16]. Currently, the shells of these mollusks are directly discarded. Thus, using these waste seashells as raw materials for catalyst preparation can not only minimize food residues, but also synthetize cost-effective catalysts. The major constituent of seashells (e.g., clams, conches) is $CaCO_3$, which can be transformed to CaO by direct calcination at appropriate temperatures. In previous work, these calcined waste-shell derived catalysts were mainly employed to produce biodiesel from transesterification of soybean oil [17], palm olein oil [18], etc. In addition, they were also used as pretreatment materials for kraft lignin pyrolysis [19]. However, there is no report about using calcined waste shells as aldol condensation catalysts of bio-derived ketones.

The aim of this paper is to demonstrate the usage of waste seashells as a bio-based and cost-effective catalyst for cyclopentanone condensation with a high activity. Multiple characterizations were also investigated to illuminate the reason for activity differences.

2. Results and Discussion

2.1. Catalytic Activity

Scapharca Broughtonii shell (S-shell), conch shell (C-shell), and oyster shell (O-shell) are three typical waste seashells in Dalian, China, which were chosen as raw materials. After simple pretreatment (calcined at 950 °C for 4 h), three kinds of seashell-derived catalysts were prepared and named as S-shell-950, C-shell-950, and O-shell-950, respectively.

Figure 1 demonstrates the conversion of cyclopentanone and carbon yield of dimer obtained under different catalysts. Among the investigated waste seashell catalysts, S-shell-950 exhibited the best catalytic activity.

From S-shell-950, an 82.2% yield of dimer was achieved, which was comparable with commercial CaO catalyst (88.2%). C-shell-950 also achieved relatively high activity (dimer yield: 82.1%). However, dimer yield of O-shell-950 catalyst was not desirable, which was only 4.6%.

Subsequently, the effect of the S-shell calcination temperature on dimer yield was also investigated (as shown in Figure 2). In the absence of catalyst, no conversion of cyclopentanone was observed. Likewise, no conversion of cyclopentanone was observed if S-shell was used as a catalyst without calcination. However, if we calcined S-shell at different temperatures (e.g., S-shell-550 represents S-shell calcined at 550 °C for 4 h), the catalytic activities showed obviously different: the dimer yield increased significantly when the calcination temperature reached 750 °C, then decreased slightly with the further increment of the calcination temperature. In contrast, trimer can only be detected above 850 °C and the yield of trimer was increased with the further increment of the calcination temperature. Among the investigated catalysts, S-shell-750, S-shell-850 and S-shell-950 exhibited a relatively good catalytic performance. In particular, S-shell-750 catalyst exhibited the best performance; a 92.1% yield of dimer was obtained with nearly 100% selectivity. Moreover, catalyst reusability of commercial CaO and S-shell-950 in this reaction was comparable (as shown in Figure S1). The evidence above indicates that commercial CaO can be substituted directly with S-shell catalysts in this condensation reaction.

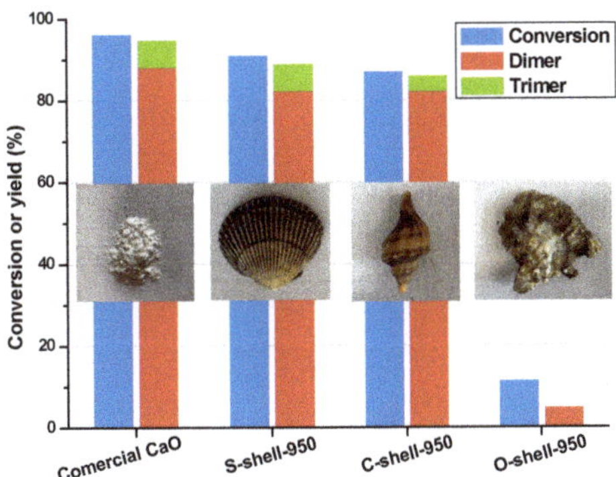

Figure 1. Effect of CaO source on carbon yield of dimer. Reaction conditions: 180 °C; 4.0 g cyclopentanone; 1.0 g catalyst; 2 h.

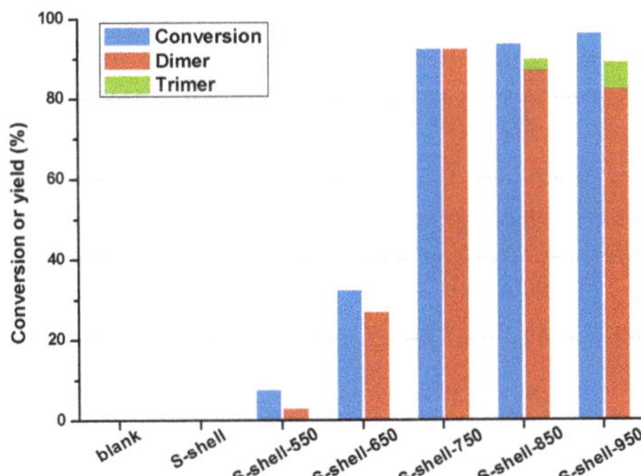

Figure 2. Effect of S-shell calcination temperature on carbon yield of dimer. Reaction conditions: 180 °C; 4.0 g cyclopentanone; 1.0 g S-shell catalyst; 2 h.

2.2. Catalyst Characterization

To get a deeper insight of activity difference of catalysts derived from different sources, multiple characterizations of the catalysts were performed. According to the results of X-ray diffraction (XRD) in Figure 3, the only CaO diffraction peaks were detected after all types of waste seashell calcined at 950 °C for 4 h. This indicates that the $CaCO_3$ ingredient in waste seashell completely converted to CaO at pretreatment conditions (calcined at 950 °C for 4 h).

The Brunauer-Emmet-Teller (BET) specific surface areas of the investigated catalysts were characterized by physical adsorption. As the results shown in Table 1, S_{BET} values of four types of catalysts showed no significant differences, which were around 10 m^2 g^{-1}.

However, the CaO content of these catalysts showed obvious variations (as shown in Table 1). There is an evident correspondence between the activity and the CaO content of catalysts. Among the investigated waste seashell catalysts, S-shell-950 exhibited the highest purity, which is comparable with commercial CaO (97.78% vs. 98.00%). The CaO content sequence for the investigated catalysts was commercial CaO > S-shell-950 > C-shell-950 > O-shell-950, which was consistent with the activity sequence of catalysts mentioned in Figure 1. This result was also confirmed by SEM-EDS (as shown in

Figures S2 and S3 in the supporting information). Compared with S-shell and C-shell, O-shell has more impurity elements (such as Na, Mg, Si, S, Cl), even if calcined at 950 °C.

Table 1. Specific Brunauer-Emmet-Teller surface areas (S_{BET}), CaO content, and contact angle of the calcined shell catalysts.

Catalyst	S_{BET} (m^2 g^{-1}) [1]	CaO Content (%) [2]	CA Mean(°) [3]
Commercial CaO	9.69	98.00	16.55
S-shell-950	8.00	97.78	10.42
C-shell-950	11.80	80.34	8.92
O-shell-950	15.73	76.14	5.13

[1]: measured by physical adsorption, [2]: measured by ICP, [3]: contact angle-measured by optical contact angle measurement.

Moreover, the contact angle (CA) of water on the catalyst surface was also measured to evaluate the hydrophilicity of the catalysts (as shown in Table 1 and Figure 4). The CA sequence of catalysts was commercial CaO > S-shell-950 > C-shell-950 > O-shell-950. This sequence was also consistent with the activity sequence of catalysts mentioned in Figure 1. It can be assumed from this result that in this reaction, the activities of CaO catalysts are also related to its hydrophilicity. The more hydrophilic a catalyst is, the lower the activity it obtains. This assumption can be supported by following the literature [20]: CaO catalysts are unavoidably poisoned by atmospheric H_2O or produced H_2O during the reaction. Therefore, if a CaO catalyst is hydrophilic, H_2O can be absorbed by catalyst more easily, which leads to bad catalytic activity. In conclusion, the excellent performance of S-shell-950 catalysts can be explained by two reasons: the high content of CaO and the relatively low hydrophilicity.

Another question to be revealed is the relationship between the calcination temperature and the catalytic activity of S-shell catalysts. As the XRD results in Figure 5a show, when the calcination temperature was below 750 °C, $CaCO_3$ residue was found in S-shell catalysts. When the calcination temperature was raised above 750 °C, no $CaCO_3$ diffraction peak was observed, which indicated $CaCO_3$ was completely transformed to CaO. Figure 5b shows the visible effect of the S-shell with different calcination temperatures. The outside surface of the S-shell transformed from black color to grey color as the calcination temperature increased, which was owing to the decomposition of organism residues. When the calcination temperature was above 750 °C, the calcined shell became fragile and easy to mill into powder. The abovementioned results indicate that $CaCO_3$ in S-shell was completely transformed into CaO after calcination at 750 °C or above.

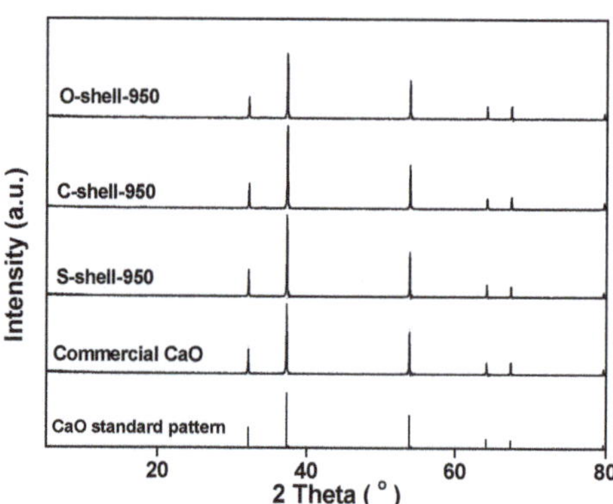

Figure 3. X-ray diffraction (XRD) pattern of commercial CaO and waste seashell after calcined at 950 °C for 4 h.

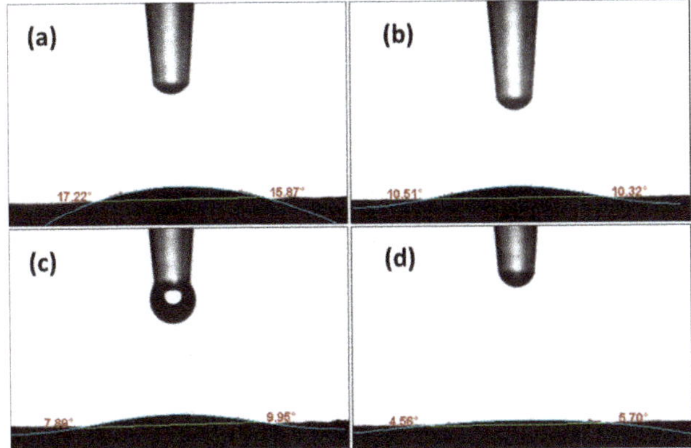

Figure 4. Contact angle of water on catalysts. (**a**) commercial CaO; (**b**) S-shell-950; (**c**) C-shell-950; (**d**) O-shell-950.

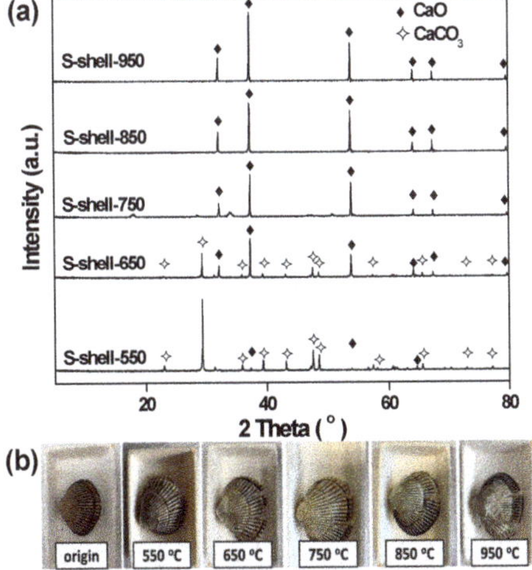

Figure 5. (**a**) XRD pattern of S-shell catalysts. (**b**) The original appearance of S-shell and visible alterations of appearance after calcination at 550–950 °C for 4 h.

In addition, CO_2 temperature-programmed desorption (CO_2-TPD) and NH_3-TPD of S-shell catalysts were also carried out. From the results shown in Figure 6 and Table 2, we can see the calcination temperature has a strong influence on the base sites amount and base strength of the catalysts. When the calcination temperature increased to 750 °C, the amount of base sites and base strength both increased significantly. Likewise, the amount of acid sites was also increased with the increment of the calcination temperature, but the amount was relatively low, which was only ~10 µmol g^{-1}. Among the investigated S-shell catalysts, S-shell-750, S-shell-850, and S-shell-950 catalysts have relatively stronger basicity and acidity. Therefore, we can attribute the excellent performance of S-shell-750/850/950 to the higher CaO content, relatively low hydrophilicity, and stronger acidity and basicity of these catalysts.

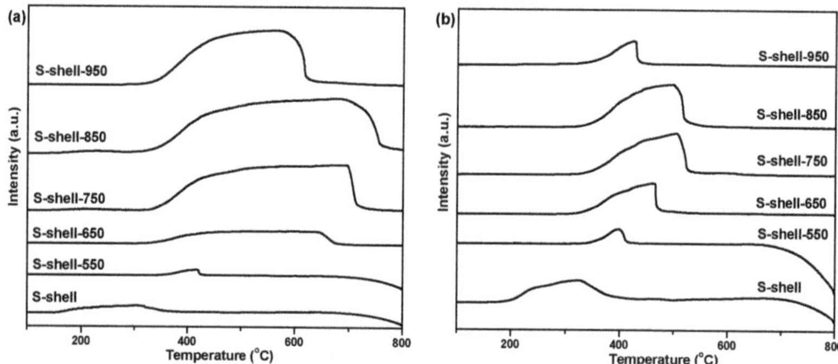

Figure 6. (**a**) CO_2 temperature-programmed desorption (CO_2-TPD) profiles of S-shell calcined at different temperatures. (**b**) NH_3-TPD profiles of S-shell calcined at different temperatures.

Table 2. The amounts of acid sites or base sites of S-shell calcined at different temperatures.

Catalysts	Base Sites Amount (μmol g^{-1})	Acid Sites Amount (μmol g^{-1})
S-shell		
S-shell-550	1.93	2.04
S-shell-650	72.79	9.47
S-shell-750	496.83	16.48
S-shell-850	744.34	14.92
S-shell-950	428.32	5.34

3. Materials and Methods

3.1. Materials and Catalyst Preparation

Waste seashells were collected from home food residue with subsequent washing and drying. Before activity testing, the seashells were calcined in a muffle furnace at the proper temperature for 4 h and milled into powders. Commercial CaO and cyclopentanone were purchased from Sinopharm Chemical Reagent Co., Ltd. (Shanghai, China).

3.2. Characterization of Catalysts

3.2.1. X-ray Diffraction (XRD)

XRD patterns of different CaO catalysts were recorded with a Shimadzu X-Ray diffractometer XRD-6100 from Shimadzu, Japan, using Cu target at a scan speed of 5° min^{-1}. Before tests, the CaO catalysts were calcined at the required temperature for 4 h in a muffle furnace.

3.2.2. Inductively Coupled Plasma Optical Emission Spectrometry (ICP-OES)

The actual CaO contents in catalysts were measured by ICP-OES (Perkin-Elmer Optima 8000, Waltham, MA, USA). Before tests, the different CaO catalysts were calcined at 950 °C for 4 h. Based on the Ca contents, the CaO contents were calculated as follow:

$$\text{CaO content} = \frac{\text{Ca content}}{\text{M(Ca)}} \times \text{M(CaO)} \times 100\%$$

M(Ca): 40.0; M(CaO): 56.0

3.2.3. Physical Adsorption

The specific Brunauer-Emmet-Teller surface areas (S_{BET}) of the CaO catalysts were measured by nitrogen physisorption at 77 K using an ASAP 2020 PLUS HD88 apparatus (Micromeritics, Norcross, Georgia, USA). Before the measurements, the samples were evacuated at 573 K for 3 h.

3.2.4. Chemi-Sorption

The basicity of S-shell catalysts was characterized by CO_2 temperature-programmed desorption (CO_2-TPD) experiments on a PCA-1200 chemi-adsorption analyzer provided by Biaode electronic technology CO.,LTD (Beijing, China). For each test, the sample was placed in a quartz reactor, pretreated in He flow at its preparation temperature for 1 h, and cooled down in He flow to 80 °C. After the saturated adsorption of CO_2, the sample was purged with He at 80 °C for 45 min to remove the physically adsorbed CO_2. The desorption of CO_2 was carried out in He flow from 80 °C to 800 °C at a heating rate of 10 °C/min.

The acidity of the solid base catalysts was measured with NH_3-TPD on the same catalyst characterization system as we used in CO_2-TPD. For each test, the sample was placed in a quartz reactor. Before the measurement, the sample was purged with He flow at 120 °C for 2 h. After the saturated adsorption of NH_3 at 120 °C, the sample was maintained at 100 °C in He flow for 45 min to remove the physically adsorbed ammonia. The desorption of NH_3 was conducted in He flow from 100 °C to 800 °C at a heating rate of 10 °C/min.

3.2.5. Contact Angle

The contact angle (CA) of water on catalyst surface was observed by Biolin Scientific Attension® Theta Flex (Gothenburg, Sweden).

3.2.6. Activity Test

The self-aldol condensation of cyclopentanone was conducted in a 20 mL stainless steel batch reactor with Teflon lining. Typically, 4.0 g cyclopentanone and 1.0 g catalyst were used. Before each reaction, the reactor was purged with nitrogen for 30 s. Then the reactor kept stirring at 180 °C for 2 h. After the reaction, the reactor was quenched to room temperature with water. The products were taken out and dissolved in 96 g 1% cyclohexanone (internal standard)—ethanol solution. The solution was filtrated, diluted with ethanol, and analyzed by a Varian 450-GC. Before the test, the catalysts were calcined at the required temperature.

4. Conclusions

CaO catalysts derived from three types of waste seashell were first reported as active catalysts in cyclopentanone self-aldol condensation. Among the investigated catalysts, S-shell-750 catalyst exhibited the highest catalytic activity and selectivity (dimer yield: 92.1%; selectivity: up to 100%), which was comparable with commercial CaO (dimer yield: 88.2%). According to the results of multiple characterizations, the excellent catalytic performance of S-shell catalysts can be rationalized by its relatively higher CaO content, low hydrophilicity, and stronger basicity/acidity. In this reaction, commercial CaO can be substituted directly by S-shell-750 catalyst. This work developed a new route for using cost-effective bio-based catalysts for biomass platform chemical conversion.

Supplementary Materials: The following are available online at http://www.mdpi.com/2073-4344/9/8/661/s1, Figure S1: Catalyst reusability of commercial CaO and S-shell-950 catalysts. Reaction conditions: 4.0 g cyclopentanone; 1.0 g catalyst; 180 °C; 2h. Catalysts were regenerated by calcination at 950 °C for 4 h. Figure S2: SEM-EDS data of catalysts before calcination. (a) commercial CaO; (b) S-shell; (c) C-shell; (d) O-shell. Figure S3: SEM-EDS data of catalysts after calcination at 950 °C. (a) commercial CaO; (b) S-shell-950; (c) C-shell-950; (d) O-shell-950.

Author Contributions: X.S. conceived and designed the experiments; Q.X. and X.W. performed the experiments; N.L., H.J., M.N. and H.S. analyzed the data; J.Z. and Q.P. contributed reagents and analysis tools; X.S. wrote the paper.

Funding: This research was funded by National Natural Science Foundation of China (No. 21808023) and Natural Science Foundation of Liaoning (No. 20180550559).

Acknowledgments: We acknowledge the Program for Scientific Research Innovation Team in Colleges and Universities of Shandong Province.

Conflicts of Interest: The authors declare no conflict of interest.

References

1. Huber, G.W.; Iborra, S.; Corma, A. Synthesis of transportation fuels from biomass: Chemistry, catalysts, and engineering. *Chem. Rev.* **2006**, *106*, 4044–4098. [CrossRef] [PubMed]
2. Mika, L.T.; Csefalvay, E.; Nemeth, A. Catalytic conversion of carbohydrates to initial platform chemicals: Chemistry and sustainability. *Chem. Rev.* **2018**, *118*, 505–613. [CrossRef] [PubMed]
3. Zang, H.; Wang, K.; Zhang, M.; Xie, R.; Wang, L.; Chen, E.Y.X. Catalytic coupling of biomass-derived aldehydes into intermediates for biofuels and materials. *Catal. Sci. Technol.* **2018**, *8*, 1777–1798. [CrossRef]
4. Wang, Y.; Peng, M.; Zhang, J.; Zhang, Z.; An, J.; Du, S.; An, H.; Fan, F.; Liu, X.; Zhai, P.; et al. Selective production of phase-separable product from a mixture of biomass-derived aqueous oxygenates. *Nat. Commun.* **2018**, *9*, 5183. [CrossRef]
5. Hu, Y.; Zhao, Z.; Liu, Y.; Li, G.; Wang, A.; Cong, Y.; Zhang, T.; Wang, F.; Li, N. Synthesis of 1,4-cyclohexanedimethanol, 1,4-cyclohexanedicarboxylic acid and 1,2-cyclohexanedicarboxylates from formaldehyde, crotonaldehyde and acrylate/fumarate. *Angew. Chem. Int. Edit* **2018**, *57*, 6901–6905. [CrossRef]
6. Deng, W.; Wang, Y.; Zhang, S.; Gupta, K.M.; Hülsey, M.J.; Asakura, H.; Liu, L.; Han, Y.; Karp, E.M.; Beckham, G.T.; et al. Catalytic amino acid production from biomass-derived intermediates. *Proc. Natl. Acad. Sci. USA* **2018**, *115*, 5093–5098. [CrossRef] [PubMed]
7. Jahromi, H.; Agblevor, F.A. Hydrodeoxygenation of Aqueous-Phase Catalytic Pyrolysis Oil to Liquid Hydrocarbons Using Multifunctional Nickel Catalyst. *Ind. Eng. Chem. Res.* **2018**, *57*, 13257–13268. [CrossRef]
8. Kaminski, T.; Sheng, Q.; Husein, M.M. Hydrocracking of Athabasca Vacuum Residue Using Ni-Mo-Supported Drill Cuttings. *Catalysts* **2019**, *9*, 216. [CrossRef]
9. Fang, R.; Liu, H.; Luque, R.; Li, Y. Efficient and selective hydrogenation of biomass-derived furfural to cyclopentanone using Ru catalysts. *Green Chem.* **2015**, *17*, 4183–4188. [CrossRef]
10. Yang, Y.; Du, Z.; Huang, Y.; Lu, F.; Wang, F.; Gao, J.; Xu, J. Conversion of furfural into cyclopentanone over Ni–Cu bimetallic catalysts. *Green Chem.* **2013**, *15*, 1932–1940. [CrossRef]
11. Yang, J.; Li, N.; Li, G.; Wang, W.; Wang, A.; Wang, X.; Cong, Y.; Zhang, T. Synthesis of renewable high-density fuels using cyclopentanone derived from lignocellulose. *Chem. Commun.* **2014**, *50*, 2572–2574. [CrossRef] [PubMed]
12. Sheng, X.; Li, G.; Wang, W.; Cong, Y.; Wang, X.; Huber, G.W.; Li, N.; Wang, A.; Zhang, T. Dual-bed catalyst system for the direct synthesis of high density aviation fuel with cyclopentanone from lignocellulose. *AIChE J.* **2016**, *62*, 2754–2761. [CrossRef]
13. Climent, M.J.; Corma, A.; Iborra, S.; Mifsud, M.; Velty, A. New one-pot multistep process with multifunctional catalysts: Decreasing the E factor in the synthesis of fine chemicals. *Green Chem.* **2010**, *12*, 99–107. [CrossRef]
14. Deng, Q.; Nie, G.; Pan, L.; Zou, J.-J.; Zhang, X.; Wang, L. Highly selective self-condensation of cyclic ketones using MOF-encapsulating phosphotungstic acid for renewable high-density fuel. *Green Chem.* **2015**, *17*, 4473–4481. [CrossRef]
15. Ngo, D.T.; Sooknoi, T.; Resasco, D.E. Improving stability of cyclopentanone aldol condensation MgO-based catalysts by surface hydrophobization with organosilanes. *Appl. Catal. B Environ.* **2018**, *237*, 835–843. [CrossRef]
16. News, C.F. Shellfish Consumption of China. Available online: http://www.shuichan.cc/news_view-375425.html (accessed on 29 July 2019).
17. Nakatani, N.; Takamori, H.; Takeda, K.; Sakugawa, H. Transesterification of soybean oil using combusted oyster shell waste as a catalyst. *Biores. Technol.* **2009**, *100*, 1510–1513. [CrossRef] [PubMed]

18. Viriya-Empikul, N.; Krasae, P.; Puttasawat, B.; Yoosuk, B.; Chollacoop, N.; Faungnawakij, K. Waste shells of mollusk and egg as biodiesel production catalysts. *Biores. Technol.* **2010**, *101*, 3765–3767. [CrossRef] [PubMed]
19. Lee, H.W.; Kim, Y.-M.; Jae, J.; Lee, S.M.; Jung, S.-C.; Park, Y.-K. The use of calcined seashell for the prevention of char foaming/agglomeration and the production of high-quality oil during the pyrolysis of lignin. *Renew. Energy* **2019**, *144*, 147–152. [CrossRef]
20. Granados, M.L.; Poves, M.D.Z.; Alonso, D.M.; Mariscal, R.; Galisteo, F.C.; Moreno-Tost, R.; Santamaría, J.; Fierro, J.L.G. Biodiesel from sunflower oil by using activated calcium oxide. *Appl. Catal. B Environ.* **2007**, *73*, 317–326. [CrossRef]

 *catalysts* 

Article

Evaluation on the Methane Production Potential of Wood Waste Pretreated with NaOH and Co-Digested with Pig Manure

Renfei Li [1,2], Wenbing Tan [1,2], Xinyu Zhao [1,2], Qiuling Dang [1,2], Qidao Song [3], Beidou Xi [1,2,*] and Xiaohui Zhang [1,2,*]

[1]   State Key Laboratory of Environmental Criteria and Risk Assessment, Chinese Research Academy of Environmental Sciences, Beijing 100012, China; lirenfei1114@163.com (R.L.); wenbingtan@126.com (W.T.); zhaoxinyu1126@126.com (X.Z.); dangling819@126.com (Q.D.)
[2]   State Environmental Protection Key Laboratory of Simulation and Control of Groundwater Pollution, Chinese Research Academy of Environmental Sciences, Beijing 100012, China
[3]   Institute of Scientific and Technical Information Catas, Chinese Academy of Tropical Agricultural Sciences, Haikou 571101, China; tanliu2009@126.com
*   Correspondence: wbtann@126.com (B.X.); baixue215@163.com (X.Z.); Tel.: +86-10-8491-3133 (B.X.)

Received: 8 June 2019; Accepted: 14 June 2019; Published: 17 June 2019

Abstract: Wood waste generated during the tree felling and processing is a rich, green, and renewable lignocellulosic biomass. However, an effective method to apply wood waste in anaerobic digestion is lacking. The high carbon to nitrogen (C/N) ratio and rich lignin content of wood waste are the major limiting factors for high biogas production. NaOH pre-treatment for lignocellulosic biomass is a promising approach to weaken the adverse effect of complex crystalline cellulosic structure on biogas production in anaerobic digestion, and the synergistic integration of lignocellulosic biomass with low C/N ratio biomass in anaerobic digestion is a logical option to balance the excessive C/N ratio. Here, we assessed the improvement of methane production of wood waste in anaerobic digestion by NaOH pretreatment, co-digestion technique, and their combination. The results showed that the methane yield of the single digestion of wood waste was increased by 38.5% after NaOH pretreatment compared with the untreated wood waste. The methane production of the co-digestion of wood waste and pig manure was higher than that of the single digestion of wood waste and had nonsignificant difference with the single-digestion of pig manure. The methane yield of the co-digestion of wood waste pretreated with NaOH and pig manure was increased by 75.8% than that of the untreated wood waste. The findings indicated that wood waste as a sustainable biomass source has considerable potential to achieve high biogas production in anaerobic digestion.

Keywords: wood waste; biofuel; lignocellulosic biomass; NaOH pretreatment; anaerobic co-digestion

1. Introduction

Wood is a natural, renewable, and recyclable green material and bioenergy source. Under the circumstance of increasing depletion of non-renewable energy and material, getting the utmost out of wood is increasingly important. In general, almost 50% of a tree can converted to the final products, and the rest remain as wood waste (WW) [1]. WW mainly consist of the residues from tree felling and processing, as well as discarded furniture and building materials [2]. Among these processes, sawmills account for 40–60% of the total WW generation [3,4]. The total amount of China's WW is estimated to be 30.28 million tons in 2013 [5]. However, to date, only a minority of WW can be used for recycling and reusing, and an effective method to fully utilize these solid wastes has not been developed. Currently, the main treating options for WW are incineration for heating, thermal power generation, and heat recovery due to its high calorific value. However, direct incineration for energy conversion is inefficient, and large amounts of greenhouse gases and volatile organic compounds can be released during the incineration, especially in small boilers or combustion chambers that often lack emission control systems. These emissions may cause serious environmental pollution.

The biofuels produced from lignocellulosic materials (e.g., wood and agricultural crops) are a green, sustainable, renewable energy, and biofuel production has become a Chinese national strategy [6]. Anaerobic digestion offers an attractive option for the utilization of WW for biogas production [7]. Nevertheless, a high carbon-to-nitrogen (C/N) ratio and high crystalline cellulosic structure make it difficult for WW to continuously and efficiently yielding biogas, which limits its large-scale application in anaerobic digestion technology [8].

Anaerobic co-digestion (AcoD) is a promising optimization technique in biogas production. This process lies in balancing the nutrients, bacterial diversity, pH, toxic compounds, and dry matter in different substrates to achieve a yield-increasing effect of CH_4 [9–11]. The methane yield of different substrates can be remarkably elevated via the AcoD technique compared with single-substrate digestion [12]. Given the ability of the intrinsic characteristics of animal manures (low C/N ratio and high NH_4–N content) to complement those of lignocellulosic biomass (high C/N ratio and rich lignocellulosic content) in digestion, mixing animal manures and lignocellulosic biomass (e.g., wood and straw) in AcoD to achieve production improvement for the biogas yield of different substrates is a common practice. For instance, Li et al. [6] recommended applying rice straw (RS) to pig manure (PM) in a (volatile solid) VS ratio of 1:1 or 1:2 can evidently increase biomethane production.

The high crystalline cellulosic structure is a substantial factor limiting the enzymatic degradation of lignocellulosic biomass during digestion [13]. To break down the linkages between lignin monomers or between lignin and polysaccharides to obtain lignocellulosic biomass, which is readily hydrolyzed, many pretreatment approaches have been established and normally categorized into three types: chemical (i.e., alkaline, acidic, and inorganic salts), physical (i.e., microwaves and liquid hot water), and biological (enzymatic and fungal). Within the pretreatment categories, NaOH pretreatment has been extensively applied to optimize biogas production of a wide range of lignocellulosic biomass [14].

Considering that the successful application of WW in anaerobic digestion to produce biomethane possibly brings considerable benefits because such feedstock is an abundant, cheap, and sustainable alternative, the utilization potentiality and optimizing strategy of WW in biogas production has to be evaluated. Although some studies have investigated the performance of biogas production from wood in anaerobic digestion [7,15,16], the information of the potential and optimizing strategy for WW to achieve high biogas production is still limited. For example, the optimizing strategy used for improving the biogas production of WW were scarce, and the comparisons of the improvements of biogas production between WW and conventional digestion substrates (e.g., lignocellulosic biomass and animal manure) under the same optimizing strategy, which can be used for more effectively assessing the optimization effect of biogas production of WW, were also lacking in previous studies. This study aimed to assess the performance and potential of methane production from WW in anaerobic digestion. The aim was achieved by (1) evaluating the disparity in biomethane productions between WW and the conventional digestion substrates (RS for lignocellulosic biomass and PM for animal manure), and (2) evaluating the improvement of biomethane production of WW under the effect of different optimizing strategies (NaOH pretreatment, AcoD technique, and their combination).

2. Results

2.1. Daily and Cumulative Methane Production from Different Substrates

Figure 1 presents the daily and cumulative methane yields with time for each digestion type. The maximum daily CH_4 yields of the single-substrate digestions of PM, WW, and RS were 14.2 (day 1), 17 (day 1), and 21.2 mL CH_4/g VS (day 2), respectively, during the 49 d of incubation (Figure 1a). After the production peak, the daily methane yield of WW was sharply decreased to approximately 2.6 mL

CH$_4$/g VS and then remained constant (Figure 1a). Although the overall trend of daily methane yields of PM and RS dropped observably during the single-substrate digestions, some brief rise and fall during the incubation period were also observed (Figure 1a). After the maximum production peak, the daily methane yield of PM temporarily increased at days 10 to 12, 14 to 17, and 22 to 27. The daily methane yield of RS temporarily increased at days 10 to 15, 24 to 25, and 26 to 27 (Figure 1a). The mean cumulative methane yield of the single-digestion of WW was 175.81 mL CH$_4$/g VS, which was significantly ($p < 0.05$) lower than that of the single-digestion of PM (245.09 mL CH$_4$/g VS) and RS (252.19 mL CH$_4$/g VS) (Figure 1b). The average pH values of the single-digestions of WW, PS, and PM ranged from 7.5 to 8.8, 6.6 to 8.7, and 5.5 to 8.1, respectively (Figure 1c).

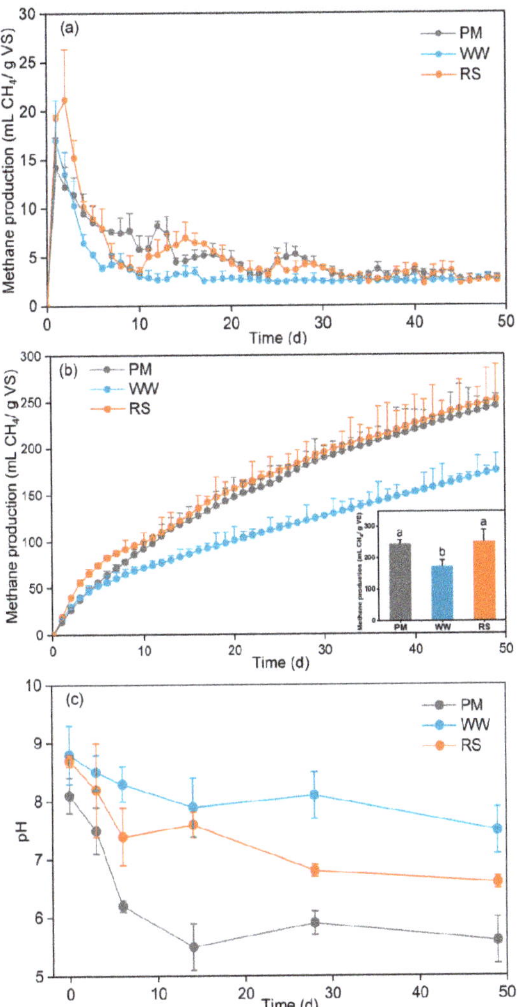

Figure 1. Daily (**a**) and cumulative (**b**) methane production and pH (**c**) of the single-digestions from different substrates. PM: pig manure, WW: wood waste, RS: rice straw. Different lowercase letters in the inset indicate a significant difference ($p < 0.05$) between the methane yields of different raw materials. For the calculation of methane volume, P and T was 1 standard atmospheric pressure and 25 °C, respectively.

2.2. Effects of NaOH Pretreatment on the Daily and Cumulative Methane Production from Different Substrates

After NaOH pretreatment, the maximum daily methane yields of WW and RS were increased from 17 CH$_4$/g VS to 19.1 mL CH$_4$/g VS (day 2) and from 21.2 CH$_4$/g VS to 22.2 mL CH$_4$/g VS (day 2), respectively (Figure 2a,c). Two mean production peaks were observed for the daily methane yield of WW on days 2 (19.1 mL CH$_4$/g VS) and 25 (6.32 mL CH$_4$/g VS) after NaOH pretreatment (Figure 2a). After the maximum production peak, the daily methane yield of WW temporarily increased at days 23 to 26 (Figure 2a). Three main production peaks were observed for the daily methane yield of RS on days 2 (22.2 mL CH$_4$/g VS), 14 (7.32 mL CH$_4$/g VS), and 30 (6.78 mL CH$_4$/g VS) after NaOH pretreatment (Figure 2a). After the maximum production peak, the daily methane yield of WW temporarily increased at days 8 to 14 and 26 to 30 (Figure 2a). Moreover, the daily methane yields of WW and RS after pretreated with NaOH were significantly ($p < 0.05$) higher than that of the untreated WW and RS, respectively (Figure 2a,c). These results showed that NaOH pretreatment could evidently increase the methane production of WW and RS in the single-substrate digestion process.

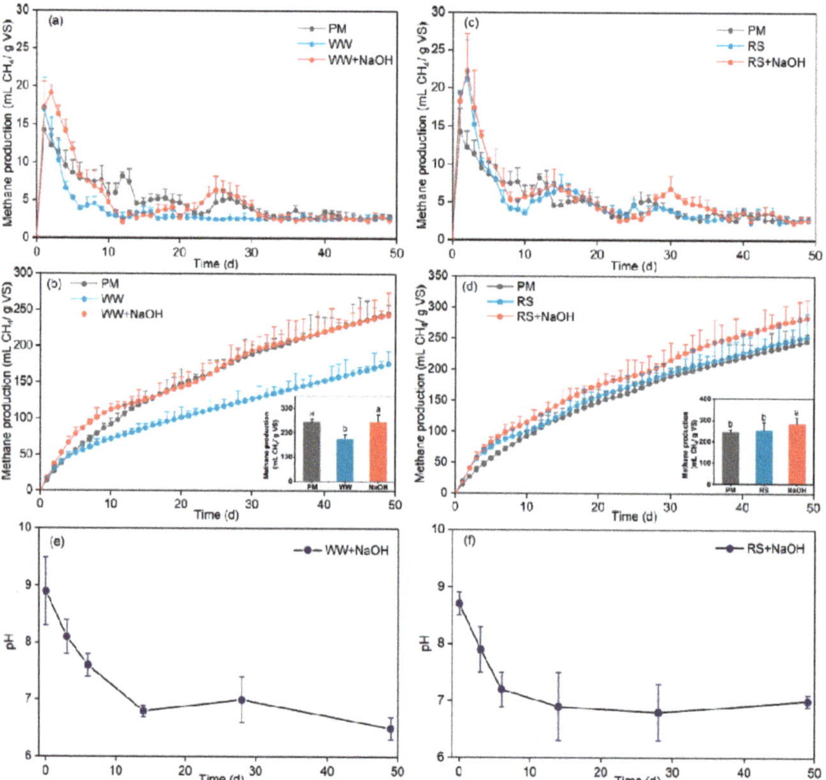

Figure 2. Daily (**a,c**) and cumulative (**b,d**) methane production and pH (**e,f**) of the single-digestions from WW and RS before and after NaOH pretreatment. PM: pig manure, WW: wood waste, RS: rice straw, WW+NaOH: WW with NaOH pretreatment, RS+NaOH: RS with NaOH pretreatment. Different lowercase letters in the inset indicate a significant difference ($p < 0.05$) between the methane yields of different raw materials.

The mean cumulative methane yield of the single digestion of WW was 243.53 mL CH$_4$/g VS after the NaOH pretreatment, which nearly reached the average cumulative methane yield of the single-digestion of PM (245.09 mL CH$_4$/g VS) (Figure 2b). When pretreated with NaOH, the mean cumulative methane yield of RS was increased from 252.19 to 282.94 mL CH_4/g VS (increased by 12.2%), which was remarkably higher than the mean cumulative methane yield of PM (Figure 2d). Moreover, the mean pH values of the single-digestions of WW and RS with NaOH pretreatment ranged from 6.5 to 8.9 and 6.8 to 8.7, respectively (Figure 2e,f).

2.3. Daily and Cumulative Methane Production from Different Co-Digestion Types

Three production peaks (17.8, 6.21, and 5.87 mL CH_4/g VS) were observed for the daily methane yield of the co-digestion of WW and PM on days 1, 17, and 29 (Figure 3a). After the maximum production peak, the daily methane yield of the co-digestion of WW and PM temporarily increased at days 6 to 7, 13 to 16, and 22 to 29 (Figure 3a). Four production peaks (23, 9.4, 5.75, and 6.34 mL CH_4/g VS) were observed for the daily methane yield of the co-digestion of RS and PM on days 2, 17, 28, and 41 (Figure 3c). After the maximum production peak, the daily methane yield of the co-digestion of RS and PM temporarily increased at days 1 to 2, 11 to 17, 24 to 28, and day 37 to 41 (Figure 3c).

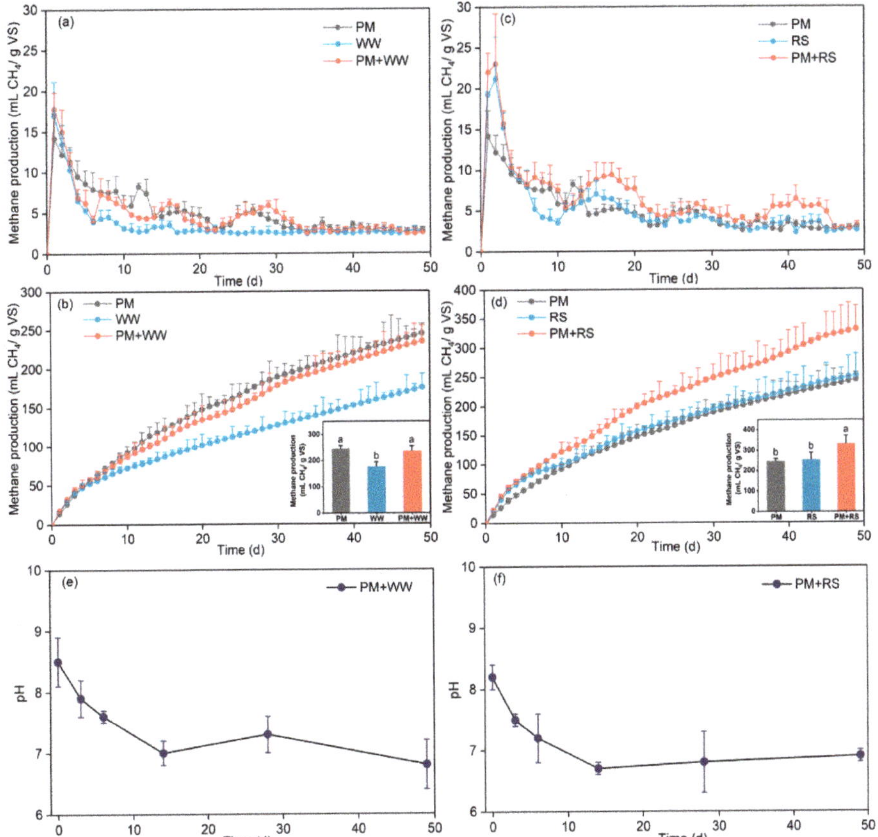

Figure 3. Daily (**a,c**) and cumulative (**b,d**) methane production and pH (**e,f**) of the co-digestions from WW and RS with PM. PM: pig manure, WW: wood waste, RS: rice straw, PM + WW: the AcoD of PM and WW, PM + RS: the AcoD of PM and RS. Different lowercase letters in the inset indicate a significant difference ($p < 0.05$) between the methane yields of different raw materials.

The average cumulative methane yield of the co-digestion of WW and PM was 234.88 mL CH_4/g VS, which was increased by 33.6% compared with the single-digestion of WW (Figure 4) and nearly reached the mean cumulative methane yield of the single-digestion of PM (Figure 3b). As shown in Figure 3d, the cumulative methane yield of the co-digestion of RS and PM was remarkably ($p <$ 0.05) higher than that of the single-digestions of both RS and PM. The average cumulative methane yield of the co-digestion of RS and PM was 331.12 mL CH_4/g VS, which was increased by 31.3% than the single-digestions of RS (Figure 4). The pH values of the co-digestion of PM and WW and the co-digestion of PM and RS ranged from 6.8 to 8.5 and 6.7 to 8.2, respectively (Figure 3e,f).

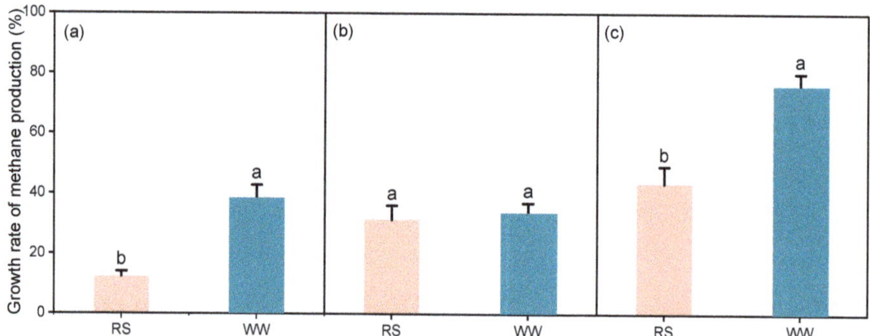

Figure 4. Growth rates of methane yield of the anaerobic digestions from WW and RS under different optimizing strategies. (**a**) NaOH pretreatment, (**b**) AcoD, (**c**) AcoD with NaOH pretreatment. Growth rate = $(X_T - X_O)/X_O$, X_T denotes the mean value of biomethane yield from the treated feedstock, X_O denotes the mean value of biomethane yield from the untreated feedstock. Different lowercase letters under the same optimizing strategy indicate a significant difference ($p < 0.05$) between the growth rates of methane production of different feedstocks.

2.4. Effect of NaOH Pretreatment on the Daily and Cumulative Methane Production from Different Co-Digestion Types

Three production peaks (20.5, 7.31, and 9.11 mL CH_4/g VS) were observed for the daily methane yield of the co-digestion of WW (pretreated with NaOH) and PM on days 2, 17, and 28 (Figure 5a). After the maximum production peak, the daily methane yield of the co-digestion of WW (pretreated with NaOH) and PM temporarily increased at days 1 to 2, 14 to 17, and 22 to 28 (Figure 5a). Four production peaks (23.9, 9.14, 7.77, and 6.82 mL CH_4/g VS) were observed for the daily methane yield of the co-digestion of RS (pretreated with NaOH) and PM on days 2, 14, 30, and 41 (Figure 5c). After the maximum production peak, the daily methane yield of the co-digestion of RS (pretreated with NaOH) and PM temporarily increased at days 1 to 2, 11 to 14, 25 to 30, and 37 to 41 (Figure 5c).

The average cumulative methane yield of the co-digestion of WW (pretreated with NaOH) and PM was 309.06 mL CH_4/g VS, which was increased by 75.8% compared with that of the single-digestion of WW (Figure 4c) and was remarkably ($p < 0.05$) higher than that of the single-digestion of WW (pretreated with NaOH) and the co-digestion of WW and PM (Figure 5b). The average cumulative methane yield of the co-digestion of RS (pretreated with NaOH) and PM was 361.73 mL CH_4/g VS, which was increased by 43.4% compared with that of the single-digestion of RS (Figure 4c) and was remarkably ($p < 0.05$) higher than that of the single-digestion of RS (pretreated with NaOH) and the co-digestion of RS and PM (Figure 5d).The pH values of the co-digestion of PM and WW (pretreated with NaOH) and the co-digestion of PM and RS (pretreated with NaOH) ranged from 6.3 to 8.6 and 6.4 to 8.2, respectively (Figure 5e,f).

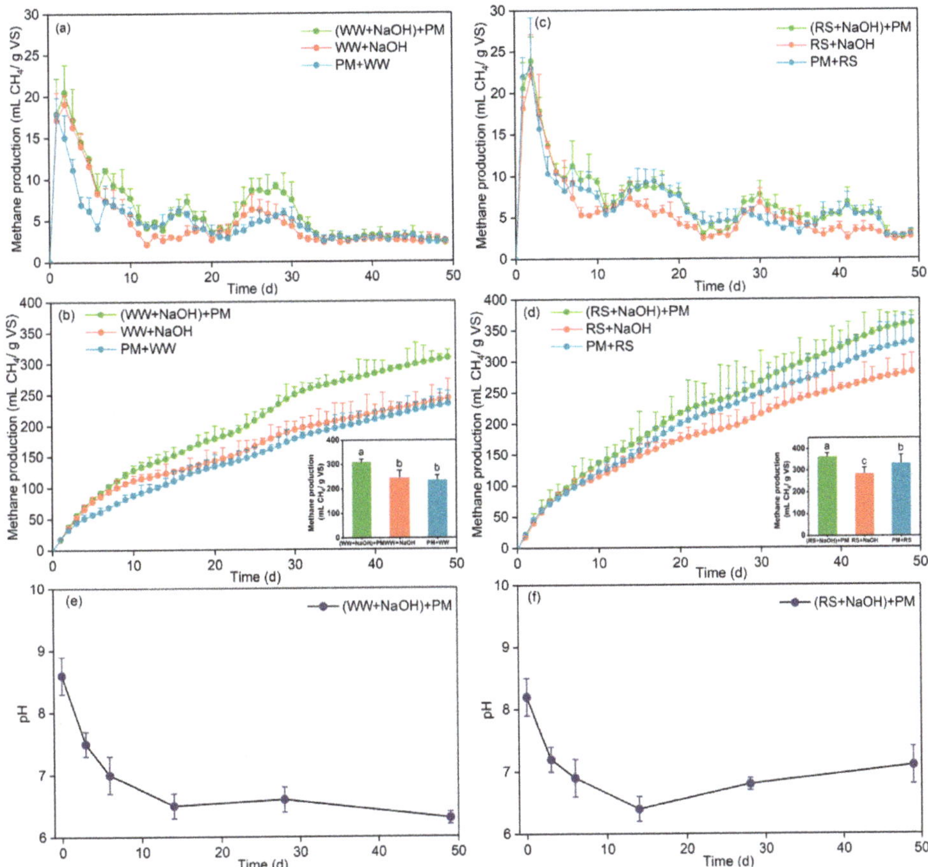

Figure 5. Daily (**a,c**) and cumulative (**b,d**) methane production and pH (**e,f**) of the co-digestions from WW and RS with PM after pretreatment with NaOH. WW + NaOH: WW with NaOH pretreatment, RS + NaOH: RS with NaOH pretreatment, (WW + NaOH) + PM: the AcoD of PM and WW (pretreated with NaOH), (RS + NaOH) + PM: the AcoD of PM and RS (pretreated with NaOH). Different lowercase letters in the inset indicate a significant difference ($p < 0.05$) between the methane yields of different raw materials.

3. Discussion

RS and WW belonging to lignocellulosic biomass are normally composed of three major components, namely polysaccharide cellulose and hemicellulose and the aromatic non-polysaccharide lignin. The higher the lignin content in WW, the tighter the physical structure of WW than RS (Table 1) [8]. Thus, the affinity of internal cellulose and hemicellulose of WW to hydrolase was relative weak. Maintaining the pH close to neutral (6.8–7.2) is preferable for methanogenesis, whereas the optimal pH for hydrolysis and acidogenesis is within the range of 5.5–6.5 [17,18]. At the initial stage of digestion, the accumulation of organic acids produced by hydrolysis and acidogenesis was difficult due to the low affinity of wood chips to hydrolase, thus the system pH cannot be lowered (Figure 1c). The high pH value in turn suppressed the activities of hydrolysis and acidogenesis, as well as methanogenesis of the single digestion of WW. Therefore, the methane production of RS was higher than WW during the single-substrate digestion process (Figure 1b).

Table 1. Characteristics of WW, RS, PM and inoculum.

Parameter	Inoculum	PM	WW	WW + NaOH	RS	RS + NaOH
TS (%)	18.9	18.1 ± 0.2	96.2 ± 0.6	NA	92.5 ± 0.5	NA
VS (%)	9.8	9.4 ± 0.1	79.8 ± 0.6	NA	84.4 ± 0.7	NA
Cellulose (%)	NA	23.6 ± 0.4	39.9 ± 0.3	41.2 ± 0.3	30.3 ± 1.2	33.7 ± 1.4
Hemicellulose (%)	NA	21.7 ± 1.2	36.7 ± 1.5	50.9 ± 1.6	29.2 ± 1.5	40.3 ± 1.8
Lignin (%)	NA	8.4 ± 0.1	13.4 ± 0.7	3.3 ± 0.2	5.4 ± 0.4	1.24 ± 0.2
Ash (%)	NA	6.9 ± 0.2	17.8 ± 0.3	NA	15.7 ± 0.3	NA
TC (%)	32.2	26.2 ± 0.3	37.1 ± 0.4	39.1 ± 0.4	34.3 ± 0.5	36.4 ± 0.4
TN (%)	1.4	1.3 ± 0.1	0.7 ± 0.1	1.4 ± 0.1	0.7 ± 0.1	1.3 ± 0.1
C/N	23.5	20	53.5	27.2	47	29.1

Notes: NA: not analyzed; PM: pig manure, WW: wood waste, RS: rice straw. WW + NaOH: WW pretreated with NaOH, RS + NaOH: RS pretreated with NaOH.

Salehian et al. [7] found that the yield of methane produced from pine wood pretreated with 8% NaOH (100 °C for 10 min) was increased from 65 mL CH_4/g VS to 178.2 mL CH_4/g VS after 45 days of incubation. Similarly, Mirahmadi et al. [15] observed a 50% enhancement in methane production after the 7% NaOH pretreatment (100 °C for 2 h) of birch. Though under different conditions of alkaline pretreatment, the yield of methane in the current work of eucalyptus was also remarkably elevated from 175.81 to 243.53 mL CH_4/g VS after NaOH pretreatment (Figure 2b). The complex structures of cellulose, hemicellulose, and lignin are difficult for microorganisms to degrade [19]. The daily and cumulative methane production of the single-digestions of WW and RS was evidently increased by NaOH pretreatment (Figure 2). This result was because alkali can break down the ether bonds between lignin and saponified the ester bonds between hemicellulose and lignin to weaken the internal hydrogen bonds within cellulose and hemicellulose [14], thus making cellulose and hemicelluloses accessible to hydrolytic enzymes. The NaOH pretreatment decreased the C/N ratio of WW and RS to a low level from 54.5 and 47 to 27.2 and 29.1, respectively, which was more acceptable for biodegradation over the first several days (Table 1) [20]. Notably, the improvement of methane production of the single digestion of WW (increased by 38.5%) by NaOH pretreatment was significantly higher ($p < 0.05$) than that of RS (increased by 12.2%) (Figure 4). This result might be occasioned by the following reasons. First, the NaOH pretreatment could increase the effective contact area between anaerobic microorganisms and substrates via reducing lignin content or breaking down lignin-hemicellulose complexes [14,21]. Therefore, WW, which contained higher amounts of lignin content than RS (Table 1) [8], had higher potential in increasing effective contact area when pretreated with NaOH. Second, the amounts of cellulose and hemicellulose that can be used to produce methane by methanogens were more contained in WW than in RS (Table 1). Thus, the improvement of the methane yield of WW could be visible when the effective contact area between anaerobic microorganisms and cellulose and hemicellulose was increased. Finally, after NaOH pretreatment, the pH value of the single digestion of WW was decreased to neutral, which was preferable for methanogenesis. However, the pH of the single-digestion of RS was almost unchanged (Figure 2e).

Compared with the single-substrate anaerobic digestions of PM and RS, the biomethane yield of the co-digestion of PM and RS was significantly ($p < 0.05$) increased (Figure 3). The methane yield of RS and PM co-digestion had increased by over 30% than both of the PM and RS single-digestions (Figures 3 and 4). High C/N ration and rich NH_4–N content are the dominant factors limiting the methane production rates of RS and PM (Table 1), respectively [22]. Mixing PM and RS could balance the C/N ratio (Table 2) and nutrition, as well as toxic compounds generated during the digestion [23]. The AcoD with different substrates could also stimulate the synergistic effects of microorganisms for achieving improved biogas production [24].

Table 2. Experimental design.

Treatments	Raw Material	VS Ratio	C/N Ratio
T1	PM		20
T2	RS		47
T3	WW		53.5
T4	WW + NaOH		27.2
T5	RS + NaOH		29.1
T6	PM/WW	2:01	31.3
T7	PM/RS	2:01	29
T8	PM/WW(NaOH)	2:01	22.4
T9	PM/WW(NaOH)	2:01	23

Notes: PM/WW indicates the AcoD of PM and WW. PM/RS indicates the AcoD of PM and RS. PM/WW(NaOH) indicates the AcoD of PM and WW (pretreated with NaOH), PM/RS(NaOH) indicates the AcoD of PM and RS (pretreated with NaOH).

Co-digestion with WW had no beneficial effect on the methane production of PM (Figure 3b), which may be ascribed to the rigid and recalcitrant lignocellulosic structure of WW. On the other hand, the methane production of the co-digestion of WW and PM had nonsignificant difference with that of the single-digestion of PM (Figure 3b), suggesting WW can be utilized as a supplementary during the PM anaerobic digestion without affecting its methane production efficiency. We showed that the WW pretreated with NaOH had a satisfactory methane production performance in anaerobic digestion (Figure 2a,b). The methane production of WW treated with AcoD (pretreated with NaOH) was increased by 75.8% compared with the untreated WW. Furthermore, the growth rates of methane production of WW treated with NaOH and AcoD (pretreated with NaOH) were signally higher than the RS that under the same optimizing strategies (Figure 4). WW is widely distributed in vast rural areas and has huge reserves. Therefore, when treated with targeted approaches, WW has a considerable potential transforming from the worthless organic waste to a promising fermentation substrate.

4. Materials and Methods

4.1. Substrates and Inoculum of Anaerobic Dry Digestion

The PM and wood chips were used as the substrate of anaerobic dry digestion. The PM was obtained from the Danzhou Pig Farm (Hainan, China) and stored in a refrigerator at 4 °C before the start of the digestion experiment. The wood chips were prepared from the branches of a eucalyptus plant. The branches were crushed to obtain a diameter of less than 6 mm, soaked in biogas slurry, and acclimated for 1 week at room temperature. The inoculum was obtained from anaerobic activated sludge (Shunyi Biogas Plant, Beijing, China) and centrifuged at 10,000 r/min for 30 min. Afterward, the precipitate (inoculum) of the anaerobic activated sludge was acclimated for 1 week at room temperature, and the supernatant was used to soak the wood chips and regulate the total solid content in the anaerobic dry digestion system. The physical and chemical properties of PM, wood chips, RS, and inoculum are shown in Table 1.

4.2. Treatment Design and Incubation Experiments

A 500-mL glass vial was used as an anaerobic dry digestion reactor in this study. Nine treatments were prepared with three replicates each (Table 2). The inoculum accounted for 40% of the digestion material based on the total solid content. The digestion system was replenished with the supernatant of the centrifuged anaerobic activated sludge to 200 g of the total mass, uniformly stirred, placed in an anaerobic dry digestion reactor, and sealed with butyl rubber with an aluminum collecting gas bag. Each anaerobic dry digestion was incubated for 49 days in a constant-temperature incubator at 35 °C in the dark.

The wood chips were chemically pretreated with NaOH solution. Specifically, they were separately mixed with distilled water and NaOH solution with a mass fraction of 2% at a solid-to-liquid ratio of 1:10, uniformly stirred, treated at 90 °C for 4 h, rinsed with deionized water until neutral, and dried. The biogas slurry was extracted through a reflux operation every 12 h and injected back into the anaerobic dry digestion reactor. In this way, the biogas slurry could be evenly dispersed on the digestion substrate. The wood chips without pretreatment and no reflux operation were also used as controls.

4.3. Sample Collection and Analysis

The biogas generated during digestion was collected in a 3 L aluminum gas collecting bag. The biogas in the gas collecting bag was obtained regularly by using a 200 mL glass syringe every day to measure the biogas production with a gas flow meter (LMF-1, Cixi Instrument Co., Ltd., Shanghai, China). In addition, a gas sample was periodically collected using a 2 mL syringe for methane determinations.

Total solid (TS) and VS contents were measured in accordance with standard methods [25]. Total carbon (TC) and total nitrogen (TN) contents were determined with a high-temperature automated elemental analyzer (Vario EL cube, Langenselbold, Germany). Cellulose, hemicellulose, and lignin were observed using a semi-automatic fiber analyzer (ANKOM A200i, Longjie Instrument Equipment Co., Ltd., Shanghai, China). Methane concentrations were identified on a gas chromatograph (Agilent 6890, Santa Clara, CA, USA) equipped with a thermal conductivity detector and a flame ionization detector.

Analysis of variance (ANOVA) was performed to compare the differences in cumulative methane productions of single and co-digestions from different substrates, and the differences in the growth rates of methane production of the anaerobic digestions from WW and RS under different optimizing strategies were determined. ANOVA was conducted with SPSS 20.0 (IBM Corporation Software Group, Somers, NY, USA). Results were considered significant at $p < 0.05$.

5. Conclusions

In this study, NaOH pretreatment, AcoD technique, and their combination were used to test the performance and potential of the methane production of WW in anaerobic digestion. After pretreatment with NaOH was administered, the mean cumulative methane yield of the single digestion of WW increased from 175.81 mL CH_4/g VS to 243.53 mL CH_4/g VS, which was equivalent to a 38.5% increase compared with that of untreated WW. The mean methane yield of the co-digestion of WW and PM was 234.88 mL CH_4/g VS, which was higher than that of the single digestion of WW (175.81 mL CH_4/g VS) and was not significantly different from that of the single digestion of PM (245.09 mL CH_4/g VS). The mean cumulative methane yield of the co-digestion of WW pretreated with NaOH and PM was 309.06 mL CH_4/g VS, which was increased by 75.8% compared with that of the untreated WW and was higher than those of the single-digestion of WW pretreated with NaOH and the co-digestion of WW and PM. The growth rates of the methane production of WW treated with NaOH and AcoD pretreated with NaOH were considerably higher than those of the RS under the same optimizing strategies. This work could provide useful insights into the development of WW as a new sustainable and efficient alternative for biogas production.

Author Contributions: X.Z. (Xiaohui Zhang) and B.X. conceived and designed the experiments; R.L. and Q.S. performed the experiments; W.T. analyzed the data with suggestions by X.Z. (Xinyu Zhao) and Q.D.; W.T. and R.L. wrote the paper; B.X. and X.Z. (Xiaohui Zhang) proofed the paper.

Funding: This research was funded by the National Key Research and Development Program of China (No. 2018YFC1900102) and the National Natural Science Foundation of China (No. 51808519).

Acknowledgments: We thank Liang Zhao for assistance in the laboratory and with data analysis.

Conflicts of Interest: The authors declare no conflicts of interest.

References

1. Turley, D.B.; Chaudhry, Q.; Watkins, R.W.; Clark, J.H.; Deswarte, F.E.I. Chemical products from temperate forest tree species—Developing strategies for exploitation. *Ind. Crop. Prod.* **2006**, *24*, 238–243. [CrossRef]
2. Souza, A.M.; Nascimento, M.F.; Almeida, D.H.; Silva, D.A.L.; Almeida, T.H.; Christoforo, A.L.; Lahr, F.A. Wood-based composite made of wood waste and epoxy based ink-waste as adhesive: A cleaner production alternative. *J. Clean. Prod.* **2018**, *193*, 549–562. [CrossRef]
3. da Silva Vieira, R.; Lima, J.T.; Moreira da Silva, J.R.; Gherardi Hein, P.R.; Baillères, H.; Pereira Barauna, E.E. Small wooden objects using eucalypt sawmill wood waste. *Bioresources* **2010**, *5*, 1463–1472.
4. Eshun, J.F.; Potting, J.; Leemans, R. Wood waste minimization in the timber sector of Ghana: A systems approach to reduce environmental impact. *J. Clean. Prod.* **2012**, *26*, 67–78. [CrossRef]
5. Wang, H.; Zuo, X.; Wang, D.; Bi, Y. The estimation of forest residue resources in China. *J. Cent. South Univ. Forest. Technol.* **2017**, *37*, 29–38.
6. Li, D.; Liu, S.; Mi, L.; Li, Z.; Yuan, Y.; Yan, Z.; Liu, X. Effects of feedstock ratio and organic loading rate on the anaerobic mesophilic co-digestion of rice straw and pig manure. *Bioresour. Technol.* **2015**, *187*, 120–127. [CrossRef]
7. Salehian, P.; Karimi, K.; Zilouei, H.; Jeihanipour, A. Improvement of biogas production from pine wood by alkali pretreatment. *Fuel* **2013**, *106*, 484–489. [CrossRef]
8. Paul, S.; Dutta, A. Challenges and opportunities of lignocellulosic biomass for anaerobic digestion. *Resour. Conserv. Recycl.* **2018**, *130*, 164–174. [CrossRef]
9. Jha, A.K.; Li, J.; Nies, L.; Zhang, L. Research advances in dry anaerobic digestion process of solid organic wastes. *Afr. J. Biotechnol.* **2011**, *10*, 14242–14253.
10. Borowski, S.; Kucner, M.; Czyżowska, A.; Berłowska, J. Co-digestion of poultry manure and residues from enzymatic saccharification and dewatering of sugar beet pulp. *Rnew. Energy* **2016**, *99*, 492–500. [CrossRef]
11. Wang, X.J.; Yang, G.H.; Feng, Y.Z.; Ren, G.X. Potential for biogas production from anaerobic co-digestion of dairy and chicken manure with corn stalks. *Adv. Mat. Res.* **2012**, *347*, 2484–2492. [CrossRef]
12. Wang, X.; Yang, G.; Feng, Y.; Ren, G.; Han, X. Optimizing feeding composition and carbon–nitrogen ratios for improved methane yield during anaerobic co-digestion of dairy, chicken manure and wheat straw. *Bioresour. Technol.* **2012**, *120*, 78–83. [CrossRef] [PubMed]
13. Monlau, F.; Barakat, A.; Trably, E.; Dumas, C.; Steyer, J.P.; Carrère, H. Lignocellulosic materials into biohydrogen and biomethane: Impact of structural features and pretreatment. *Crit. Rev. Environ. Sci. Technol.* **2013**, *43*, 260–322. [CrossRef]
14. Sambusiti, C.; Ficara, E.; Malpei, F.; Steyer, J.P.; Carrère, H. Benefit of sodium hydroxide pretreatment of ensiled sorghum forage on the anaerobic reactor stability and methane production. *Bioresour. Technol.* **2013**, *144*, 149–155. [CrossRef] [PubMed]
15. Mirahmadi, K.; Kabir, M.M.; Jeihanipour, A.; Karimi, K.; Taherzadeh, M. Alkaline pretreatment of spruce and birch to improve bioethanol and biogas production. *Bioresources* **2010**, *5*, 928–938.
16. Nieves, D.C.; Karimi, K.; Horváth, I.S. Improvement of biogas production from oil palm empty fruit bunches (OPEFB). *Ind. Crop. Prod.* **2011**, *34*, 1097–1101. [CrossRef]
17. Park, S.; Li, Y. Evaluation of methane production and macronutrient degradation in the anaerobic co-digestion of algae biomass residue and lipid waste. *Bioresour. Technol.* **2012**, *111*, 42–48. [CrossRef] [PubMed]
18. Lemmer, A.; Merkle, W.; Baer, K.; Graf, F. Effects of high-pressure anaerobic digestion up to 30 bar on pH-value, production kinetics and specific methane yield. *Energy* **2017**, *138*, 659–667. [CrossRef]
19. Jaffar, M.; Pang, Y.; Yuan, H.; Zou, D.; Liu, Y.; Zhu, B.; Li, X. Wheat straw pretreatment with KOH for enhancing biomethane production and fertilizer value in anaerobic digestion. *Chin. J. Chen. Eng.* **2016**, *24*, 404–409. [CrossRef]
20. Siddique, M.N.I.; Wahid, Z.A. Achievements and perspectives of anaerobic co-digestion: A review. *J. Clean. Prod.* **2018**, *194*, 359–371. [CrossRef]
21. Li, X.; Li, L.; Zheng, M.; Fu, G.; Lar, J.S. Anaerobic co-digestion of cattle manure with corn stover pretreated by sodium hydroxide for efficient biogas production. *Energy Fuels* **2009**, *23*, 4635–4639. [CrossRef]
22. Jantrania, A.R.; White, R.K. High-solids anaerobic fermentation of poultry manure. In Proceedings of the Fifth International Symposium on Agricultural Waste, St. Joseph, MI, USA, 16–17 December 1985; pp. 73–80.

23. Li, Y.; Chen, Y.; Wu, J. Enhancement of methane production in anaerobic digestion process: A review. *Appl. Energy* **2019**, *240*, 120–137. [CrossRef]
24. Ağdağ, O.N.; Sponza, D.T. Co-digestion of mixed industrial sludge with municipal solid wastes in anaerobic simulated landfilling bioreactors. *J. Hazard. Mater.* **2007**, *140*, 75–85. [CrossRef] [PubMed]
25. American Public Health Association (APHA). *Standard Methods for the Examination of Water and Wastewater*; American Public Health Association: Washington, DC, USA, 2005.

Article

Effect of Dilute Acid and Alkali Pretreatments on the Catalytic Performance of Bamboo-Derived Carbonaceous Magnetic Solid Acid

Yikui Zhu [1,2], Jiawei Huang [1,2], Shaolong Sun [3], Aimin Wu [1,2,*] and Huiling Li [1,2,*]

[1] State Key Laboratory for Conservation and Utilization of Subtropical Agro-bioresources, South China Agricultural University, Guangzhou 510642, China; ZhuYKWerid@163.com (Y.Z.); jiaweihuangkawy@gmail.com (J.H.)

[2] Guangdong Key Laboratory for Innovative Development and Utilization of Forest Plant Germplasm, South China Agricultural University, Guangzhou 510642, China

[3] College of Natural Resources and Environment, South China Agricultural University, Guangzhou 510642, China; sunshaolong328@scau.edu.cn

* Correspondence: wuaimin@scau.edu.cn (A.W.); lihl@scau.edu.cn (H.L.); Tel.: +020-85280259 (A.W.)

Received: 26 January 2019; Accepted: 1 March 2019; Published: 7 March 2019

Abstract: Lignocellulose is a widely used renewable energy source on the Earth that is rich in carbon skeletons. The catalytic hydrolysis of lignocellulose over magnetic solid acid is an efficient pathway for the conversion of biomass into fuels and chemicals. In this study, a bamboo-derived carbonaceous magnetic solid acid catalyst was synthesized by $FeCl_3$ impregnation, followed by carbonization and –SO_3H group functionalization. The prepared catalyst was further subjected as the solid acid catalyst for the catalytic conversion of corncob polysaccharides into reducing sugars. The results showed that the as-prepared magnetic solid acid contained –SO_3H, –COOH, and polycyclic aromatic, and presented good catalytic performance for the hydrolysis of corncob in the aqueous phase. The concentration of H^+ was in the range of 0.6487 to 2.3204 mmol/g. Dilute acid and alkali pretreatments of raw material can greatly improve the catalytic activity of bamboo-derived carbonaceous magnetic solid acid. Using the catalyst prepared by 0.25% H_2SO_4-pretreated bamboo, 6417.5 mg/L of reducing sugars corresponding to 37.17% carbohydrates conversion could be obtained under the reaction conditions of 120 °C for 30 min.

Keywords: bamboo; pretreatment; magnetic solid acid; corncob; reducing sugar

---

1. Introduction

The depletion of fossil fuel reserves and climate change issues have raised concerns about renewable petroleum alternatives with the increment of global energy demand [1,2]. Biomass has received increasing attention in recent decades due to it being widespread, abundant, diverse, and inexpensive. Moreover, it has been intensively investigated as a highly sustainable carbon-containing source for the production of bioplatform molecules and biochemicals. The unique property of lignocellulosic biomass as the only renewable carbon carrier makes it an attractive source for bioenergy production. The conversion of lignocellulosic biomass to useful chemicals and biofuels via green and efficient approaches is one of the most popular topics in recent years [3]. However, problems such as its high pretreatment cost and difficulty in catalyst recovery hinder their utilization. Therefore, the development of new reaction techniques, including novel catalysts, novel pretreatment methods, or reaction media, is crucial for biomass-to-bioenergy industries [4].

The cell wall of lignocellulose mainly consists of lignin, cellulose, and hemicellulose [5]. The amount of each constituent is related to the type of plant species and their age [6]. Although lignocellulose is rich in cellulose and hemicellulose, its high lignin content, and tough and strong physical structure make it difficult to convert into chemicals and biofuels. Chemical methods such as acid pretreatment are commonly used to prepare reducing sugars. Compared to concentrated acid pretreatment, dilute acid pretreatment needs a higher reaction temperature and longer reaction time. Although the required reaction time and temperature of concentrated acid pretreatment are milder than that of dilute acid pretreatment, disadvantages such as a high equipment loss rate and environmental pollution impede its competitiveness [7]. In order to achieve low-cost and green sustainable production, solid acid catalysts had been proposed and applied to the production of reducing sugars [8].

The most common used solid acid catalysts include silica solid acids [9], biopolymer-based solid acids [10], ion-exchange resin solid acids [11], zirconia solid acids [12], and hydroxyapatite solid acids [13]. Nowadays, green catalysts, which are based on the use of renewable raw materials, are getting more and more attention [14]. Hence, more environmentally and economically-friendly solid acids were proposed. Biomass-based magnetic solid acid is prepared by using lignocellulose as a carbon carrier. It was regarded as a porous solid with a large surface area. Biomass-based magnetic solid acid was widely used in a large amount of reactions, such as the separation and purification of gases, and the removal of organic pollutants from water, refrigeration, and electrochemical devices [15]. Compared to traditional heterogeneous acid catalysts, the biomass-based magnetic solid acid has the characteristics of simple preparation, better recovery, convenient material selection, and low cost. Li et al. used a corn straw biomass-based solid acid to catalyze the hydrolysis of corn straw. The results showed that the prepared catalyst exhibited high catalytic activity for the conversion of corn straw into levulinic acid, and the most favorable values of catalyst dosage, hydrolysis temperature, hydrolyzation duration, and the maximum yield of LA were 3 g, 249.66 °C, 67.3 min, and 23.17%, respectively [15]. Chen et al. prepared a series of carbonaceous solid acids from biorenewable feedstock and used them as catalysts for the direct conversion of carbohydrates into 5-ethoxymethylfurfural (EMF) [16]. The results showed that the prepared catalysts presented a porous structure, high acid density, and easy separation. An EMF yield of 63.2% could be obtained from fructose at 120 °C. Lignocellulose contains a large amount of hemicellulose. It has been reported that the yield of xylan-based activated carbon (mainly based on hemicellulose) was much lower than that of cellulose and lignin due to the instability of hemicellulose [17]. Moreover, the presence of hemicellulose decreases the specific surface area of the solid acid when the carbonized temperature was up to 400 °C, which may reduce the catalytic performance of the solid acid. Simultaneously, hemicelluloses decompose at low temperatures and produce waste gas, which lead to environmental pollution. Therefore, in order to achieve a better carbon precursor, the biomass needs to be treated with suitable pretreatment methods [18]. Dilute acid and dilute alkali pretreatments are usually employed to destroy the complex structure of biomass. Dilute acid pretreatment can not only effectively remove hemicellulose, but can also minimize the damage of lignin and cellulose [19]. Meanwhile, dilute alkali pretreatment can effectively expand the biomass, leaving carbohydrates (cellulose and hemicelluloses) behind, thus increasing the contact area of the solid acid [20].

Bamboo is a fast-growing perennial herbaceous plant with a large phytomass, which is widely distributed in China [21]. Bamboo has many excellent properties that make it a suitable carbonized material for catalyst preparation, such as its porous structure and high thermal stability. Corncob is one of the abundant lignocellulose sources in China, and the annual global corncob production exceeds 1.03 billion metric tons, which can be used as a substrate for the production of platform products [8]. In this study, a bamboo-derived carbonaceous magnetic solid acid with a unique magnetic core–shell and high acid content was prepared by the impregnation-incomplete carbonization–sulfonation method and used as a magnetic solid acid to catalyze the hydrolysis of corncob to produce reducing sugar. The effects of dilute acid and dilute alkaline pretreatments on the catalytic performance of the as-prepared catalysts were investigated. The pretreated bamboo-derived carbonaceous support is expected with a porous structure for the introducing of –SO$_3$H, forming layers of adsorbate molecules that can interact with reactants. This process mainly comprises two steps: the preparation of bamboo-based magnetic solid acid and the catalytic hydrolysis of corncob by solid acid to prepare reducing sugars.

2. Results and Discussion

2.1. Characterization of the Catalyst

The Fourier transform infrared (FT-IR) spectrum of bamboo-derived carbonaceous magnetic solid acid prepared by different pretreatment methods is shown in Figure 1. The peak of 1560 cm^{-1} that appeared in all the samples corresponds to the C=C stretching vibration in aromatic carbons [22]. The typical band that appeared at 585 cm^{-1} represents the Fe–O vibration in Fe_3O_4, indicating that the solid acid was successfully magnetized during magnetization [23]. Moreover, strong bands such as 1041 cm^{-1} and 1380 cm^{-1} (O=S=O asymmetric) indicate that the –SO$_3$H group was successfully introduced into the bamboo-derived magnetic solid materials [24]. However, the strength of 1137 cm^{-1} (O=S=O symmetric stretching vibrations) was varied between solid acids prepared by different pretreatment methods. The order of the strength of the O=S=O peak at 1137 cm^{-1} was: dilute acid pretreated > dilute alkali pretreated > deionized (DI) water pretreated. The activated carbon prepared by dilute acid pretreatment was better combined with sulfuric acid, which was consistent with the subsequent catalytic hydrolysis results. In addition, a band at 3423 cm^{-1} corresponds to the O–H stretching vibration in –COOH, suggesting that phenolic –OH groups were also successfully generated to the prepared solid acid during the carbonization and sulfonation processes [25].

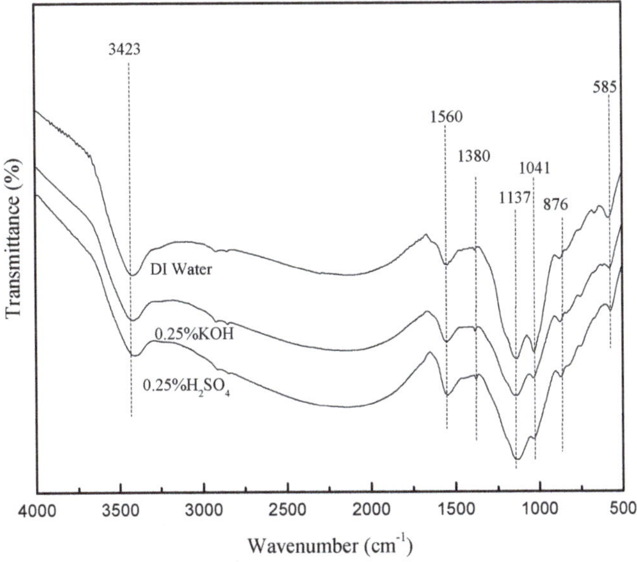

Figure 1. Fourier transform infrared (FT-IR) spectrum of bamboo-derived carbonaceous magnetic solid acid.

In order to understand the H$^+$ concentration on the surface of the prepared materials, an acid/base titration test was performed, and the results are shown in Figure 2. H$^+$ concentration on the surface of bamboo-derived carbonaceous magnetic solid acid prepared by different pretreatment methods was in the range of 0.6487 mmol/g to 2.3204 mmol/g. The acidity of the catalyst increased with the increment of acid/alkaline concentration, which may be related to the better interaction between the substrate and chemicals [18]. Compared with dilute alkali pretreatment, the acid density of the solid acids prepared by dilute acid pretreatment is higher, except at high concentration (2%). This phenomenon was consistent with the above FT-IR results. However, the acid density of the sample prepared by 2% KOH was much higher than that of the sample prepared by 2% H_2SO_4. Furthermore, 2.3204 mmol/g of H^+ concentration could be obtained from the DI-pretreated samples. These results indicated that the surface of a bamboo-derived carbonaceous magnetic solid had been successfully acidified.

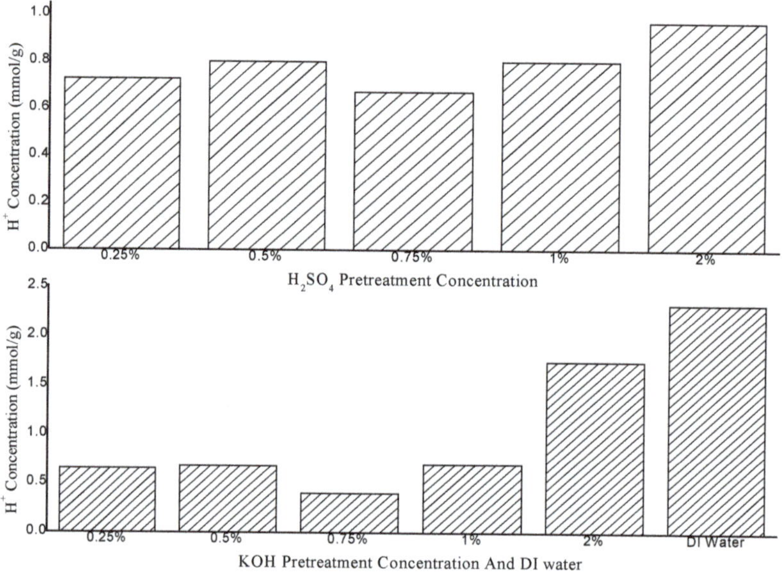

Figure 2. H^+ concentration of the bamboo-derived magnetic solid acid prepared by different pretreatment conditions.

During the carbonization and sulfonation processes, changes occurred in the crystallinity of the material. The XRD spectras of bamboo-derived magnetic solid acid prepared at different pretreatment conditions are shown in Figure 3. The observed sharp peaks at 2θ values of 35.61°, 57°, and 62° were assigned to the (220), (311), and (440) lattice planes of Fe_3O_4 (JCPDS19-629) [22,24], which indicated that the carbon coated by Fe_3O_4 existed in the form of polycyclic aromatic hydrocarbons, and their structures were kept stable during the sulfonation process. Moreover, material prepared by 0.5% H_2SO_4 pretreated bamboo showed the strongest intensity at 35.61°, which may lead to the better catalytic performance for the reducing production of sugars. This speculation was also verified by the following catalysis experiments. Compared with the solid acid prepared by dilute alkali pretreated under the same concentration, the peak intensity of the dilute acid pretreated sample was higher, which was consistent with the results of the subsequent catalytic output. In addition, both patterns exhibited a broad and weak diffraction peak at $2\theta = 20$–30°, which was due to the presence of amorphous carbon, which was composed of aromatic carbon sheets oriented in a random fashion [26].

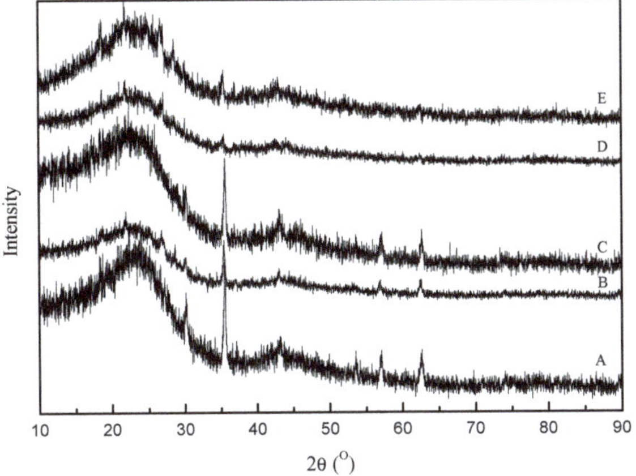

Figure 3. XRD spectra of bamboo-derived carbonaceous magnetic solid acid prepared with different pretreatment conditions (A: 0.5% H_2SO_4 pretreated sample; B: 2% H_2SO_4 pretreated sample; C: 0.5% KOH pretreated sample; D: 2% KOH pretreated sample; E: Deionized (DI) water pretreated sample).

The thermal behavior of the bamboo-derived magnetic solid acid (pretreated by 0.25% H_2SO_4) was shown in Figure 4. The thermal decomposition of the prepared catalyst can be divided into three stages. The weight loss which occurred before 150 °C was mainly because of the evaporation of water adsorbed on the surface of the material [27]. The weight loss stage from 180 °C to 420 °C corresponded to the decomposition of –SO_3H groups [3]. The last stage that occurred above 400 °C was attributed to the further condensation of amorphous carbons [25]. About an 11% weight loss of the Fe precursor may be occurred between 180–500 °C, which may be ascribed to the decomposition and conversion reaction of the Fe species of the catalyst at high temperatures [24].

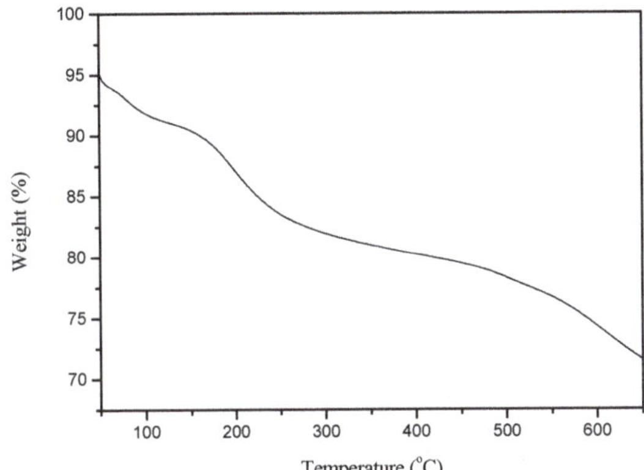

Figure 4. Thermal behavior of the bamboo-derived magnetic solid acid (pretreated by 0.25% H_2SO_4).

2.2. Effects of Pretreatment Conditions on the Chemical Composition of Bamboo

The detection of chemical composition of lignocelluloses can visually verify their structure in different conditions. In order to evaluate the effects of dilute acid and alkali pretreatments on the major sugar compositions of bamboo, dewaxed bamboo was treated by dilute H_2SO_4 and KOH in various concentrations (0.25%, 0.5%, 0.75%, 1%, and 2%) in a solid ratio of 1:10 (g/mL) at 120 °C for 30 min, respectively. The major sugar components of hydrolysates and solid residues are shown in Tables 1 and 2.

Table 1. Concentrations of major sugars in bamboo hydrolysates (mg/L).

Entry	Sample	Xylose	Arabinose	Glucose
1	0.25% H_2SO_4	1995.35	510.50	166.85
2	0.5% H_2SO_4	2157.45	582.70	243.45
3	0.75% H_2SO_4	2805.00	605.05	266.75
4	1% H_2SO_4	6278.50	665.25	417.15
5	2% H_2SO_4	8258.90	711.00	613.30
6	0.25% KOH	ND [a]	ND	ND
7	0.5% KOH	ND	ND	ND
8	0.75% KOH	ND	ND	ND
9	1% KOH	ND	ND	ND
10	2% KOH	ND	ND	ND
11	DI Water	ND	ND	ND

[a] ND: Undetectable.

Table 2. Major sugar composition of bamboo residues (g/g).

Entry	Sample	Xylose	Arabinose	Glucose
1	0.25% H_2SO_4	9.61	0.058	38.02
2	0.5% H_2SO_4	6.08	0.051	39.93
3	0.75% H_2SO_4	5.71	ND [a]	42.05
4	1% H_2SO_4	4.17	ND	44.81
5	2% H_2SO_4	4.83	ND	48.31
6	0.25% KOH	9.69	0.69	40.61
7	0.5% KOH	12.14	0.84	40.55
8	0.75% KOH	11.79	0.88	43.14
9	1% KOH	14.74	1.15	50.23
10	2% KOH	12.05	0.74	51.82
11	DI Water	10.93	0.74	33.94

[a] ND: Undetectable.

In the absence of a catalyst (H_2SO_4 and KOH), the dissolution of reducing sugars (xylose, arabinose, and glucose) was not observed under the investigated conditions (Table 1, entry 11). However, after dilute H_2SO_4 was added as the catalyst, the contents of reducing sugars in the hydrolysates enhanced sharply (Table 1, entries 1–5). The productions of xylose, arabinose, and glucose increased from 1995.35 mg/L to 8258.9 mg/L, 510.5 mg/L to 711.0 mg/L, and 166.85 mg/L to 613.3 mg/L with the acid concentration increased from 0.25% to 2.0%, respectively. This result indicated that about 74.74% of hemicellulose and 1.56% of cellulose could be extracted by 2.0% H_2SO_4 at 120 °C for 30 min. Therefore, the major compounds in the acid-treated bamboo residues may be cellulose and lignin. This phenomenon illustrated that a higher acid concentration benefited the dissolution of bamboo hemicellulose, especially for xylose during the pretreatment process. In addition, the hemicellulose structure of bamboo was elucidated as arabinoxylan oligosaccharides with xylose as the backbone and arabinose as the side chain [27]. Compared with xylose (16.61–68.83%), about 25.52–35.55% of arabinose was removed during dilute H_2SO_4 pretreatment, which suggested that the main chain of bamboo hemicellulose breaks faster than the side chain. However, glucose, xylose, and arabinose could not be detected in the KOH-pretreated hydrolysates (Table 1, entries 6–10), which suggested that dilute alkali pretreatment showed less power for the dissolution of hemicellulose and cellulose from bamboo than dilute acid pretreatment, which was consistent with the previous reports [2]. As shown in Table 1, xylose was the dominating sugar in the liquid fractions, and its content increased gradually as the pretreatment acid concentration increased. Moreover, it was found that arabinose was the secondary sugar constituent in the liquid fractions, and the yield of glucose was much lower than that of the former two. These results indicated that the dilute acid pretreatment mainly promotes the dissolution of hemicellulose rather than cellulose.

The hydrolyzed residues were simultaneously subjected to sugar analysis to verify the effect of pretreatments on the change of bamboo composition (Table 2). The contents of xylose and arabinose decreased gradually as the concentration of dilute acid increased (Table 2, entries 1–5), which was consistent with the results in Table 1. Moreover, a sugar composition analysis of dilute alkali-pretreated bamboo showed that the contents of major monosaccharides were kept stable in different KOH concentrations. This phenomenon suggested that cellulose and hemicellulose did not change much after the dilute alkali pretreatment. No monosaccharides could be detected in the hydrolysates after the dilute alkali and DI water pretreatments, which may be due to their further conversion into formic acid, acetic acid, furfural, etc. [17].

An FT-IR spectra of the bamboo residues obtained from dilute acid and dilute alkali pretreatments was performed to complete the study of all the samples from the experimental design (Figure 5). A band near 1733 cm^{-1} was ascribed to the stretching of C=O in hemicelluloses [28]. However, the absorption of this band decreased with the increment of sulfuric acid concentration, which suggested that the hemicellulose partly dissolved during the pretreatment process. Additionally, strong bonds at 1205 cm^{-1} (OH in-plane bending in cellulose I and cellulose II), 1160 cm^{-1} (C–O–C stretching asymmetric), 1100 cm^{-1} (glucose ring-stretching asymmetric), 898 cm^{-1} (β-glycosidic linkages between glucose unites in cellulose) could be observed both in dilute acid and alkali pretreated samples [27]. The strong bands at 1596 cm^{-1} and 1267 cm^{-1} corresponded to the aromatic skeletal vibration breathing with C=O stretching and vibration of the guaiacyl ring of lignin, respectively [16], which suggested that hemicelluloses could be removed effectively during the pretreatment process, resulting in the increment of the relative content of lignin in the bamboo residues. Therefore, the FT-IR results showed that the relative amount of the lignin and cellulose increased during the dilute acid/alkaline pretreatment process, and the change of the chemical composition of bamboo may affect the catalytic ability of the following prepared bamboo-derived carbonaceous magnetic solid acid.

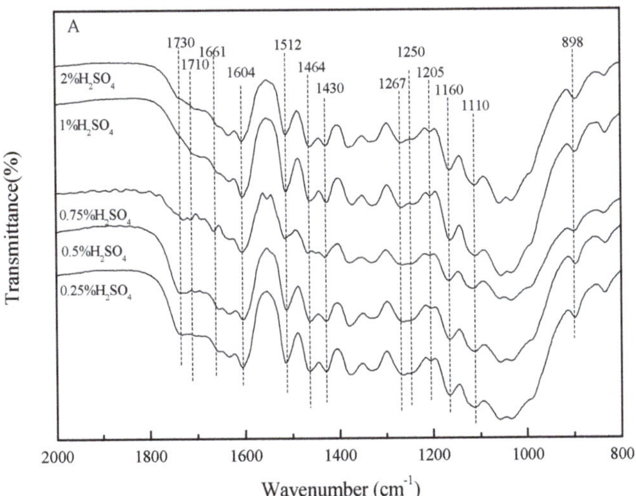

Figure 5. *Cont.*

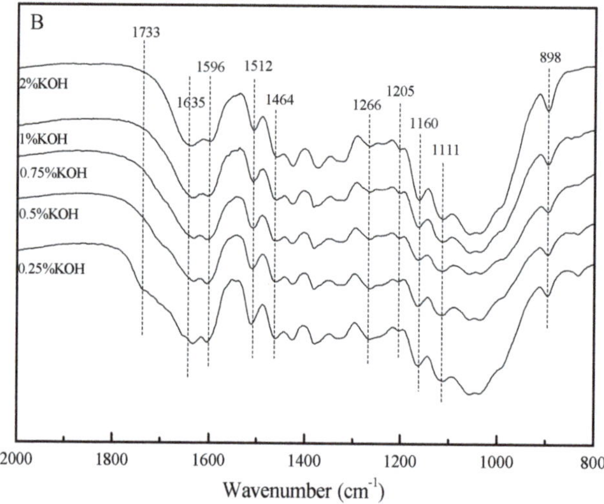

Figure 5. FT-IR spectra of dilute acid and alkali pretreated bamboo ((**A**) spectra of dilute acid pretreated bamboo; (**B**) spectra of dilute alkali pretreated bamboo).

2.3. Hydrolysis of Corncob with Bamboo-Derived Carbonaceous Magnetic Solid Acid

It should be noted that the pretreatment conditions had a great influence on the catalytic performance of the prepared bamboo-derived carbonaceous magnetic solid acid. In this study, experiments were executed for the catalytic hydrolysis of corncob at 120 °C for 30 min via the prepared magnetic solid acids. The main components of the hydrolysates and corncob solid residues were shown in Tables 3 and 4. The basic trend of xylose, arabinose, and glucose yields decreased when the pretreatment concentration of H_2SO_4 rose from 0.25% to 2%, which may be due to the reduction of the carbohydrates contents in the H_2SO_4-pretreated bamboo residues [29]. A higher carbohydrate content can achieve higher surface areas of specific materials during carbonization, which may be because an incomplete carbonization process is primarily the process of carbohydrates dehydrating and becoming volatile [18]. Moreover, the acidity of solid acids enhanced when prepared by a higher H_2SO_4 concentration. During the hydrolysis process, the obtained reducing sugar will be further catalyzed to form other products, such as furfural, 5-hydroxymethylfurfural, and so on [30]. However, the opposite phenomenon was observed for bamboo-derived magnetic solid acid prepared by dilute alkali pretreatment. When the KOH concentration was raised from 0.25% to 2%, the yields of xylose, arabinose, and glucose increased from 1856.2 mg/L to 2601.0 mg/L, 27.4 mg/L to 402.7 mg/L, and 672.5 mg/L to 1226.2 mg/L, respectively. This phenomenon may be because lignin was partly removed during the dilute alkali pretreatment, leaving most of the carbohydrates behind [31]. Therefore, higher reducing sugars yields could be obtained from the solid acid prepared by high concentration alkali-pretreated bamboo. In addition, solid acid prepared by authohydrolysis presented the lowest reducing sugars productivity, which may be ascribed to the weak damage strength of the bamboo structure under mild conditions. The highest reducing sugars yield of 37.17% could be obtained from bamboo-derived carbonaceous magnetic solid acid prepared by 0.25% H_2SO_4-pretreated bamboo. The prepared catalyst possessed a good catalytic performance for the conversion of hemicellulose into xylose and arabinose without further dehydration to form other products, which may be due to its moderate acidity and crystallinity.

Table 3. Effect of the pretreatment conditions on the catalytic performance of bamboo-derived magnetic solid acid (mg/L).

Entry	Sample	Xylose	Arabinose	Glucose
1	0.25% H_2SO_4	4643.40	662.80	1111.30
2	0.5% H_2SO_4	3100.05	492.15	1100.90
3	0.75% H_2SO_4	2284.05	360.20	683.6
4	1% H_2SO_4	3356.75	497.80	991.40
5	2% H_2SO_4	2259.40	318.40	219.50
6	0.25% KOH	1856.20	278.40	672.50
7	0.5% KOH	2642.85	366.55	668.50
8	0.75% KOH	2358.30	320.90	562.90
9	1% KOH	2034.75	302.4	850.80
10	2% KOH	2601.00	402.70	1226.20
11	DI Water	286.90	25.55	386.45

Reaction conditions: 1 g of corncob, 0.2 g of bamboo-derived magnetic solid acid, 20.0 mL of DI water, 300 rpm, 120 °C and 30 min.

Table 4. Analysis of corncob solid residue after catalysis (mg/L).

Entry	Sample	Xylose	Arabinose	Glucose
1	0.25% H_2SO_4	22.35	0.00	1172.05
2	0.5% H_2SO_4	46.30	0.00	1288.85
3	0.75% H_2SO_4	205.85	0.10	874.35
4	1% H_2SO_4	39.20	0.00	1150.00
5	2% H_2SO_4	63.55	0.00	1193.85
6	0.25% KOH	42.00	0.00	1038.50
7	0.5% KOH	44.30	0.00	1073.10
8	0.75% KOH	105.30	0.65	959.90
9	1% KOH	197.15	0.50	817.05
10	2% KOH	0.00	0.00	1004.10
11	DI Water	0.00	0.00	999.05

Determining the recyclability of a catalyst is important to evaluate its efficiency based on economic and environmental factors. The solid acids prepared in this study had the characteristic of being easily separated due to their magnetic properties. In this study, the recyclability of the prepared bamboo-derived magnetic solid acid (pretreated by 0.25% H_2SO_4) was performed, and the results are shown in Figure 6. After three recycle runs, the prepared catalyst remained active for the conversion of corncob into reducing sugar. In comparison to the fresh catalyst, the yields of xylose, arabinose, and glucose decreased slightly from 4643.4 mg/L to 3902.5 mg/L, 662.8 mg/L to 455.5 mg/L, and 1111.3 mg/L to 805.4 mg/L after three recycles, respectively, which suggested that the prepared catalyst had good reusability. Compared with conventional solid acids, the biomass-based magnetic solid acid used in this study has the advantages of easy recycling, moderate acidity, and widely resourced. Moreover, it possessed good catalytic performance for the conversion of lignocellulosic raw materials into reducing sugars.

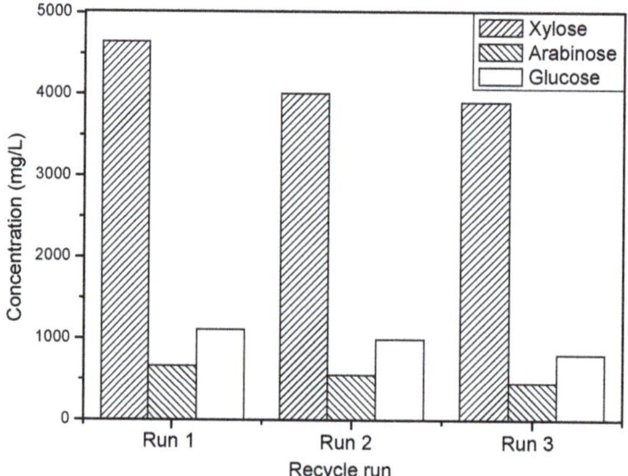

Figure 6. Recyclability of the bamboo-derived magnetic solid acid (pretreated by 0.25% H_2SO_4).

3. Materials and Methods

3.1. Materials

The bamboo used in the experiment was collected from Hunan Province, China. Prior to the experiments, the bamboo was decorticated, ground, and sieved to about 40-mesh size. The obtained particles were oven-dried at 55 °C to a constant weight. The compositional analysis (glucose, xylose, arabinose, and lignin) of the bamboo was performed according to an established National Renewable Energy Laboratory procedure [32] with a resulting composition of 49.1 wt.% glucose, 12.0 wt.% xylose, 2.0 wt.% arabinose, and 37.1 wt.% lignin.

The corncob used in this study was obtained from Shandong Province, China. Before the experiments, corncob was ground to pass through a 40-mesh screen, and then dewaxed with a 2:1 (v/v) acetone/ethanol mixture in a Soxhlet extractor (1000 mL, Synthware Glass Co. Ltd., Beijing, China) for 6 h. The dewaxed corncob was oven-dried at 60 °C to constant weight and milled for 6 h (300 rpm) with a ball-milling machine (DECO-PBM-V, Deco Co. Ltd., Hunan, China). The composition of the dewaxed corncob was 38.1 wt.% glucose, 31.9 wt.% xylose, 4.2 wt.% arabinose, and 25.8 wt.% lignin.

Chemicals $FeCl_3 \bullet 6H_2O$ (AR, $\geq$99.0%), H_2SO_4 (AR, $\geq$98.0%), ethanol (AR, $\geq$99.0%), phenolphthalein (AR, $\geq$99.0%), and NaOH (AR, $\geq$99.0%) were purchased from Kermel Co. Ltd. (Tianjin, China). Standard reagents, including glucose (HPLC, $\geq$99.0%), xylose (HPLC, $\geq$99.0%), arabinose (HPLC, $\geq$98.0%) were purchased from Shanghai Sigma-Aldrich Trading Co. Ltd. (Shanghai, China). Deionized (DI) water was used to prepare all of the solutions. All of the chemicals were used as received.

3.2. Methods

3.2.1. Catalyst Characterization

The Fourier transform infrared (FT-IR) spectrum of the prepared bamboo-derived magnetic solid acid was recorded on a spectrophotometer (Tensor 27, Bruker Optics, Karlsruhe, Germany). An acid-base titration test was performed to verify the free hydrogen ion (H^+) concentration in the bamboo-derived magnetic solid acid. In briefly, 0.1 g of catalyst was added into 20 mL of NaCl solution (20 mmol/L), and then stirred for 24 h (100 rpm) at room temperature. Subsequently, the mixture was filtered, and a liquid fraction (5 mL) was titrated with 50 mmol/L of NaOH solution. Phenolphthalein was used as the indicator. Experiments were done in triplicate, and the average value was obtained. The X-ray diffraction (XRD) patterns of the prepared catalysts were detected in the 2θ range of 10° to 90° on a Bruker D8 ADVANCE X-ray diffractometer (Karlsruhe, Germany) with Cu $K\alpha$ radiation. The thermostability of the samples was tested with thermogravimetric analysis (TA Q200, New Castle, DE, USA).

3.2.2. Two-Step Preparation of Bamboo-Derived Magnetic Carbonaceous Solid Acid

The hydrothermal pretreatment of bamboo was carried out in a high-pressure reactor (SLM-100, Shenlang Co. Ltd., Beijing, China). In this study, a mixture of 10 g of feedstock and 100 mL of liquid (KOH/H_2SO_4/water) was added into the reactor. The concentrations of the KOH and H_2SO_4 solution were 0.25%, 0.5%, 0.75%, 1%, and 2% respectively. A control group was performed using DI water to compare with dilute acid and alkali pretreatments. The pretreatment time, temperature, and the agitation rate were 120 °C, 30 min, and 300 rpm, respectively. Once the reaction finished, flowing water was used to cool the reactor quickly. After that, the mixture was separated by filtration. Solid residue was washed with DI water several times and dried at 60 °C overnight for the following preparation of bamboo-derived magnetic solid acid. The liquid products were determined by high-performance liquid chromatography (HPLC, Waters 2414, America) coupled with a refractive index detector (RID) and a Bio-rad Aminex HPX-87H (300 × 7.8 mm) column. Five mM of H_2SO_4 was employed as the eluent with a flow rate of 0.5 mL/min at 60 °C.

Five g of pretreated residue was dispersed in 500 mL of the $FeCl_3 \bullet 6H_2O$ solution (10 mmol/L), and then stirred at room temperature (150 rpm) for 5 h. After impregnation, bamboo was carbonized at 500 °C for 1 h under nitrogen atmosphere to produce the bamboo carbon. The bamboo-derived carbon was mixed with a concentrated sulfuric acid at a ratio of 1:10 (g/mL) and sonicated for 15 min. Subsequently, the mixture was heated under vigorous at 90 °C for 10 h to introduce the sulfo-group ($-SO_3H$) to the surface of the bamboo-derived magnetic solid precursor. After the reaction, the mixture was diluted with DI water and dried in a vacuum oven at 80 °C for 12 h.

3.2.3. Hydrolysis of Corncob by Bamboo-Derived Magnetic Solid Acid

One g of ball-milled corncob, 0.2 g of bamboo-derived magnetic solid acid, and 20 mL of DI water were mixed first and ultrasonicated for 30 min, and then added into the reactor. The reaction time, temperature, and the agitation was 120 °C, 30 min, and 300 rpm, respectively. At the same time, a catalyst-free reaction including 1 g of ball-milled corncob and 20 mL of DI water was also conducted at the same reaction conditions. After the reaction, the product was filtered with a 0.22-μm syringe. Liquid fractions were stored in the fridge prior to HPLC analysis, as mentioned above. Solid residues were washed with DI water several times and oven-dried at 60 °C to a constant weight.

3.2.4. Recyclability of Catalyst

After reaction, the bamboo-derived magnetic solid acid catalyst was separated by a magnet. The obtained catalyst was washed with DI water and oven-dried at 70 °C for 6 h for the next catalytic run.

4. Conclusions

The conversion of corncob into reducing sugars was achieved using bamboo-derived carbonaceous magnetic solid acid as the catalyst. The prepared catalysts contained $-SO_3H$, $-COOH$, and polycyclic aromatic, and the number of H^+ sites was from 0.6487 mmol/g to 2.3204 mmol/g. Comparing with the catalysts prepared by dilute alkali and DI water pretreatments, bamboo-derived carbonaceous magnetic solid acid prepared by dilute acid-pretreated bamboo showed better catalytic activity for the production of reducing sugars. The highest reducing sugars yield of 37.17% could be obtained using 0.25% of H_2SO_4-pretreated bamboo-derived carbonaceous solid acid as catalyst at 120 °C for 30 min.

Author Contributions: Formal analysis, Y.Z., J.H., and S.S.; Methodology, A.W. and H.L.; Supervision, A.W. and H.L.; Writing—review and editing, Y.Z., J.H. and H.L.

Funding: This research was funded by National Natural Science Foundation of China (NO. 31700506), Natural Science Foundation of Guangdong Province, China (NO. 2017A030310550) and Guangdong Province Science and Technology Project (NO. 2016A010104012).

Conflicts of Interest: The authors declare no conflict of interest.

References

1. Schmitt, C.; Belén, M.; Reolon, G.; Zimmermann, M.; Raffelt, K.; Grunwaldt, J.-D.; Dahmen, N. Synthesis and regeneration of nickel-based catalysts for hydrodeoxygenation of beech wood fast pyrolysis bio-oil. *Catalysts* **2018**, *8*, 449–477. [CrossRef]

2. Xin, D.; Yang, Z.; Liu, F.; Xu, X.; Zhang, J. Comparison of aqueous ammonia and dilute acid pretreatment of bamboo fractions: Structure properties and enzymatic hydrolysis. *Bioresour. Technol.* **2015**, *175*, 529–536. [CrossRef] [PubMed]

3. Li, H.; Deng, A.; Ren, J.; Liu, C.; Lu, Q.; Zhong, L.; Peng, F.; Sun, R. Catalytic hydrothermal pretreatment of corncob into xylose and furfural via solid acid catalyst. *Bioresour. Technol.* **2014**, *158*, 313–320. [CrossRef] [PubMed]

4. Gupta, P.; Paul, S. Solid acids: Green alternatives for acid catalysis. *Catal. Today* **2014**, *236*, 153–170. [CrossRef]

5. Lee, J. Biological conversion of lignocellulosic biomass to ethanol. *J. Biotechnol.* **1997**, *56*, 1–24. [CrossRef]

6. Börjesson, M.; Larsson, A.; Westman, G.; Ström, A. Periodate oxidation of xylan-based hemicelluloses and its effect on their thermal properties. *Carbohydr. Polym.* **2018**, *202*, 280–287. [CrossRef] [PubMed]

7. Pellera, F.-M.; Gidarakos, E. Chemical pretreatment of lignocellulosic agroindustrial waste for methane production. *Waste Manag.* **2018**, *71*, 689–703. [CrossRef] [PubMed]

8. Li, H.; Ren, J.; Zhong, L.; Sun, R.; Liang, L. Production of furfural from xylose, water-insoluble hemicelluloses and water-soluble fraction of corncob via a tin-loaded montmorillonite solid acid catalyst. *Bioresour. Technol.* **2015**, *176*, 176–183. [CrossRef] [PubMed]

9. Zaccheria, F.S.F.; Iftitah, E.D.; Ravasio, N. Brønsted and Lewis Solid Acid Catalysts in the Valorization of Citronellal. *Catalysts* **2018**, *8*, 410–411. [CrossRef]

10. Solinas, A.; Taddei, M. Solid-supported reagents and catch-and-release techniques in organic synthesis. *Synthesis* **2007**, *38*, 2409–2453. [CrossRef]

11. Sebti, S.D.; Tahir, R.; Nazih, R.; Boulaajaj, S.D. Comparison of different Lewis acid supported on hydroxyapatite as new catalysts of Friedel-Crafts alkylation. *Appl. Catal. A Gen.* **2001**, *218*, 25–30. [CrossRef]

12. Tanabe, K.; Yamaguchi, T. Acid-base bifunctional catalysis by ZrO_2 and its mixed oxides. *Catal. Today* **1994**, *20*, 185–197. [CrossRef]

13. Hara, M.; Yoshida, T.; Takagaki, A.; Takata, T.; Kondo, J.; Hayashi, S.; Domen, K. Carbon material as a strong protonic acid. *Angew. Chem. Int. Ed.* **2004**, *43*, 2955–2958. [CrossRef] [PubMed]

14. Oregui, M.; Miletic, N.; Hao, W.; Björnerbäck, F.; Rosnes, M.; Garitaonandia, J.; Hedin, N.; Arias, P.L.; Barth, T. High-performance magnetic activated carbon from solid waste from lignin conversion processes. Part II: Their use as NiMo catalyst supports for lignin conversion. *Energy Procedia* **2017**, *114*, 6272–6296.

15. Liu, F.; Rotaru, A.E.; Shrestha, P.M.; Malvankar, N.S.; Nevin, K.P.; Lovley, D.R. Promoting direct interspecies electron transfer with activated carbon. *Energy Environ. Sci.* **2012**, *10*, 8982–8989. [CrossRef]

16. Li, X.; Lei, T.; Wang, Z.; Li, X.; Wen, M.; Yang, M.; Chen, G.; He, X.; Xu, H.; Guan, Q.; et al. Catalytic pyrolysis of corn straw with magnetic solid acid catalyst to prepare levulinic acid by response surface methodology. *Ind. Crop. Prod.* **2018**, *116*, 73–80. [CrossRef]

17. Chen, T.; Peng, L.; Yu, X.; He, L. Magnetically recyclable cellulose-derived carbonaceous solid acid catalyzed the biofuel 5-ethoxymethylfurfural synthesis from renewable carbohydrates. *Fuel* **2018**, *219*, 344–352. [CrossRef]

18. Guo, Y.; Rockstraw, D.A. Physical and chemical properties of carbons synthesized from xylan, cellulose, and Kraft lignin by H_3PO_4 activation. *Carbon* **2006**, *44*, 1464–1475. [CrossRef]

19. Zhu, M.Q.; Wang, Z.W.; Wen, J.L.; Qiu, L.; Zhu, Y.H.; Su, Y.Q.; Wei, Q.; Sun, R.C. The effects of autohydrolysis pretreatment on the structural characteristics, adsorptive and catalytic properties of the activated carbon prepared from Eucommia ulmoides Oliver based on a biorefinery process. *Bioresour. Technol.* **2017**, *232*, 159–167. [CrossRef] [PubMed]

20. Wei, H.; Chen, X.; Shekiro, J.; Kuhn, E.; Wang, W.; Ji, Y.; Kozliak, E.; Himmel, M.E.; Tucker, M.P. Kinetic modelling and experimental studies for the effects of Fe^{2+} ions on xylan hydrolysis with dilute-acid pretreatment and subsequent enzymatic hydrolysis. *Catalysts* **2018**, *8*, 39–57. [CrossRef]

21. Safari, A.; Karimi, K.; Shafiei, M. Dilute alkali pretreatment of softwood pine: A biorefinery approach. *Bioresour. Technol.* **2017**, *234*, 67–76. [CrossRef] [PubMed]

22. Xu, L.; Shi, Y.; Zhou, G.; Xu, X.; Liu, E.; Zhou, Y.; Zhang, F.; Li, C.; Fang, H.; Chen, L. Structural development and carbon dynamics of Moso bamboo forests in Zhejiang Province, China. *For. Ecol. Manag.* **2018**, *409*, 479–488. [CrossRef]

23. Sluiter, J.B.; Ruiz, R.O.; Scarlata, C.J.; Sluiter, A.D.; Templeton, D.W. Compositional analysis of lignocellulosic feedstocks: Review and description of methods. *J. Agric. Food Chem.* **2010**, *58*, 9043–9053. [CrossRef] [PubMed]

24. Qi, X.; Watanabe, M.; Aida, T.M.; Smith, R.L., Jr. Sulfated zirconia as a solid acid catalyst for the dehydration of fructose to 5-hydroxymethylfurfural. *Catal. Commun.* **2009**, *10*, 1771–1775. [CrossRef]

25. Li, H.; Deng, A.; Ren, J.; Liu, C.; Wang, W.; Peng, F.; Sun, R. A modified biphasic system for the dehydration of d-xylose into furfural using $SO_4{}^{2-}/TiO^{2-}ZrO_2/La^{3+}$ as a solid catalyst. *Catal. Today* **2014**, *234*, 251–256. [CrossRef]

26. Hu, L.; Tang, X.; Wu, Z.; Lin, L.; Xu, J.; Xu, N.; Dai, B. Magnetic lignin-derived carbonaceous catalyst for the dehydration of fructose into 5-hydroxymethylfurfural in dimethylsulfoxide. *Chem. Eng. J.* **2015**, *263*, 299–308. [CrossRef]

27. Jia, X.; Dai, R.; Sun, Y.; Song, H.; Wu, X. One-step hydrothermal synthesis of $Fe_3O_4/g\text{-}C_3N_4$ nanocomposites with improved photocatalytic activities. *J. Mater. Sci.* **2016**, *27*, 3791–3798. [CrossRef]

28. Ishii, T.; Hiroi, T.; Thomas, J.R. Feruloylated xyloglucan and p-coumaroyl arabinoxylan oligosaccharides from bamboo shoot cell-walls. *Phytochemistry* **1990**, *29*, 1999–2003. [CrossRef]

29. Xu, F.; Yu, J.; Tesso, T.; Dowell, F.; Wang, D. Qualitative and quantitative analysis of lignocellulosic biomass using infrared techniques: A mini-review. *Appl. Energ.* **2013**, *104*, 801–809. [CrossRef]

30. Faix, O. Classification of lignins from different botanical origins by FT-IR Spectroscopy. *Holzforschung* **1991**, *45*, 21–28. [CrossRef]

31. Li, J.; Li, K.; Zhang, T.; Wang, S.; Jiang, Y.; Bao, Y.; Tie, M. Development of activated carbon from windmill palm sheath fiber by KOH activation. *Fiber. Polym.* **2016**, *17*, 880–887. [CrossRef]

32. He, C.; Sasaki, T.; Shimizu, Y.; Koshizaki, N. Synthesis of ZnO nanoparticles using nanosecond pulsed laser ablation in aqueous media and their self-assembly towards spindle-like ZnO aggregates. *Appl. Surf. Sci.* **2008**, *254*, 2196–2202. [CrossRef]

 *catalysts*

Review

Catalytic Thermochemical Conversion of Algae and Upgrading of Algal Oil for the Production of High-Grade Liquid Fuel: A Review

Yingdong Zhou and Changwei Hu *

Key Laboratory of Green Chemistry and Technology, Ministry of Education, College of Chemistry, Sichuan University, Chengdu 610064, Sichuan, China; yingdong_zhou@163.com
* Correspondence: changweihu@scu.edu.cn; Tel.: +86-28-85411105

Received: 27 December 2019; Accepted: 18 January 2020; Published: 21 January 2020

Abstract: The depletion of fossil fuel has drawn growing attention towards the utilization of renewable biomass for sustainable energy production. Technologies for the production of algae derived biofuel has attracted wide attention in recent years. Direct thermochemical conversion of algae obtained biocrude oil with poor fuel quality due to the complex composition of algae. Thus, catalysts are required in such process to remove the heteroatoms such as oxygen, nitrogen, and sulfur. This article reviews the recent advances in catalytic systems for the direct catalytic conversion of algae, as well as catalytic upgrading of algae-derived oil or biocrude into liquid fuels with high quality. Heterogeneous catalysts with high activity in deoxygenation and denitrogenation are preferable for the conversion of algae oil to high-grade liquid fuel. The paper summarized the influence of reaction parameters and reaction routes for the catalytic conversion process of algae from critical literature. The development of new catalysts, conversion conditions, and efficiency indicators (yields and selectivity) from different literature are presented and compared. The future prospect and challenges in general utilization of algae are also proposed.

Keywords: algae; bio-oil; thermochemical conversion; catalytic upgrading; high-grade liquid fuel

---

1. Introduction

The development of renewable and sustainable fuels is heavily required worldwide due to the depletion of finite fossil fuels [1,2]. In addition, the combustion of fossil fuels has caused several environmental problems, including global warming and environmental pollution [3]. Consequently, much of the attention has been drawn to alternatives such as biofuels from renewable biomass resources, which are able to mitigate CO_2 emission [4,5]. Usually, sugar or oil crops, and non-edible lignocellulosic feedstocks are applied for the production of the first- and second-generation liquid biofuels [6–9]. Different from terrestrial crops, algae are also attractive feedstock due to the advantages such as high lipid accumulation, short growth cycle, and the ability to grow in aquatic environments [5,10,11]. Generally, algal biomass accumulates about 20–50 wt % lipids based on the dry weight of biomass, and some species even have more than 60% lipids content [12–15]. In addition, some kind of natural algae can fix nitrogen, phosphorus, and heavy metals in waste water and polluted lakes [12,16,17]. Thus, algae cultivation cooperating with waste water treatment not only relieves the environmental pollution, but also offers the resource for the generation of renewable energy.

Varieties of technologies for algal biofuel production have been developed in recent decades. As shown in Figure 1, the methods of utilization of algae generally include bio-chemical conversion, lipid extraction, transesterification, and thermochemical conversion. Biochemical conversion mainly produces bioethanol from the fermentation of algal sugar [18,19]. Macroalgae with high saccharides content are preferred in such process [20]. However, different kinds of algae need different enzymes for their conversion. Since algae contain considerable amount of lipids, extraction techniques are used to produce algal oil, which mainly consists of triglycerides and free fatty acids [21]. The extracted oil can be transesterified to biodiesel (defined as fatty acid alkyl esters, FAAE) over acid or base catalysts [22,23]. The traditional acid/base catalysts might corrode the equipment and cost a lot. Another way to produce biodiesel from algae is through *in situ* transesterification of algal feedstock with the aid of solvent. Technologies such as microwave, ultrasound, or supercritical fluid are usually applied in this process to enhance the biodiesel yield [24–28]. Biodiesel has the advantages of high biodegradability and renewability. Nonetheless, the extraction–transesterification process only utilizes the lipid fraction of algae, leaving the other parts (i.e., sugars and proteins) still remaining in algal cells [29]. In addition, not all kinds of algae have high lipid content. Some species of low-lipid algae contain lipid less than 15 wt % [5,11,16,17,30]. Moreover, in order to increase the content of lipids in algae, many researchers focused on the cultivation of algae, which required suitable temperature, pH, nutrient, some rigorous conditions and so on, resulting in extra expense [31–33].

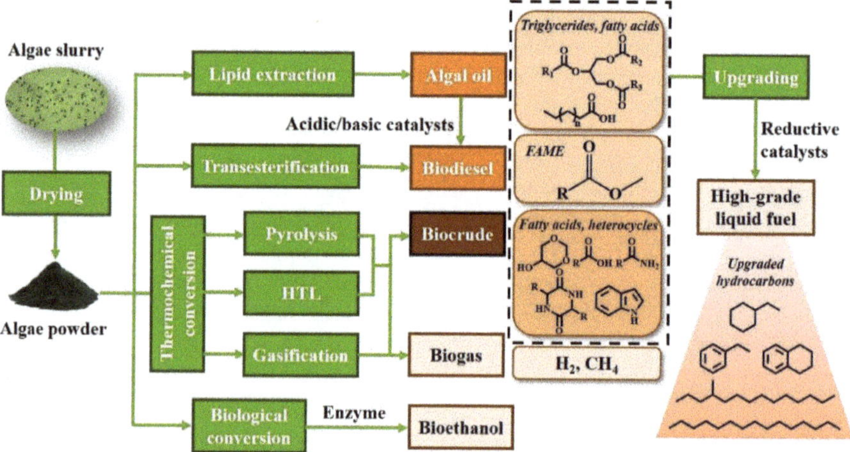

Figure 1. Schematic diagram for conversion of algae into biofuel.

Researchers have been seeking for the method to fully utilize the components in algae. Among the technologies, thermochemical routes are viable for the utilization of the entire algal cell, including lipids, sugars, and proteins. Thermochemical conversion can be classified to pyrolysis, hydrothermal liquefaction (HTL) and gasification [10,34]. Gasification produces mainly gas fuels (CH_4 and H_2) at high temperatures, while liquid bio-oil is obtained via pyrolysis and HTL [35]. Pyrolysis of dried algae produces bio-oil, biochar, and gas products via thermal degradation of feedstock at relatively high temperatures (300–700 °C) [36]. The HTL process usually conducts at mild temperatures (200–450 °C) but high pressure (4–22 MPa), which requires high-pressure resistant equipment [37,38]. Since dehydration of algae feedstock requires energy, hydrothermal liquefaction is favorable for wet biomass conversion. Thus, a great deal of studies focused on the thermochemical conversion of algae, in an attempt to convert the whole algae into a mixture of gases, liquid, and solids. The higher heating value (HHV) of bio-oil obtained from thermochemical conversion is around 30 MJ/kg, higher than the algae material itself [39–42].

However, the problem is that the composition of bio-oil obtained from direct thermochemical conversion of algae is complex, containing hydrocarbons, aromatics, organic oxygenates, and nitrogenous compounds, because of the complexity of algae [5,43–45]. The high oxygen content reduces the heating value of bio-oil, while the high nitrogen content makes the bio-oil not suitable for combustion. In addition, biodiesel obtained from transesterification and bio-oil from thermochemical conversion have drawbacks such as poorer chemical stability, higher viscosity, and lower energy density compared with petroleum-based liquid fuel [46–48]. In order to improve the selectivity of target products and the quality of biofuel, thermochemical conversion of algae and the upgrading process in the presence of catalysts are developed. On the one hand, one-step catalytic thermochemical conversion of algae is able to increase the bio-oil yield and reduce the content of oxygen and heteroatoms with a proper catalyst. Usually, heterogeneous catalysts such as zeolite are employed in the thermochemical process. The obtained bio-oil has higher energy density and lower oxygen content, as well as a much higher aromatic hydrocarbon content [49–51]. On the other hand, the extracted algal oil, biodiesel, and biocrude obtained from direct thermochemical conversion can go through an upgrading process (hydroprocessing) for obtaining high-grade liquid fuels (mainly hydrocarbons) [52–54]. Therefore, catalysts with high activity in deoxygenation and denitrogenation are highly preferred in algae biorefinery.

This article focuses on the catalytic thermochemical conversion (pyrolysis and HTL) of algae as well as the catalytic upgrading of algal oil or biocrude to high-grade liquid fuels. In this review, the characteristics of biofuels from direct thermochemical conversion in recent research, the catalytic performances of catalysts on algae feedstock and bio-oil conversion are overviewed systematically, and the selection of catalysts is summarized in detail.

2. Production of Biofuel from Direct Conversion of Algae

2.1. Algal Oil from Lipid Extraction and Transformation to Biodiesel

Algal lipids are regarded as a renewable source for the production of the third-generation biofuel [55]. The methods of lipid extraction from algae have been studied for decades, which are classified into mechanical and chemical methods [56]. Chemical methods are generally solvent assisted extraction, including Soxhlet extraction and supercritical fluid extraction, while mechanical methods include grinding, bead beating, ultrasound, and microwave [57–60]. Organic solvents such as chloroform and methanol (Bligh and Dyer method) are frequently used [61], most of which are harmful to the environment and human health [62]. In addition, the lipids in algal cells can hardly be fully extracted when a single method is applied. Since algae have multi-layered cell walls, the mechanical methods combined with chemical methods are used in the extraction process for the disruption of algal cell walls and enhancing the lipid yield. However, the cost and energy-intensive mechanical methods for cell disruption and lipids extraction make the process less attractive and limit its industrialization.

Nevertheless, the selectivity of extracted lipids becomes poor when the lipid yield enhances, due to the fact that the solvents with high extraction efficiency usually extract more compositions from algae, containing neutral lipids (triglycerides), glycolipids, phospholipids, chlorophyll, carotenoid, and sterol [13,63]. For example, lipids extracted from algae by hexane contained a high fraction of neutral lipids, while chloroform/methanol and hexane/isopropanol mixture gave a higher composition of polar lipids [63]. According to the US Department of energy's standard, the N, O and S content in upgraded algal fuels should be <0.05%, <1% and 0, respectively [64]. Due to the high oxygen content and existence of nitrogen and sulfur, algal derived oil is unsuitable to directly use in diesel engines. Consequently, the extracted lipids need to be hydrotreated to high-grade hydrocarbons. Technologies developed in the petroleum refinery for deoxygenation, denitrogenation, and desulfurization can be directly applied in hydrotreatment of algal lipids [12].

Furthermore, triglycerides and free fatty acid in algal lipids can be transformed into biodiesel via transesterification with alcohol, which exhibits better performance in diesel engines [22]. However, the oxygen content of biodiesel is still high, resulting in many drawbacks such as low stability and poor flow property at low temperature. Thus, biodiesel itself is only used as an additive in diesel engines [2,12]. In order to get algae-derived high-grade liquid fuels, which can be directly combusted in diesel engines, the upgrading process is needed to remove the heteroatoms in algal oil or algae derived biodiesel, and convert them into diesel-range hydrocarbons.

2.2. Bio-Oil from Direct Pyrolysis

Heating biomass in the absence of air or oxygen for thermal degradation at relatively high temperatures is called pyrolysis. Usually, temperature between 300–700 °C is applied [2]. In some cases, the pyrolysis temperature is over 800 °C or below 300 °C [65]. Bio-oil, biochar, and gaseous products are obtained via pyrolysis. Direct pyrolysis can be classified into slow pyrolysis (SP), fast pyrolysis (FP), and microwave assisted pyrolysis (MAP). In slow pyrolysis, the heating rate is relatively slow (between 0.1–1 °C·s^{-1}) so that the process of heating cannot be ignored [65]. It is reported that the heating rate had an influence on the yields and distribution of oil products from the slow pyrolysis of lignocellulosic biomass [66,67]. In fast pyrolysis, the temperature rises to the designated temperature within seconds (>100 °C·s^{-1}). The bio-oil yields from fast pyrolysis are usually higher than slow pyrolysis because the high heating rate makes the vapor products stay in the reactor within only seconds [65,68]. Recently, microwave pyrolysis is considered as an efficient way for biomass conversion. Microwaves can be controlled easily with instantaneous start-up and shut-off with high heating efficiency [17]. The pyrolysis atmosphere is reported to have an influence on the pyrolytic products [17,69]. Because algae are poor absorbers of microwaves, suitable absorbers such as activated carbon are mixed with the feedstock [65]. There are several parameters influencing the yields and properties of pyrolysis products. Expected heating rate and pyrolysis atmosphere, pyrolysis temperature, time, and particle size of feedstock have shown influence on pyrolysis products [65]. For example, high temperature and residence time usually resulted in high bio-oil yield and complex composition of bio-oil, while higher yield of oil was obtained from microalgae with a larger particle size in MAP [17,70].

However, the residue (%) from pyrolysis of algae is usually over 30%, which indicates that pyrolysis of algae is not able to convert the whole algal cell. Even at 700 °C, fast pyrolysis of *Chlorella vulgaris*, *Schizochytrium limacinum*, *Arthrospira platensis*, and *Nannochloropsis oculate* yielded residue of 39%, 48%, 36%, and 48%, respectively [71]. Since pyrolysis at a relatively high temperature cannot selectively degrade only one component (i.e., lipids, sugars or proteins) from algae, the composition of bio-oil is complex. Recent research on direct (non-catalytic) pyrolysis of algae are summarized and listed in Table 1. It can be seen that, whatever the method of pyrolysis is applied, the oxygen and nitrogen content of bio-oil are too high to be directly used in diesel engines. The high oxygen content of pyrolytic bio-oil results in low HHV and poor stability, and the combustion of the high nitrogen content bio-oil may generate NO$_x$ and cause air pollution [2,5]. Therefore, developing strategies for improving the quality of pyrolytic bio-oil becomes a hot topic. Since different components of algae have different thermal stability, the degradation temperature influences the selectivity of products in bio-oil. Recently, researchers have developed fractional pyrolysis of algae, which is separate conversion of the three main components (lipids, carbohydrates and proteins) by controlling the pyrolytic temperature and to realize the multistep conversion. Fractional pyrolysis of cyanobacteria from water booms was studied, and it was found that fractional pyrolysis separated the degradation of different components in algae and improved the selectivity of products in bio-oil [72,73]. However, the stepwise pyrolysis made the process complex and might require extra energy consumption, which increased the cost. Thus, one-step catalytic pyrolysis to remove O and N during the thermal degradation process is favorable for the production of high-grade green diesel.

Table 1. Properties of bio-oil from non-catalytic pyrolysis.

Feedstock	Method	T (°C)	Bio-Oil Yield (wt %)	Elemental Composition (wt %)				HHV (MJ·kg⁻¹)	Ref.
				C	H	O	N		
Nannochloropsis oculata	FP	500	-	-	-	-	-	33.4	[74]
Nannochloropsis sp. residue	SP	400	31.1	56.1	7.6	30.1	5.3	24.4	[75]
Scenedesmus dimorphus	SP	500	40.0	74.7	10.6	8.3	5.8	28.5	[76]
Cyanobacteria	SP	300	9.8	66.0	8.6	14.8	10.7	32.0	[73]
C. vulgaris remnants	FP	500	53	51.4	8.3	27.5	12.8	24.6	[77]
C. vulgaris	FP	400	72	61.0	8.2	24.8	6.0	28.9	[78]
Chlorella sp.	MAP	490	28.6	65.4	7.8	16.5	10.3	30.7	[79]
Chlorella sp.	MAP	550	57	65.7	9.3	15.8	8.5	32.4	[80]

2.3. Biocrude from Direct Hydrothermal Liquefaction

Hydrothermal liquefaction (HTL) of algae is usually conducted in a pressurized water environment (4-28 MPa) at a relatively moderate temperature (200–450 °C) [3,37,38,81,82]. At elevated temperature and pressure, the properties (e.g., solubility) of water change, which promote the degradation of macromolecules in algae to small molecular compounds [5,37]. HTL is considered to be a favorable technique for the conversion of wet algae due to no need for the dewatering process. Like pyrolysis, HTL also has the ability to utilize all of the algae including lipids, carbohydrates, and proteins. After HTL, liquid products, solid residues and gas products are obtained. The water-insoluble phase of liquid products can be recovered by extraction using organic solvents, and the obtained oil-like products are called biocrude. The water-soluble phase separated from liquid products containing nutrients (e.g., N, P, Mg and K) makes up a large fraction, which can be recycled for microalgae cultivation and anaerobic digestion [81,83–86].

Compared with pyrolysis, the yield of solid residue from HTL of algae was around 20% at 350 °C [87–89]. In other words, the conversion of algae via HTL is higher than pyrolysis even when HTL is conducted at relative mild temperatures. This is probably due to the high solubility of some components in water. A summary of the yields and properties of biocrude from direct (non-catalytic) HTL of algae is given in Table 2. In general, the biocrude obtained from HTL has higher HHV compared with pyrolysis oil. The nitrogen content of biocrude is relatively low because some fraction of nitrogen remains in aqueous products. Despite the advantages of HTL, the biocrude composition from HTL of algae is also complex. To gain a high biocrude yield, HTL treatment usually conducts at high temperatures (>300 °C) [89,90]. The dissolution of proteins or saccharides which contain a high proportion of N and O makes the quality of biocrude reduced and limits its commercialization [91]. Therefore, strategies for improving the biocrude yield and quality have drawn lots of researchers' attention. With a proper catalyst, the yield of biocrude can be increased to a certain amount, whereas the content of heteroatoms decreases via denitrogenation or deoxygenation [92,93]. In addition, HTL of algae usually couples with the catalytic upgrading process for the production of high-quality biofuel containing low oxygen and nitrogen.

Table 2. Properties of biocrude from non-catalytic HTL.

Feedstock	Condition	Biocrude Yield (wt %)	Elemental Composition (wt %)				HHV (MJ·kg^{-1})	Ref.
			C	H	O	N		
Scenedesmus sp.	250 °C, 7 min	20	66.1	8.6	20.8	4.5	31.2	[90]
Scenedesmus sp.	350 °C, 30 min	36	72.5	9.1	12.9	5.5	35.0	[90]
G. sulphuraria	350 °C, 6 min	28.1	78.2	11.4	4.2	4.6	38.2	[94]
L. digitata	350 °C, 15 min	17.6	70.5	7.8	17	4.0	32	[88]
L. hyperborea	350 °C, 15 min	9.8	72.8	7.7	14.9	3.7	33	[88]
Kirchneriella sp.	300 °C, 30 min	45.4	76.6	9.0	7.9	5.2	37.5	[89]
Algal blooms	300 °C, 60 min	18.4	73.5	9.0	10.9	6.6	35.5	[95]
Tetraselmis sp.	350 °C, 30 min	31.0	75.6	9,9	12.7	5.2	33.3	[96]

3. Catalytic Thermochemical Conversion of Algae

3.1. Catalytic Pyrolysis of Algae

The quality of bio-oil can be improved by using catalysts in pyrolysis. The oxygen content of pyrolytic bio-oil needs to be reduced in order to improve the stability and heating value. In addition, the high proportion of N atoms in the bio-oil should also be removed to meet the standard of combustion fuel. With the aid of an appropriate catalyst, the pyrolytic pathway can be changed, and consequently the selectivity of products is influenced [97]. Generally, catalysts such as zeolite are frequently applied in catalytic pyrolysis, due to the high activity in deoxygenation, cracking, and dehydration [68,98]. Moreover, metal oxides and supported metal catalysts have also been explored for catalytic pyrolysis of algae. This part overviews the performance of different types of catalysts on catalytic pyrolysis of algae.

3.1.1. Zeolites

Zeolites are regarded as highly efficient catalysts for upgrading bio-oil from algae, due to the suitable acidity, resistance of carbon deposition, and the ability to eliminate oxygen atoms without hydrogen [98–100]. Of the zeolite catalysts, ZSM-5 is commonly used because of its adjustable acidity and high performance in deoxygenation, decarboxylation, and decarbonylation [68]. Its acidity can be controlled by varying the Si/Al ratio. High Si/Al ratio results in low acidity of the zeolite [68,101]. The acid sites on zeolites make the macromolecules of algae degrade to compounds with small molecular size. Subsequently, the formed compounds pass through deoxygenation or aromatization forming reduced compounds such as hydrocarbons.

The performance of a catalyst on catalytic pyrolysis of algae can be evaluated by the bio-oil yield, the oxygen and nitrogen content of bio-oil. Pyrolysis of macroalgae *Enteromorpha clathrata* over metal modified Mg-Ce/ZSM-5 catalysts at 550 °C produced bio-oil with high quality [99]. The 1 mmol Mg-Ce/ZSM-5 showed the ability to increase bio-oil yield from 33.77% (without catalyst) to 37.45% and decrease the acid content. In addition, the average molecular weight of bio-oil obtained over such catalyst seemed to decrease, with the content of gasoline-like (C_5-C_7) compounds increased. Primary cracking and decarboxylation might occur due to the presence of 1 mmol Mg-Ce/ZSM-5. Catalytic fast pyrolysis of *spirulina* sp. over different types of zeolites (ZSM-5, zeolite-β and zeolite-Y) was performed [45]. The HHV of pyrolysate ranged over 30-37 MJ·kg^{-1}, which was much higher than that of algae feedstock. All types of zeolites facilitated the formation of aromatics (monoaromatics, PAHs, and indoles). Cycloalkanes were formed over ZY and Zβ, while C_2-C_4 nitriles formed over high acidity zeolites. Anash et al. studied the pyrolysis behavior of *Chlamydomonas debaryana* with and without β-zeolite or activated carbon (AC) [102]. The yields of total hydrocarbons were highest over β-zeolite than that over AC and without catalyst. The combination of hydrothermally carbonized pretreatment and catalytic pyrolysis could effectively reduce nitrogen content of bio-oil, and produce more hydrocarbons, including aromatics. It was found that AC catalyst was more likely to form coke than β-zeolite. More detailed results of catalytic pyrolysis over zeolites were listed in Table 3.

Table 3. Catalytic pyrolysis of algae over zeolites.

Feedstock	Catalyst	Condition	Catalytic Performance	Ref.
Spirulina	Fe/HMS-ZSM5	500 °C, catalyst/algae = 0.5	Highest bio-oil yield: 37.8 %, hydrocarbon yield: 33.04 %	[98]
Tetraselmis suecica	CBV zeolite	400 °C, catalyst/algae = 0.2	Zeolite (Si/Al=30) showed good activity in deoxygenation.	[100]
Cyanobacteria	MgAl4-LDO/ZSM-5	550 °C, catalyst/algae = 0.75	Maximum liquid yield: 41.1%, O/C: 0.09, HHV: 37.164 MJ·kg^{-1}	[51]
Spirulina	HZSM-5 and H-β with different Si/Al ratio	450 °C, 30 s catalyst/algae = 0.75	Maximum of aromatic (6.54%) hydrocarbons were obtained over HZSM-5 (Si/Al=23), but lower acidity catalysts increased the production of aliphatic hydrocarbons.	[101]
Isochrysis sp.	Li-LSX-zeolite	500 °C, catalyst/algae = 0.5	Bio-oil yield: 29%11.8% aromatics and 23.1% aliphatics	[103]

To summarize, catalysts applied for pyrolysis of algae usually have high activity in deoxygenation. The deoxygenation performance of zeolites can be adjusted by changing the Si/Al ratio of the zeolite. In addition, aromatization of pyrolytic products can be observed, generating abundant aromatics in bio-oil. The increasing acidity of catalysts (low Si/Al ratio) results in promotion of aromatization [98,101]. After catalytic pyrolysis with zeolites, bio-oil with high HHV and low O/C ratio was obtained, but the nitrogen content could hardly be reduced to meet the standard of commercial transportation fuel. The bio-oil obtained from catalytic pyrolysis contained about 5% of N content [51,98,100,101,103]. To produce high-quality biofuel with low N content, catalysts with the ability to remove N in bio-oil need to be developed. The bio-oil obtained from catalytic pyrolysis needs to be further upgraded by catalysts such as sulfide CoMo/Al$_2$O$_3$ or NiMo/Al$_2$O$_3$, which have high activity in hydrodenitrogenation (HDN) [2].

3.1.2. Other Catalysts

Except zeolites, other catalysts such as metal oxides and supported metal catalysts are applied in catalytic pyrolysis. Transition metal such as nickel has high activity in C-C and C-O bonds cleavage, resulting in high performance for decarbonylation and decarboxylation [12]. Furthermore, reducible metal oxides are considered as a favorable support or catalyst because of their superior redox properties and low carbon deposition rate [104]. It was found that the pyrolysis of *Tetraselmis* sp. and *Isochrysis* sp. over Ni-Ce/Al$_2$O$_3$ and Ni-Ce/ZrO$_2$ produced a higher yield of bio-oil (26 wt %) [105]. The catalysts exhibited strong deoxygenation and denitrogenation ability, with only 9–15% oxygen remained and removal of 15–20% nitrogen from bio-oil. In addition, pyrolysis of Pavlova sp. over Ce/Al$_2$O$_3$-based catalysts produced bio-oil with a low O/C ratio (0.1–0.15). MgCe/Al$_2$O$_3$ exhibited the best performance on the reduction of oxygen content from 14.1 to 9.8 wt %, while NiCe/Al$_2$O$_3$ produced the highest hydrocarbon fraction [106]. Catalytic fast pyrolysis of *Nannochloropsis oculate* over Co-Mo/γ-Al$_2$O$_3$ was carried out in an analytical micropyrolyzer coupled with a gas chromatograph/mass spectrometer (py-GC-MS) [74]. It was found that aliphatic alkanes and alkenes, aromatic hydrocarbons, and long-chain nitriles were the main products in bio-oil. Co-Mo/γ-Al$_2$O$_3$ catalyst could promote the formation of 1-isocyanobutane and dimethylketene with 35% selectivity. In addition, catalytic pyrolysis over Co-Mo/γ-Al$_2$O$_3$ produced pyrolysates with higher calorific value (33–39 MJ·kg^{-1}) compared with that of algal feedstock (18 MJ·kg^{-1}).

3.2. Catalytic HTL of Algae

3.2.1. Catalytic HTL with Homogeneous Catalysts

Because HTL is conducted in aqueous phase, water-soluble homogeneous catalysts can be used. They commonly have the ability to improve biocrude yield or produce target compounds with high selectivity. Generally, the homogeneous catalysts include acid catalysts (HCl, H_2SO_4 and other organic acids), alkali catalysts (Na_2CO_3), or other inorganic salts. These catalysts usually promote bonds cleavage, so the degradation of components in algae to small molecular compounds is facilitated. In the presence of homogeneous catalysts, the dissolution of components from algae is facilitated and hence the biocrude yield enhances. However, the homogeneous catalysts can hardly improve the quality of biocrude. In other words, they have a weak influence on deoxygenation and denitrogenation of algal biocrude [3]. A summary of recent works on catalytic HTL on homogeneous catalysts is displayed in Table 4.

In general, the homogeneous catalysts used in HTL are catalysts with proper acidity or basicity. Koley et al. studied the catalytic and non-catalytic HTL of wet *Scenedesmus obliquus* (Table 4) [107]. The optimizations of HTL temperature, pressure, and residence time were conducted on both catalytic and non-catalytic HTL. They found that 300 °C, 200 bar, and 60 min was the optimum condition for obtaining the maximum yield of biocrude (35.7 wt %). By adding catalysts, the biocrude yield increased and followed the order: acidic catalyst CH_3COOH (45%) > HCOOH (40%) > HCl (39%) > H_2SO_4 (38%) > H_3BO_3 (37%), and basic catalyst Na_2CO_3 (40%) > NaOH (38%) > Ca(OH)$_2$ (37%) > KOH (37%) > K_2CO_3 (36%). In addition, the acetic acid had the ability to reduce the oxygen content of biocrude, resulting in a considerable HHV of 40.2 MJ·kg^{-1}. However, the composition of biocrude in the presence of CH_3COOH was still complex, containing fatty acids, phenols, indoles, monoaromatics, and N-heterocycles.

The effect of acidic, neutral and basic catalysts on the conversion of microalgae (*Spirulina platensis*) by HTL at various temperatures was studied by Zhang et al. [108]. HCl and acetic acid were used to create acidic condition, while KCl for neutral, and K_2CO_3 and KOH for basic conditions. Among these catalysts, only acetic acid and KOH are found to positively influence the biocrude yield, which was more obvious at lower temperatures. The acid and base catalysts promoted the degradation of components in microalgae and suppressed the condensation reaction, resulting in lower average molecular weight of biocrude. However, the distribution of compounds detected by GC-MS showed little change even with the aid of acid or base catalysts.

Typically, Na_2CO_3 is a commonly-used homogeneous catalyst for HTL because of its ability to enhance the biocrude yield. Shakya et al. studied the effect of temperature on the HTL of three kinds of algal species *Nannochloropsis*, *Pavlova*, and *Isochrysis* over Na_2CO_3 [109]. It can be concluded that the biocrude yield increased with the rise of temperature from 250 to 350 °C. The maximum biocrude yields from HTL of three algae species followed the order: *Nannochloropsis* (48.67 wt %) > *Isochrysis* (40.69 wt %) > *Pavlova* (39.96 wt %). When using Na_2CO_3 as the catalyst, the biocrude yield changed to the order of *Pavlova* > *Isochrysis* > *Nannochloropsis*. The biocrude yields of algae with high carbohydrates content (*Pavlova* and *Isochrysis*) increased at higher temperatures (300–350 °C) with the aid of Na_2CO_3, whereas the high-protein-containing algae (*Nannochloropsis*) showed higher yield of biocrude at lower temperature (i.e., 250 °C). However, the conversion of algae with Na_2CO_3 did not significantly improve biocrude properties. The biocrudes obtained were still not suitable for application in transportation.

Overall, homogeneous catalysts have a strong impact on the products yields, especially enhancing the biocrude yields. This is probably ascribed to the degradation ability of these catalysts to create an acid or basic condition. There were few reports about the deoxygenation and denitrogenation ability of homogeneous catalysts. Thus, the quality of biocrude from HTL can hardly be improved by homogeneous catalysts. The biocrude obtained cannot meet the standard for transportation fuel. Furthermore, the acidity and basicity of catalysts might influence the pH value of biocrude and lead to corrosion to the equipment. The homogeneous catalysts can hardly be recovered after reaction, and this leads to further expense [3,82]. Therefore, finding catalysts with good reusability and highly efficient for deoxygenation and denitrogenation is pressingly required.

Table 4. Catalytic HTL of algae with homogeneous catalysts.

Catalyst	Algae Feedstock	Condition	Biocrude Yield (wt %)	Performance	Ref.
Without catalyst	*Scenedesmus obliquus*	300 °C, 200 bar, 1 h	35.7	HHV: 39.4 MJ·kg^{-1}	[107]
CH$_3$COOH	*Scenedesmus obliquus*	300 °C, 200 bar, 1 h	45.1	HHV: 39.1 MJ·kg^{-1}, O%: 8.9%	[107]
Na$_2$CO$_3$	*Scenedesmus obliquus*	300 °C, 200 bar, 1 h	40.2	HHV: 40.2 MJ·kg^{-1}, N%: 4.7%	[107]
KOH	*Spirulina platensis*	300 °C, 35 min	30.1	Positive effect on biocrude yield, more ketone and amides formed, lighter volatiles generated	[108]
K$_2$CO$_3$	defatted *Cryptococcus curvatus*	350 °C, 1 h	68.9	The highest biocrude yield, HHV: 38.2 MJ·kg^{-1}	[110]
K$_2$CO$_3$	defatted *Cryptococcus curvatus*	300 °C, 1 h	63.9	The best biocrude quality, HHV: 36.9 MJ·kg^{-1}, lowest nitrogen (0.77%)	[110]
KOH	*Cyanidioschyzon merolae*	300 °C, 0.5 h, 120 bar	22.67	HHV: 33.66 MJ·kg^{-1}	[111]
CH$_3$COOH	*Cyanidioschyzon merolae*	300 °C, 0.5 h, 120 bar	21.23	HHV: 33.36 MJ·kg^{-1}	[111]
KOH	*Ulva prolifera*	290 °C, 10 min	26.7	Biocrude yield increased from 12.0 wt % to 26.7 wt % with KOH; Higher HHV: 33.6 MJ·kg^{-1}	[112]
Na$_2$CO$_3$	Green macroalgal blooms	270 °C, 45 min	20.1	The highest bio-oil yield was achieved over Na$_2$CO$_3$	[113]

3.2.2. Catalytic HTL with Heterogeneous Catalysts

The heterogeneous catalysts, which exist in different phases with the reaction media, are usually solid catalysts [114]. Heterogeneous catalysts can be easily recovered after reaction, therefore reducing the cost of the process [115,116]. Conversion of algae in the presence of heterogeneous catalysts might cover the deficiency of homogeneous catalysts. Commonly, heterogeneous catalysts include zeolites (e.g., H-ZSM-5), supported metal catalysts (Pt/C), and other metal oxide supported catalysts (e.g., sulfide CoMo/Al$_2$O$_3$ and Ni/TiO$_2$) [117–119]. These materials have strong activity for bonds cleavage, resulting in facilitation of macromolecule degradation and conversion of oxygenates and nitrogenates to high-grade hydrocarbons. As a result, biocrude with low viscosity, high HHV, and low N content is produced in the presence of heterogeneous catalysts.

The effects of heterogeneous catalysts on the yield and quality of biocrude are summarized in Table 5. After screening, the majority of the catalysts listed in the table are supported metal catalysts and can be recycled several times. For example, magnetic nanoparticles (MNPs) were synthesized for microalgae separation and catalytic HTL by Egesa et al. [115]. Firstly, the MNPs were used for separation of algae from the culture medium, with a separation efficiency of 99% achieved. Then, the MNPs were applied in catalytic HTL for the production of biocrude from microalgae. The biocrude yield significantly increased from 23.2% (without catalyst) to 37.1% in the presence of Zn/Mg-ferrite MNPs. Moreover, the percentage of hydrocarbons increased by 26.4%, and the percentage of heptadecane increased by 27.8%, while the percentage of oxygenates and N-containing compounds decreased. This indicated the catalysts had activity in deoxygenation and denitrogenation. In addition, the MNPs could be easily recovered and recycled several times.

Additionally, noble metal catalysts such as commercial Pd/C, Ru/C and Pt/C are widely used in algae conversion and show excellent catalytic performance. Liu et al. reported a two-step catalytic conversion of algae (*Spirulina*) via solvent extraction followed by catalytic HTL of the extracted residue [120]. In the extraction process, ethanol was found to be the best solvent with the highest extraction efficiency, and the introduction of MgSO$_4$ could produce ethyl esters from fatty acids. The lipid extracted residue was treated by HTL in the presence of commercial Pd/C, Pt/C, Ru/C, Rh/C, and Pd/HZSM-5. Among all catalysts, Rh/C exhibited the best performance in catalytic conversion of algae, producing 50.98% yield of biocrude with 30.7 MJ·kg^{-1} HHV. The O and N content of biocrude obtained from HTL over Rh/C decreased significantly from 32.2% and 7.1% to 23.6% and 4.4%, respectively. In addition, the percentage of hydrocarbons in biocrude obtained over Rh/C based on GC-MS results was 55.7%. Xu et al. studied the catalytic effects of Pt/C, Ru/C and Pt/C + Ru/C on the HTL of *Chlorella* in the presence of H$_2$ (Table 5) [117]. They divided the biocrude into water-soluble biocrude (WSB) and water-insoluble biocrude (WISB). The addition of catalysts could decrease the fraction of WSB but increase WISB fraction. At optimized conditions, Pt/C and Ru/C led to the highest carbon (63.6% and 74.2%) and hydrogen (7.3% and 8.4%) contents but lowest oxygen (14.1% and 9.2%) and nitrogen (12.2% and 7.1%) contents, and the highest HHV (29.7 and 35.6 MJ·kg^{-1}) for WSB and WISB fraction, respectively. The water insoluble biocrude obtained from HTL of algae over Pt/C contained amides (48.2%), hydrocarbons (17.7%), acids (12.8%), and phenols (7.7%). In addition, catalytic HTL produced biocrude with more low boiling point fractions.

Apart from noble metal catalysts, the application of non-noble metal catalysts in algae HTL has drawn lots of attention due to their low cost and high activity in bonds cleavage. Among the non-noble metals, nickel, cobalt, iron, and molybdenum are proved to be active in deoxygenation and denitrogenation [2]. Kohansal et al. conducted the HTL of *Scenedesmus obliquus* in the presence of Ni-based catalysts (Ni/AC, Ni/AC-CeO$_2$ nanorods and Ni/CeO$_2$ nanorods) [121]. The optimum condition for the catalytic HTL of microalgae over the catalysts was to set at 324.12 °C, 43.52 min, and 19.90 wt % feedstock. With the addition of heterogeneous catalysts, the biocrude yields over three catalysts were higher than that from a non-catalytic process. The highest biocrude yield of 41.87% was achieved over Ni/AC-CeO$_2$ nanorods, with the HHV of 38.57 MJ·kg^{-1}. In the presence of Ni-based catalysts, the percentage of hydrocarbons in biocrude was higher than that of the non-catalytic biocrude, but the content of nitrogen-containing compounds was also higher.

Overall, heterogeneous catalysts perform better than homogeneous catalysts in terms of the improvement of algal biocrude quality. In addition, the yield of biocrude can be improved in the presence of heterogeneous catalysts. The catalysts can be easily recovered and reused after HTL, but the coke formation is still the major problem during catalytic HTL. However, in the previous literature, the contents of oxygen and nitrogen in the obtained biocrude are still too high to satisfy the standard of transportation fuel. Therefore, the technologies for production of high-quality liquid fuel with extremely low O and N content from algae need to be further developed.

Table 5. Catalytic HTL of algae with heterogeneous catalysts.

Catalyst	Algae Feedstock	Condition	Biocrude Yield (wt %)	Performance	Ref.
Co/CNTs	*Dunaliella tertiolecta*	320 °C, 30 min, catalyst/algae = 0.1	40.25	Higher percentage of hydrocarbons, lower content of fatty acid and lower N-compounds	[122]
nano-Ni/SiO$_2$	*Nannochloropsis* sp.	250 °C, 60 min, catalyst/algae = 0.05	30.5	Bio-oil with lower O and N content, catalyst recovery 2–3 times	[123]
Pd/HZSM-5@MS	*Spirulina*	380 °C, 2 h, HCOOH, catalyst/algae = 0.02	37.3	Promotion of HDO for bio-oil, HHV: 32.65 MJ·kg^{-1}, catalyst recovery for 5 times	[116]
Ni/TiO$_2$	*Spirulina*	250 °C, 30 min, catalyst/algae = 0.05	43.1	Promoting the formation of hydrocarbons (14%) and esters (15%), and decreasing oxygenates and nitrogenates.	[124]
Ru/C	*Nannochloropsis* sp.	350 °C, 20 min, H$_2$, catalyst/algae = 0.2	43.5	Increasing the yield of water-insoluble biocrude, decreasing O (7.95%) and N (4.95%) content	[125]
Ru/C	*Chlorella*	350 °C, 30 min, 0.3 MPa H$_2$, catalyst/algae = 0.2	27	Water insoluble biocrude with highest C (74.2%) and H (8.4%) contents, the lowest O (9.15%) and N (7.1%) and highest the HHV (35.6 MJ·kg^{-1})	[117]
ZSM-5	*Ulva prolifera*	280 °C, 10 min, catalyst/algae = 0.15	29.3	HHV of biocrude was 34.8 MJ·kg^{-1} (non-catalytic was 21.2 MJ·kg^{-1});	[118]
Ni/TiO$_2$	*Nannochloropsis*	300 °C, 30 min, catalyst/algae = 0.1	48.23	Biocrude with lower viscosity, more light fractions, HHV: 35 MJ·kg^{-1}, reproduction for at least 10 times	[119]
Fe/HZSM-5	*Nannochloropsis*	365 °C, 60 min	38.1	Increase of carbon into biocrude, nitrogen into aqueous phase	[126]
Zn/Mg-ferrite MNPs	*Scenedesmus obliquus*	320 °C, 60 min, catalyst/algae = 0.12	37.1	The percentage of hydrocarbons (46.5%), heptadecane (37.8%), HHV: 35.4 MJ·kg^{-1}	[115]

4. Catalytic Conversion of Oil Derived from Algae

4.1. Catalytic Hydroprocessing of Extracted Algal Oil

Generally, the oil recovered from algal cell consists of different types of triglycerides. The fatty acids fraction of triglycerides usually contains palmitic, palmitoleic acid, stearic acid, and oleic acid. Some algal species also contain polyunsaturated fatty acids (PUFA) such as eicosapentaenoic acid (EPA), arachidonic acid (AA), and docosahexaenoic acid (DHA), which are value-added health care products [12,127]. The algae derived triglycerides can be hydrotreated by catalysts to fuel-like hydrocarbons with the aid of a proper catalyst. Since hydrogen could be obtained from a wealth of sources, including water splitting, especially by electrolysis of water with renewable electricity such as wind power or solar power, the consumption of hydrogen in the hydroprocessing can be ignored. A summary of recent works on hydroprocessing of algal oil is listed in Table 6.

Conventional NiMo sulfide catalyst has been widely used in hydrogenation of natural oil into fuel-ranged hydrocarbons. Liu et al. investigated the hydrocracking of algal oil from *Botryococcus braunii* (C$_n$H$_{2n-10}$, n = 29–34) with sulfide NiMo into fuel-ranged hydrocarbons [128]. The support effect on the selectivity of products was studied under the conditions of 300 °C for 6 h under 4 MPa H$_2$. For the hydrotreating of the model compound (squalene C$_{30}$H$_{50}$), the main product was squalane (C$_{30}$H$_{62}$) over NiMo/SiO$_2$, C$_1$-C$_4$ gas hydrocarbons over NiMo/HZSM-5, C$_5$-C$_9$ gasoline-ranged hydrocarbons over NiMo/HY and NiMo/SiO$_2$-Al$_2$O$_3$, and C$_{10}$-C$_{15}$ aviation fuel-ranged hydrocarbons over NiMo/Al$_{13}$-Mont, respectively. The hydrocracking of algal oil over NiMo/Al$_{13}$-Mont gave aviation fuel-ranged hydrocarbons (C$_{10}$-C$_{15}$) with a yield of 52%. The sulfide NiMo catalyst acted as a bifunctional catalyst for hydrogenation of squalene to squalane followed by cracking of the formed squalane to shorter-chain hydrocarbons. Zhao et al. explored the hydrotreating of extracted algal lipids from *Nannochloropsis* for the production of aviation fuel [129]. The effect of hydrotreating temperature (270–350 °C) and catalyst loading (10-30%) was investigated. The optimum condition for hydrotreating reaction was 350 °C and 30% catalyst loading. The main components of biofuel were C_8-C_{16} hydrocarbons and aromatics. The two-step hydrotreating process obtained biofuels with oxygen content below 0.3% and nitrogen content below 0.007% and HHV of 46.24 MJ·kg^{-1}.

In addition, noble metal catalysts such as Pt, Pd and Ru, which have high activity in hydrodeoxygenation, are also suitable for conversion of algal oil into high-grade hydrocarbons [130,131]. Xu et al. explored a technology for selective extraction of neutral lipid from algae *Scendesmus dimorphus* and subsequently conversion into jet fuel [132]. Hexane and ethanol solvent mixture was used for selective extraction of neutral lipids. Then, the extracted lipid was hydrogenated over the Pt/Meso-ZSM-5 catalyst. The obtained product oil (38%) mainly contained branched paraffin with C_9-C_{15} chain length. The jet fuel product satisfied the ASTM 7566 standard with the desired freeze point (−57 °C), flash point (42 °C), heating value (45 MJ·kg^{-1}), and aromatics content (<1%).

Although metal sulfides and noble metals are highly active in deoxygenation of algal lipid/oil, the sulfur leaching of metal sulfides and high-cost noble metal make these process not environmentally and economically friendly [47,133]. Except for metal sulfides and noble metal catalysts, the sulfur-free non-noble metal catalysts are promising in heterogeneous catalysis. It is found that non-noble metals (e.g., Ni, Co, Cu) are active in deoxygenation of fatty acids and natural oil to hydrocarbons [47,134,135]. Santillan-Jimenez et al. investigated the continuous catalytic hydrogenation of model compound and algal lipids to fuel-like hydrocarbons using Ni-Al layered double hydroxide [136]. In addition, Ni/Al_2O_3, Ni/ZrO_2, and Ni/La-CeO_2 were applied for the comparison experiments. Of all Ni-based catalysts, Ni-Al LDH showed the best results for conversion of tristearin to C_{10}-C_{17} hydrocarbons at 260 °C. Higher temperatures favored the cracking reaction to form lighter alkanes, while lower H_2 pressure favored the formation of heavier hydrocarbons. For the hydrogenation of algal oil, ~50% yield of hydrocarbons was obtained over Ni-Al LDH.

Table 6. Catalytic hydroprocessing of algal oil.

Catalyst	Feedstock	Condition	Conversion (%)	Performance	Ref.
Ni/meso-Y	Microalgae biodiesel	275 °C, injection rate: 0.02 mL·min^{-1}	91.5	Hydrocarbon selectivity: 56.2%, isomer ratio: 46.4%	[137]
Ni-Cu/Al$_2$O$_3$	Extracted algal lipids	260 °C and 580 psi	-	Diesel-ranged hydrocarbons content: 83%	[138]
Ni-Al LDH	Extracted algal lipids	300 °C, 4 h stream	100	Hydrocarbon content: 99 wt %, diesel-like hydrocarbons: 76 wt %	[139]
Co/clay	Algae DHA oil	260 °C, 8 h, 40 bar H$_2$	100	Hydrocarbon yield: 85.5 wt %	[47]
Ce/Zeo-β	Algal oil	400 °C, 6 h	98	Selectivity for C$_{10}$-C$_{14}$ hydrocarbons: 85%	[140]

Generally, extracted algal oil/lipids have the potential to be converted to high-grade, fuel-like hydrocarbons. The nitrogen and oxygen content of algal-lipid derived fuel are low enough to satisfy the standard for transportation fuel due to the low nitrogen content and easily editable oxygen of algal lipids. However, due to the limited lipid content of algae, the yield of algal lipids derived green fuel based on the whole algal cell is also low. Based on the concept of "waste-free biorefinery", the utilization of other components of algae except for lipids needs to be explored.

4.2. Catalytic Upgrading of Biocrude Oil from Thermochemical Conversion of Algae

Another way for production of fuel-ranged hydrocarbons from algae is the upgrading of the biocrude oil from thermochemical conversion. The biocrude oil obtained at high temperatures contains the components derived from saccharides and proteins apart from lipid. Some of the oxygenates and nitrogenates cause the undesired properties of biocrude oil [141]. Therefore, the large proportion of oxygen and nitrogen of bio-oil needs to be removed for obtaining high-quality biofuel. This process involved the use of a proper heterogeneous catalyst with high activity in deoxygenation and denitrogenation. Generally, the upgrading process also needs H_2 to remove the heteroatoms (O, N, and S), for improving the heating value and reducing the O, N, and S content of product oil. The resulted biofuel should have low viscosity, high stability, and high HHV [68,82]. A summary of recent works on catalytic upgrading of algal biocrude oil is presented in Table 7.

4.2.1. Catalytic Upgrading of Pyrolysis Bio-Oil

As mentioned in the previous section, the bio-oil from direct pyrolysis of algae contains high proportion of O and N due to the degradation of the whole algae cell at high temperatures. The quality of pyrolytic bio-oil needs to be upgraded via catalytic process before utilization as transportation fuel. Elkasabi et al. investigated one-step hydrotreating and aqueous extraction of O and N-containing compounds, for the production of fuel-range hydrocarbons from *Spirulina* pyrolysis bio-oil [142]. The catalytic hydrodeoxygenation (HDO) and hydrodenitrogenation (HDN) were conducted over commercial Ru/C catalyst. The upgrading at 385 °C resulted in organic oil with low N and O content (<1 wt %). The selectivity of products could be controlled by varying reaction conditions. More paraffins were obtained at higher temperature (~400 °C), while lower temperature (350 °C) resulted in more phenolics. The remaining oxygen and nitrogen-containing compounds in the upgraded oil could be removed through aqueous extraction with HCl.

Guo et al. reported an approach to get high-quality liquid fuel from catalytic HDO of pyrolysis oil from *Chlorella* and *Nannochloropsis* by using Ni-Cu/ZrO$_2$ bimetallic catalysts [143]. The highest HDO activity was obtained over 15.71 wt % Ni 6.29 wt % Cu supported on ZrO$_2$, with the HDO efficiency of 82% for the upgrading of bio-oil from *Chlorella*. The Ni-Cu/ZrO$_2$ catalyst showed excellent stability after reaction, with low sintering and coking. In addition, the heating value, viscosity, and the water content of the upgraded oil were improved. Particularly, the cetane number of the product oil from *Nannochloropsis* satisfied the standard of EN 590-09.

Overall, catalytic upgrading of pyrolytic oil from algae can successfully remove the O and N content to a low level. In addition, the remaining N and O can possibly be removed using physical adsorption or the extraction method. The resulting liquid biofuel can meet the standard of transportation fuel. However, the mentioned works did not study much on the recyclability and regeneration of the catalysts, and the mechanism of upgrading process for a better understanding of catalyst design.

4.2.2. Catalytic Upgrading of HTL Biocrude

Like pyrolysis, direct HTL of algae produces biocrude oil with poor quality, especially for its high O and N content, and the physical properties that do not suit the fuel standards [141]. Patel et al. explored a method for catalytic upgrading of biocrude from fast HTL of algae over Pt, Pd, Ru supported on C and Al$_2$O$_3$, and sulfide NiMo/Al$_2$O$_3$ [144]. The highest oil yield (60 wt %) and highest denitrogenation ability (2.05 wt %) were obtained over NiMo/Al$_2$O$_3$, but the effect of deoxygenation was poor. The oxygen content of the upgraded biocrude ranged from 1.60-6.07 wt %, while the nitrogen content ranged from 2.05-3.47 wt %. The decrease of O content resulted in the increase of HHV to 38.36-45.40 MJ·kg^{-1}. The boiling point distribution of upgraded biocrude decreased from the gas oil fraction (271–343 °C) to the kerosene fraction (<271 °C). In addition, the abundant components in upgraded biocrude were branched alkanes and straight-chain alkanes.

Shakya et al. studied the catalytic upgrading of biocrude oil from HTL of *Nannochloropsis* [145]. Five different catalysts (Pt/C, Ru/C, Ni/C, ZSM-5 and Ni/ZSM-5) were used as the upgrading catalysts. Upgrading at 300 °C showed higher oil yield, while higher temperature at 350 °C resulted in bio-oil with higher quality. The maximum upgraded oil yield was obtained over Ni/C at 350 °C, whereas upgraded oil with higher HHV, lower acidity, and nitrogen content was achieved over Ru/C and Pt/C. The HHVs of upgraded biocrude ranged from 40–44 MJ·kg^{-1}, which were highly improved compared with biocrude feedstock (36.44 MJ·kg^{-1}). The catalytic upgrading produced upgraded oil with a 65–75% decrease in nitrogen content, and 95-98% decrease in oil acidity.

Biller et al. investigated the hydroprocessing of biocrude on sulfide CoMo and NiMo catalysts from continuous HTL of Chlorella [146]. In the non-catalytic HTL step, 40 wt % yield of biocrude was obtained with 6% nitrogen, 11% oxygen, and HHV of 35 MJ·kg^{-1}. The upgrading of biocrude over both NiMo and CoMo catalysts was conducted at 405 °C and 350 °C. The two catalysts showed similar performance on the improvement of hydroprocessed oil. The upgraded oil with the highest HHV (45.4 MJ·kg^{-1}) was achieved over CoMo catalyst at 405 °C, with the oil yield of 69.4 wt %, nitrogen content of 2.7%, and oxygen content of 1.0%. Hydroprocessing at high temperature (405 °C) resulted in upgraded oil with higher gasoline and diesel #1 fractions. Moreover, hydrocarbons (C_9-C_{26}) were the main products in upgraded bio-oil.

Table 7. Catalytic upgrading of biocrude oil from thermochemical conversion.

Catalyst	Feedstock	Condition	Yield (wt %)	Performance	Ref.
Hβ	Pretreated algal bio-oil	400 °C, 4 h, 6 MPa H_2	44.8	HHV: 44.3 MJ·kg^{-1}, O: 2.3% N: 2.1%, total hydrocarbons: 75.4%	[147]
MCM-41 (100%Si)	Pretreated algal bio-oil	400 °C, 4 h, 6 MPa H_2	54.5	HHV: 45.2 MJ·kg^{-1}, O: 1.9% N: 1.7%, total hydrocarbons: 81.5%	[147]
HY (5%Na_2O)	Pretreated algal bio-oil	400 °C, 4 h, 6 MPa H_2	48.1	HHV: 44.9 MJ·kg^{-1}, O: 1.9% N: 1.7%, total hydrocarbons: 80.5%	[147]
HZSM-5 (Si/Al=25)	Pretreated algal bio-oil	400 °C, 4 h, 6 MPa H_2	52.3	HHV: 44.5 MJ·kg^{-1}, O: 2.0% N: 1.8%, total hydrocarbons: 60.8%	[147]
NiMo/Al_2O_3	HTL biocrude from *Spirulina*	400 °C, 4 h, 8 MPa H_2	70.4	Hydrocarbons: 78%, reduced boiling point	[148]
Ru/C+alumina	Pretreated algal oil	400 °C, 4 h, 6 MPa H_2	70	HHV: 47.0 MJ·kg^{-1}, O: 0.4% N: 2.5%, total hydrocarbons: 43.2%	
Ru/C+Mo_2C	Pretreated algal oil	400 °C, 4 h, 6 MPa H_2	77	HHV: 46.8 MJ·kg^{-1}, O: 0.1% N: 3.1%, total hydrocarbons: 36.7%	[149]

In the previous section, catalytic HTL can improve the quality of biocrude, but the characteristics of biocrude still cannot meet the ideal standard. In comparison to one-step catalytic HTL, the upgrading of biocrude from HTL obtained biofuel with higher quality, with low boiling point, low viscosity, high HHV over 40 MJ·kg^{-1}, and low oxygen and nitrogen content, which is more preferable than the one-step catalytic HTL.

5. Conclusions and Outlook

Algae is considered as a potential feedstock for the production of biofuel. The common approaches for utilization of algae are lipid extraction and transesterification, pyrolysis, and hydrothermal liquefaction. The extraction–transesterification of algae only utilizes the lipid fraction of algae, leaving other parts (i.e., saccharides and proteins) as waste. Thermochemical conversion of algae is a viable method to fully utilize algae. However, the percentage of solid residue via pyrolysis of algae is usually high (30–50%) even at high temperatures (e.g., 700 °C), indicating the low conversion of algae [71]. The conversion of algae via HTL is usually high, with about 10% yield of solid residue at relatively mild temperatures [87–89]. However, the yield of biocrude oil recovered from the organic fraction is relatively low, leaving much of the water-soluble fraction remaining in aqueous products. In addition, the produced bio-oil or biocrude has some unwanted properties such as high viscosity, high acidity, high oxygen content, high nitrogen content, and low energy density. In order to improve the yield and quality of algal biofuel, catalysts are employed in the conversion process. Homogeneous catalysts (e.g., HCOOH, KOH and Na_2CO_3) enhance the biocrude yield via HTL, but these catalysts show little effect on improving the biocrude quality. In the presence of heterogeneous catalysts (zeolites, metal oxides, supported metal catalysts, etc.), the thermochemical conversion produces bio-oil with higher yield, lower oxygen content, and higher energy density, but the nitrogen content (~5%) and some properties (such as viscosity and boiling point) can hardly be improved by the one-step catalytic conversion.

Another way to improve algal bio-oil quality is upgrading of the obtained bio-oil with heterogeneous catalysts and hydrogen. Catalysts with high activity in deoxygenation and denitrogenation are employed in this process. Generally, commercial metal sulfides (sulfide CoMo and NiMo), commercial noble metal catalysts (Pt/C, Ru/C), and non-noble metal catalysts (Ni, Co, Fe) show excellent performance in the upgrading of algae-based biocrude oil. Usually in this process, higher temperature results in higher yield of hydrocarbons. The upgrading process reduces the oxygen and nitrogen content (<1%), oil acidity, boiling point, and enhances the hydrocarbons content up to 80%. The remaining nitrogen can be removed by using physical adsorption or extraction. Possibly, by catalytic pyrolysis or HTL of algae cooperating with catalytic upgrading of the obtained oil product, the oxygen and nitrogen content might be reduced to a very low level.

According to the recent advances on algal biorefinery, we have proposed the following six aspects of the future prospect and challenges in utilization of algae.

(1) The denitrogenation of algal bio-oil should be further studied. There are many research works focusing on the catalytic deoxygenation of algal based molecules. The technology of deoxygenation process is relatively mature. However, few studies have focused on the denitrogenation of algal biocrude oil. The N-containing compounds in bio-oil might cause the emission of NO_X during combustion. Hence, developing technology for lowering the N-containing compounds is essential. In general, denitrogenation is more difficult than deoxygenation due to the less reactive organic nitrogenates than oxygenates [68]. Denitrogenation of algal oil requires more energy, making the process costly. The mechanism needs to be further studied, for better understanding of the insight into denitrogenation and design of the denitrogenation catalysts. Therefore, searching for an effective catalyst, which is highly active in denitrogenation, is pressingly required.

(2) The characterization of oil products from algae conversion, including quantification of monomers and identification of oligomers should be improved and further developed. Until now, the composition of products obtained by thermochemical conversion is usually analyzed by GC-MS because of the lack of authentic samples. The percentage of compounds in biocrude oil is determined by the relative area of total ion peaks. Based on the fundamental knowledge of GC, the response factors of different compounds are different, so the response factors must be introduced to quantify the absolute concentration of each compound accurately. However, the exact structure of the compounds should be accurately analyzed for quantification, so some other techniques should be involved. In addition, the polysaccharides and proteins from algae are macromolecules, and the full degradation of them to monomers or compounds with small molecular size is difficult. The existence of oligomers in the liquid products from the thermochemical conversion is unavoidable. Biocrude with high fraction of oligomers usually has undesired properties such as high viscosity and boiling point, so the content of oligomers in algal biofuel should be minimized, for obtaining high-quality bio-oil. However, oligomeric products usually cannot be detected by GC-MS, so the development of other techniques, such as ESI-MS and MALDI-TOF MS, is needed. By identifying the structure of oligomers, it is easier to design a specific catalyst for efficient degradation of oligomers or even algae-derived polymers (polysaccharides and proteins) and their further conversion to small-molecule fuel chemicals.

(3) Converting algae through fractional route might be preferable. For improving the selectivity of target products from starting materials with complex components, the concept of fractional conversion has been proposed and accepted for lignocellulosic biomass, which is defined as

a selective in situ conversion of biomacromolecules to desired products [5,11,72,73]. Due to the complexity of algae, the fractional conversion, in other words, selective conversion of one specific component in algae, needs to be developed for the full utilization of algae. With the multi-step conversion, the feedstock of each step is simplified and it becomes possible to obtain high-grade biofuel or high selectivity to target products. For example, the lipid fraction can be extracted and hydrogenated to hydrocarbons. The saccharides in lipid extracted algae could be converted to value-added chemicals, and the protein component might be converted to amino acids or N-containing compounds. In this way, the algae feedstock can be fully utilized to avoid generating waste, which might do harm to the environment. In addition, the strategy for algae fractionation by simultaneous separating and converting needs to be developed—for example, simultaneously separating algal lipid and catalytic converting it to hydrocarbons. Thus, finding a method to selectively convert one component of algae without destroying the others is also a way to obtain products with high selectivity.

(4)   The utilization of algal sugar to produce valuable chemicals might be a potential research goal. It is well-known that polysaccharides can be hydrolyzed and further converted to valuable platform molecules—for example, monosaccharides, 5-HMF, levulinic acid, lactic acid, etc. However, only a few research works have focused on the conversion of sugars in algae to value-added chemicals, taking advantage of the oxygen content of the algae feedstock itself. Generally, carbohydrates from algae are suitable for the conversion into platform molecules. Hydrolysis of algal polysaccharides is usually carried out in the presence of acidic catalyst, such as Brønsted acid, metal chlorides, and heterogeneous catalysts such as HZSM-5 and Sn-Beta. Over these catalysts, highly selective chemicals such as rhamnose, 5-HMF, levulinic acid or lactic acid are formed [150–154].

(5)   The research of algal biorefinery should also focus on improvement of the process, in order to realize the commercialization and industrialization. Although high-quality algae-based liquid fuel can be obtained via catalytic upgrading process, the operating conditions are severe. Usually, high temperature (400 °C), high pressure (6 MPa), and long reaction time (4 h) are required in the upgrading of algal oil. This causes high costs in energy consumption and equipment. In addition, the selection of the catalyst should not only focus on the activity, but also on the reusability, and the selectivity of products. Except for activity in deoxygenation and denitrogenation, the catalyst should also be active in degradation of the oligomers from bio-oil. The catalyst with good reusability should have high thermal stability and resistance to coke in the upgrading process. Due to catalytic HTL conducts in aqueous environments, the catalyst should also be resistant to water. In addition, the upgrading process also needs to be performed on a large scale with low costs, to make it possible for commercialization and industrialization.

Author Contributions: Conceptualization, Y.Z. and C.H.; methodology, Y.Z.; formal analysis, Y.Z.; investigation, Y.Z.; resources, C.H.; writing—original draft preparation, Y.Z.; writing—review and editing, Y.Z. and C.H.; supervision, C.H.; project administration, C.H.; funding acquisition, C.H. All authors have read and agreed to the published version of the manuscript.

Funding: This research was funded by the National Natural Science Foundation of China (No. 21536007) and the 111 project (B17030).

Acknowledgments: This work was financially supported by the National Natural Science Foundation of China (No. 21536007) and the 111 project (B17030).

References

1.   Hermida, L.; Abdullah, A.Z.; Mohamed, A.R. Deoxygenation of fatty acid to produce diesel-like hydrocarbons: A review of process conditions, reaction kinetics and mechanism. *Renew. Sustain. Energy Rev.* **2015**, *42*, 1223–1233. [CrossRef]

2.   Yang, C.; Li, R.; Cui, C.; Liu, S.; Qiu, Q.; Ding, Y.; Wu, Y.; Zhang, B. Catalytic hydroprocessing of microalgae-derived biofuels: A review. *Green Chem.* **2016**, *18*, 3684–3699. [CrossRef]

3. Galadima, A.; Muraza, O. Hydrothermal liquefaction of algae and bio-oil upgrading into liquid fuels: Role of heterogeneous catalysts. *Renew. Sustain. Energy Rev.* **2018**, *81*, 1037–1048. [CrossRef]

4. Xu, D.; Savage, P.E. Characterization of biocrudes recovered with and without solvent after hydrothermal liquefaction of algae. *Algal Res.* **2014**, *6*, 1–7. [CrossRef]

5. Zhou, Y.; Chen, Y.; Li, M.; Hu, C. Production of high-quality biofuel via ethanol liquefaction of pretreated natural microalgae. *Renew. Energy* **2020**, *147*, 293–301. [CrossRef]

6. Menon, V.; Rao, M. Trends in bioconversion of lignocellulose: Biofuels, platform chemicals & biorefinery concept. *Prog. Energy Combust. Sci.* **2012**, *38*, 522–550. [CrossRef]

7. Van den Bosch, S.; Schutyser, W.; Vanholme, R.; Driessen, T.; Koelewijn, S.F.; Renders, T.; De Meester, B.; Huijgen, W.J.J.; Dehaen, W.; Courtin, C.M.; et al. Reductive lignocellulose fractionation into soluble lignin-derived phenolic monomers and dimers and processable carbohydrate pulps. *Energy Environ. Sci.* **2015**, *8*, 1748–1763. [CrossRef]

8. Varakin, A.N.; Salnikov, V.A.; Nikulshina, M.S.; Maslakov, K.I.; Mozhaev, A.V.; Nikulshin, P.A. Beneficial role of carbon in Co(Ni)MoS catalysts supported on carbon-coated alumina for co-hydrotreating of sunflower oil with straight-run gas oil. *Catal. Today* **2017**, *292*, 110–120. [CrossRef]

9. Pattanaik, B.P.; Misra, R.D. Effect of reaction pathway and operating parameters on the deoxygenation of vegetable oils to produce diesel range hydrocarbon fuels: A review. *Renew. Sustain. Energy Rev.* **2017**, *73*, 545–557. [CrossRef]

10. Brennan, L.; Owende, P. Biofuels from microalgae—A review of technologies for production, processing, and extractions of biofuels and co-products. *Renew. Sustain. Energy Rev.* **2010**, *14*, 557–577. [CrossRef]

11. Zhou, Y.; Li, L.; Zhang, R.; Hu, C. Fractional conversion of microalgae from water blooms. *Faraday Discuss.* **2017**, *202*, 197–212. [CrossRef] [PubMed]

12. Zhao, C.; Brück, T.; Lercher, J.A. Catalytic deoxygenation of microalgae oil to green hydrocarbons. *Green Chem.* **2013**, *15*, 1720–1739. [CrossRef]

13. Enamala, M.K.; Enamala, S.; Chavali, M.; Donepudi, J.; Yadavalli, R.; Kolapalli, B.; Aradhyula, T.V.; Velpuri, J.; Kuppam, C. Production of biofuels from microalgae—A review on cultivation, harvesting, lipid extraction, and numerous applications of microalgae. *Renew. Sustain. Energy Rev.* **2018**, *94*, 49–68. [CrossRef]

14. Mata, T.M.; Martins, A.A.; Caetano, N.S. Microalgae for biodiesel production and other applications: A review. *Renew. Sustain. Energy Rev.* **2010**, *14*, 217–232. [CrossRef]

15. Song, M.; Pei, H.; Hu, W.; Ma, G. Evaluation of the potential of 10 microalgal strains for biodiesel production. *Bioresour. Technol.* **2013**, *141*, 245–251. [CrossRef]

16. Zeng, Y.; Zhao, B.; Zhu, L.; Tong, D.; Hu, C. Catalytic pyrolysis of natural algae from water blooms over nickel phosphide for high quality bio-oil production. *RSC Adv.* **2013**, *3*, 10806. [CrossRef]

17. Zhang, R.; Li, L.; Tong, D.; Hu, C. Microwave-enhanced pyrolysis of natural algae from water blooms. *Bioresour. Technol.* **2016**, *212*, 311–317. [CrossRef]

18. Xia, A.; Jacob, A.; Tabassum, M.R.; Herrmann, C.; Murphy, J.D. Production of hydrogen, ethanol and volatile fatty acids through co-fermentation of macro- and micro-algae. *Bioresour. Technol.* **2016**, *205*, 118–125. [CrossRef]

19. Alfonsin, V.; Maceiras, R.; Gutierrez, C. Bioethanol production from industrial algae waste. *Waste Manag.* **2019**, *87*, 791–797. [CrossRef]

20. Jambo, S.A.; Abdulla, R.; Mohd Azhar, S.H.; Marbawi, H.; Gansau, J.A.; Ravindra, P. A review on third generation bioethanol feedstock. *Renew. Sustain. Energy Rev.* **2016**, *65*, 756–769. [CrossRef]

21. Williams, P.J.l.B.; Laurens, L.M.L. Microalgae as biodiesel & biomass feedstocks: Review & analysis of the biochemistry, energetics & economics. *Energy Environ. Sci.* **2010**, *3*, 554–590. [CrossRef]

22. Chisti, Y. Biodiesel from microalgae. *Biotechnol. Adv.* **2007**, *25*, 294–306. [CrossRef] [PubMed]

23. Tran, D.-T.; Chang, J.-S.; Lee, D.-J. Recent insights into continuous-flow biodiesel production via catalytic and non-catalytic transesterification processes. *Appl. Energy* **2017**, *185*, 376–409. [CrossRef]

24. Wahlen, B.D.; Willis, R.M.; Seefeldt, L.C. Biodiesel production by simultaneous extraction and conversion of total lipids from microalgae, cyanobacteria, and wild mixed-cultures. *Bioresour. Technol.* **2011**, *102*, 2724–2730. [CrossRef] [PubMed]

25. Martinez-Guerra, E.; Gude, V.G.; Mondala, A.; Holmes, W.; Hernandez, R. Microwave and ultrasound enhanced extractive-transesterification of algal lipids. *Appl. Energy* **2014**, *129*, 354–363. [CrossRef]

26. Ali, M.; Watson, I.A. Microwave treatment of wet algal paste for enhanced solvent extraction of lipids for biodiesel production. *Renew. Energy* **2015**, *76*, 470–477. [CrossRef]

27. Zhou, D.; Qiao, B.; Li, G.; Xue, S.; Yin, J. Continuous production of biodiesel from microalgae by extraction coupling with transesterification under supercritical conditions. *Bioresour. Technol.* **2017**, *238*, 609–615. [CrossRef]

28. Mohamadzadeh Shirazi, H.; Karimi-Sabet, J.; Ghotbi, C. Biodiesel production from Spirulina microalgae feedstock using direct transesterification near supercritical methanol condition. *Bioresour. Technol.* **2017**, *239*, 378–386. [CrossRef]

29. Du, Z.; Hu, B.; Ma, X.; Cheng, Y.; Liu, Y.; Lin, X.; Wan, Y.; Lei, H.; Chen, P.; Ruan, R. Catalytic pyrolysis of microalgae and their three major components: Carbohydrates, proteins, and lipids. *Bioresour. Technol.* **2013**, *130*, 777–782. [CrossRef]

30. Chen, Y.; Wu, Y.; Hua, D.; Li, C.; Harold, M.P.; Wang, J.; Yang, M. Thermochemical conversion of low-lipid microalgae for the production of liquid fuels: Challenges and opportunities. *RSC Adv.* **2015**, *5*, 18673–18701. [CrossRef]

31. Xu, H.; Miao, X.; Wu, Q. High quality biodiesel production from a microalga Chlorella protothecoides by heterotrophic growth in fermenters. *J. Biotechnol.* **2006**, *126*, 499–507. [CrossRef]

32. Demirbas, A. Use of algae as biofuel sources. *Energy Convers. Manag.* **2010**, *51*, 2738–2749. [CrossRef]

33. Zhu, L.; Hiltunen, E.; Shu, Q.; Zhou, W.; Li, Z.; Wang, Z. Biodiesel production from algae cultivated in winter with artificial wastewater through pH regulation by acetic acid. *Appl. Energy* **2014**, *128*, 103–110. [CrossRef]

34. Suali, E.; Sarbatly, R. Conversion of microalgae to biofuel. *Renew. Sustain. Energy Rev.* **2012**, *16*, 4316–4342. [CrossRef]

35. Beneroso, D.; Bermudez, J.M.; Arenillas, A.; Menendez, J.A. Microwave pyrolysis of microalgae for high syngas production. *Bioresour. Technol.* **2013**, *144*, 240–246. [CrossRef] [PubMed]

36. Wan, Y.; Chen, P.; Zhang, B.; Yang, C.; Liu, Y.; Lin, X.; Ruan, R. Microwave-assisted pyrolysis of biomass: Catalysts to improve product selectivity. *J. Anal. Appl. Pyrolysis* **2009**, *86*, 161–167. [CrossRef]

37. Guo, Y.; Yeh, T.; Song, W.; Xu, D.; Wang, S. A review of bio-oil production from hydrothermal liquefaction of algae. *Renew. Sustain. Energy Rev.* **2015**, *48*, 776–790. [CrossRef]

38. Gollakota, A.R.K.; Kishore, N.; Gu, S. A review on hydrothermal liquefaction of biomass. *Renew. Sustain. Energy Rev.* **2018**, *81*, 1378–1392. [CrossRef]

39. Shakya, R.; Adhikari, S.; Mahadevan, R.; Shanmugam, S.R.; Nam, H.; Hassan, E.B.; Dempster, T.A. Influence of biochemical composition during hydrothermal liquefaction of algae on product yields and fuel properties. *Bioresour. Technol.* **2017**, *243*, 1112–1120. [CrossRef]

40. Xu, D.; Savage, P.E. Effect of reaction time and algae loading on water-soluble and insoluble biocrude fractions from hydrothermal liquefaction of algae. *Algal Res.* **2015**, *12*, 60–67. [CrossRef]

41. Cheng, F.; Cui, Z.; Mallick, K.; Nirmalakhandan, N.; Brewer, C.E. Hydrothermal liquefaction of high- and low-lipid algae: Mass and energy balances. *Bioresour. Technol.* **2018**, *258*, 158–167. [CrossRef] [PubMed]

42. Parsa, M.; Jalilzadeh, H.; Pazoki, M.; Ghasemzadeh, R.; Abduli, M. Hydrothermal liquefaction of Gracilaria gracilis and Cladophora glomerata macro-algae for biocrude production. *Bioresour. Technol.* **2018**, *250*, 26–34. [CrossRef] [PubMed]

43. Budarin, V.L.; Zhao, Y.; Gronnow, M.J.; Shuttleworth, P.S.; Breeden, S.W.; Macquarrie, D.J.; Clark, J.H. Microwave-mediated pyrolysis of macro-algae. *Green Chem.* **2011**, *13*, 2330–2333. [CrossRef]

44. Babich, I.V.; van der Hulst, M.; Lefferts, L.; Moulijn, J.A.; O'Connor, P.; Seshan, K. Catalytic pyrolysis of microalgae to high-quality liquid bio-fuels. *Biomass Bioenergy* **2011**, *35*, 3199–3207. [CrossRef]

45. Anand, V.; Sunjeev, V.; Vinu, R. Catalytic fast pyrolysis of Arthrospira platensis (spirulina) algae using zeolites. *J. Anal. Appl. Pyrolysis* **2016**, *118*, 298–307. [CrossRef]

46. Chen, J.; Xu, Q. Hydrodeoxygenation of biodiesel-related fatty acid methyl esters to diesel-range alkanes over zeolite-supported ruthenium catalysts. *Catal. Sci. Technol.* **2016**, *6*, 7239–7251. [CrossRef]

47. Soni, V.K.; Sharma, P.R.; Choudhary, G.; Pandey, S.; Sharma, R.K. Ni/Co-Natural Clay as Green Catalysts for Microalgae Oil to Diesel-Grade Hydrocarbons Conversion. *ACS Sustain. Chem. Eng.* **2017**, *5*, 5351–5359. [CrossRef]

48. Kukushkin, R.G.; Bulavchenko, O.A.; Kaichev, V.V.; Yakovlev, V.A. Influence of Mo on catalytic activity of Ni-based catalysts in hydrodeoxygenation of esters. *Appl. Catal. B Environ.* **2015**, *163*, 531–538. [CrossRef]

49. Arun, J.; Shreekanth, S.J.; Sahana, R.; Raghavi, M.S.; Gopinath, K.P.; Gnanaprakash, D. Studies on influence of process parameters on hydrothermal catalytic liquefaction of microalgae (Chlorella vulgaris) biomass grown in wastewater. *Bioresour. Technol.* **2017**, *244*, 963–968. [CrossRef]

50. Casoni, A.I.; Zunino, J.; Piccolo, M.C.; Volpe, M.A. Valorization of Rhizoclonium sp. algae via pyrolysis and catalytic pyrolysis. *Bioresour. Technol.* **2016**, *216*, 302–307. [CrossRef]

51. Gao, L.; Sun, J.; Xu, W.; Xiao, G. Catalytic pyrolysis of natural algae over Mg-Al layered double oxides/ZSM-5 (MgAl-LDO/ZSM-5) for producing bio-oil with low nitrogen content. *Bioresour. Technol.* **2017**, *225*, 293–298. [CrossRef] [PubMed]

52. Yang, W.; Li, X.; Zhang, D.; Feng, L. Catalytic upgrading of bio-oil in hydrothermal liquefaction of algae major model components over liquid acids. *Energy Convers. Manag.* **2017**, *154*, 336–343. [CrossRef]

53. Fu, J.; Yang, C.; Wu, J.; Zhuang, J.; Hou, Z.; Lu, X. Direct production of aviation fuels from microalgae lipids in water. *Fuel* **2015**, *139*, 678–683. [CrossRef]

54. Duan, P.; Bai, X.; Xu, Y.; Zhang, A.; Wang, F.; Zhang, L.; Miao, J. Catalytic upgrading of crude algal oil using platinum/gamma alumina in supercritical water. *Fuel* **2013**, *109*, 225–233. [CrossRef]

55. Alaswad, A.; Dassisti, M.; Prescott, T.; Olabi, A.G. Technologies and developments of third generation biofuel production. *Renew. Sustain. Energy Rev.* **2015**, *51*, 1446–1460. [CrossRef]

56. Mubarak, M.; Shaija, A.; Suchithra, T.V. A review on the extraction of lipid from microalgae for biodiesel production. *Algal Res.* **2015**, *7*, 117–123. [CrossRef]

57. Kim, D.Y.; Vijayan, D.; Praveenkumar, R.; Han, J.I.; Lee, K.; Park, J.Y.; Chang, W.S.; Lee, J.S.; Oh, Y.K. Cell-wall disruption and lipid/astaxanthin extraction from microalgae: Chlorella and Haematococcus. *Bioresour. Technol.* **2016**, *199*, 300–310. [CrossRef]

58. Pan, J.; Muppaneni, T.; Sun, Y.; Reddy, H.K.; Fu, J.; Lu, X.; Deng, S. Microwave-assisted extraction of lipids from microalgae using an ionic liquid solvent [BMIM][HSO4]. *Fuel* **2016**, *178*, 49–55. [CrossRef]

59. Obeid, S.; Beaufils, N.; Camy, S.; Takache, H.; Ismail, A.; Pontalier, P.-Y. Supercritical carbon dioxide extraction and fractionation of lipids from freeze-dried microalgae Nannochloropsis oculata and Chlorella vulgaris. *Algal Res.* **2018**, *34*, 49–56. [CrossRef]

60. Yamamoto, K.; King, P.M.; Wu, X.; Mason, T.J.; Joyce, E.M. Effect of ultrasonic frequency and power on the disruption of algal cells. *Ultrason. Sonochem.* **2015**, *24*, 165–171. [CrossRef]

61. Bligh, E.G.; Dyer, W.J. A Rapid Method of Total Lipid Extraction and Purification. *Can. J. Biochem. Physiol.* **1959**, *37*, 911–917. [CrossRef]

62. Jeevan Kumar, S.P.; Vijay Kumar, G.; Dash, A.; Scholz, P.; Banerjee, R. Sustainable green solvents and techniques for lipid extraction from microalgae: A review. *Algal Res.* **2017**, *21*, 138–147. [CrossRef]

63. Ryckebosch, E.; Bruneel, C.; Termote-Verhalle, R.; Muylaert, K.; Foubert, I. Influence of extraction solvent system on extractability of lipid components from different microalgae species. *Algal Res.* **2014**, *3*, 36–43. [CrossRef]

64. Jones, S.B.; Zhu, Y.; Anderson, D.B.; Hallen, R.T.; Elliott, D.C.; Schmidt, A.J.; Albrecht, K.O.; Hart, T.R.; Butcher, M.G.; Drennan, C. *Process Design and Economics for the Conversion of Algal Biomass to Hydrocarbons: Whole Algae Hydrothermal Liquefaction and Upgrading*; Pacific Northwest National Lab.(PNNL): Richland, WA, USA, 2014.

65. Azizi, K.; Keshavarz Moraveji, M.; Abedini Najafabadi, H. A review on bio-fuel production from microalgal biomass by using pyrolysis method. *Renew. Sustain. Energy Rev.* **2018**, *82*, 3046–3059. [CrossRef]

66. Li, Q.; Xing, F.; Li, J.; Lv, X.; Hu, C. The effect of heating rate on the yields and distribution of oil products from the pyrolysis of pubescen. *Energy Technol.* **2018**, *6*, 366–378. [CrossRef]

67. Mundike, J.; Collard, F.X.; Gorgens, J.F. Torrefaction of invasive alien plants: Influence of heating rate and other conversion parameters on mass yield and higher heating value. *Bioresour. Technol.* **2016**, *209*, 90–99. [CrossRef]

68. Li, F.; Srivatsa, S.C.; Bhattacharya, S. A review on catalytic pyrolysis of microalgae to high-quality bio-oil with low oxygeneous and nitrogenous compounds. *Renew. Sustain. Energy Rev.* **2019**, *108*, 481–497. [CrossRef]

69. Fang, S.; Gu, W.; Dai, M.; Xu, J.; Yu, Z.; Lin, Y.; Chen, J.; Ma, X. A study on microwave-assisted fast co-pyrolysis of chlorella and tire in the N_2 and CO_2 atmospheres. *Bioresour. Technol.* **2018**, *250*, 821–827. [CrossRef]

70. Chang, Z.; Duan, P.; Xu, Y. Catalytic hydropyrolysis of microalgae: Influence of operating variables on the formation and composition of bio-oil. *Bioresour. Technol.* **2015**, *184*, 349–354. [CrossRef]

71. Debiagi, P.E.A.; Trinchera, M.; Frassoldati, A.; Faravelli, T.; Vinu, R.; Ranzi, E. Algae characterization and multistep pyrolysis mechanism. *J. Anal. Appl. Pyrolysis* **2017**, *128*, 423–436. [CrossRef]

72. Li, H.; Li, L.; Zhang, R.; Tong, D.; Hu, C. Fractional pyrolysis of Cyanobacteria from water blooms over HZSM-5 for high quality bio-oil production. *J. Energy Chem.* **2014**, *23*, 732–741. [CrossRef]

73. Li, L.-l.; Zhang, R.; Tong, D.-m.; Hu, C.-w. Fractional Pyrolysis of Algae and Model Compounds. *Chin. J. Chem. Phys.* **2015**, *28*, 525–532. [CrossRef]

74. Gautam, R.; Vinu, R. Non-catalytic fast pyrolysis and catalytic fast pyrolysis of Nannochloropsis oculata using Co-Mo/γ-Al$_2$O$_3$ catalyst for valuable chemicals. *Algal Res.* **2018**, *34*, 12–24. [CrossRef]

75. Pan, P.; Hu, C.; Yang, W.; Li, Y.; Dong, L.; Zhu, L.; Tong, D.; Qing, R.; Fan, Y. The direct pyrolysis and catalytic pyrolysis of Nannochloropsis sp. residue for renewable bio-oils. *Bioresour. Technol.* **2010**, *101*, 4593–4599. [CrossRef] [PubMed]

76. Bordoloi, N.; Narzari, R.; Sut, D.; Saikia, R.; Chutia, R.S.; Kataki, R. Characterization of bio-oil and its sub-fractions from pyrolysis of Scenedesmus dimorphus. *Renew. Energy* **2016**, *98*, 245–253. [CrossRef]

77. Wang, K.; Brown, R.C.; Homsy, S.; Martinez, L.; Sidhu, S.S. Fast pyrolysis of microalgae remnants in a fluidized bed reactor for bio-oil and biochar production. *Bioresour. Technol.* **2013**, *127*, 494–499. [CrossRef]

78. Belotti, G.; de Caprariis, B.; De Filippis, P.; Scarsella, M.; Verdone, N. Effect of Chlorella vulgaris growing conditions on bio-oil production via fast pyrolysis. *Biomass Bioenergy* **2014**, *61*, 187–195. [CrossRef]

79. Du, Z.; Li, Y.; Wang, X.; Wan, Y.; Chen, Q.; Wang, C.; Lin, X.; Liu, Y.; Chen, P.; Ruan, R. Microwave-assisted pyrolysis of microalgae for biofuel production. *Bioresour. Technol.* **2011**, *102*, 4890–4896. [CrossRef]

80. Borges, F.C.; Xie, Q.; Min, M.; Muniz, L.A.; Farenzena, M.; Trierweiler, J.O.; Chen, P.; Ruan, R. Fast microwave-assisted pyrolysis of microalgae using microwave absorbent and HZSM-5 catalyst. *Bioresour. Technol.* **2014**, *166*, 518–526. [CrossRef]

81. Hu, Y.; Gong, M.; Feng, S.; Xu, C.; Bassi, A. A review of recent developments of pre-treatment technologies and hydrothermal liquefaction of microalgae for bio-crude oil production. *Renew. Sustain. Energy Rev.* **2019**, *101*, 476–492. [CrossRef]

82. Xu, D.; Lin, G.; Guo, S.; Wang, S.; Guo, Y.; Jing, Z. Catalytic hydrothermal liquefaction of algae and upgrading of biocrude: A critical review. *Renew. Sustain. Energy Rev.* **2018**, *97*, 103–118. [CrossRef]

83. Hu, Y.; Feng, S.; Yuan, Z.; Xu, C.C.; Bassi, A. Investigation of aqueous phase recycling for improving bio-crude oil yield in hydrothermal liquefaction of algae. *Bioresour. Technol.* **2017**, *239*, 151–159. [CrossRef] [PubMed]

84. Shanmugam, S.R.; Adhikari, S.; Shakya, R. Nutrient removal and energy production from aqueous phase of bio-oil generated via hydrothermal liquefaction of algae. *Bioresour. Technol.* **2017**, *230*, 43–48. [CrossRef] [PubMed]

85. Maddi, B.; Panisko, E.; Wietsma, T.; Lemmon, T.; Swita, M.; Albrecht, K.; Howe, D. Quantitative characterization of the aqueous fraction from hydrothermal liquefaction of algae. *Biomass Bioenergy* **2016**, *93*, 122–130. [CrossRef]

86. Leng, L.; Li, J.; Wen, Z.; Zhou, W. Use of microalgae to recycle nutrients in aqueous phase derived from hydrothermal liquefaction process. *Bioresour. Technol.* **2018**, *256*, 529–542. [CrossRef]

87. Nava Bravo, I.; Velásquez-Orta, S.B.; Cuevas-García, R.; Monje-Ramírez, I.; Harvey, A.; Orta Ledesma, M.T. Bio-crude oil production using catalytic hydrothermal liquefaction (HTL) from native microalgae harvested by ozone-flotation. *Fuel* **2019**, *241*, 255–263. [CrossRef]

88. Anastasakis, K.; Ross, A.B. Hydrothermal liquefaction of four brown macro-algae commonly found on the UK coasts: An energetic analysis of the process and comparison with bio-chemical conversion methods. *Fuel* **2015**, *139*, 546–553. [CrossRef]

89. Dandamudi, K.P.R.; Muppaneni, T.; Markovski, J.S.; Lammers, P.; Deng, S. Hydrothermal liquefaction of green microalga Kirchneriella sp. under sub- and super-critical water conditions. *Biomass Bioenergy* **2019**, *120*, 224–228. [CrossRef]

90. Wądrzyk, M.; Janus, R.; Vos, M.P.; Brilman, D.W.F. Effect of process conditions on bio-oil obtained through continuous hydrothermal liquefaction of Scenedesmus sp. microalgae. *J. Anal. Appl. Pyrolysis* **2018**, *134*, 415–426. [CrossRef]

91. López Barreiro, D.; Prins, W.; Ronsse, F.; Brilman, W. Hydrothermal liquefaction (HTL) of microalgae for biofuel production: State of the art review and future prospects. *Biomass Bioenergy* **2013**, *53*, 113–127. [CrossRef]

92. Yang, L.; Li, Y.; Savage, P.E. Catalytic Hydrothermal Liquefaction of a Microalga in a Two-Chamber Reactor. *Ind. Eng. Chem. Res.* **2014**, *53*, 11939–11944. [CrossRef]

93. Li, H.; Hu, J.; Zhang, Z.; Wang, H.; Ping, F.; Zheng, C.; Zhang, H.; He, Q. Insight into the effect of hydrogenation on efficiency of hydrothermal liquefaction and physico-chemical properties of biocrude oil. *Bioresour. Technol.* **2014**, *163*, 143–151. [CrossRef] [PubMed]

94. Cheng, F.; Jarvis, J.M.; Yu, J.; Jena, U.; Nirmalakhandan, N.; Schaub, T.M.; Brewer, C.E. Bio-crude oil from hydrothermal liquefaction of wastewater microalgae in a pilot-scale continuous flow reactor. *Bioresour. Technol.* **2019**, *294*, 122184. [CrossRef] [PubMed]

95. Tian, C.; Liu, Z.; Zhang, Y.; Li, B.; Cao, W.; Lu, H.; Duan, N.; Zhang, L.; Zhang, T. Hydrothermal liquefaction of harvested high-ash low-lipid algal biomass from Dianchi Lake: Effects of operational parameters and relations of products. *Bioresour. Technol.* **2015**, *184*, 336–343. [CrossRef] [PubMed]

96. Han, Y.; Hoekman, S.K.; Cui, Z.; Jena, U.; Das, P. Hydrothermal liquefaction of marine microalgae biomass using co-solvents. *Algal Res.* **2019**, *38*, 101421. [CrossRef]

97. Yu, J.; Maliutina, K.; Tahmasebi, A. A review on the production of nitrogen-containing compounds from microalgal biomass via pyrolysis. *Bioresour. Technol.* **2018**, *270*, 689–701. [CrossRef]

98. Jafarian, S.; Tavasoli, A. A comparative study on the quality of bioproducts derived from catalytic pyrolysis of green microalgae Spirulina (Arthrospira) plantensis over transition metals supported on HMS-ZSM5 composite. *Int. J. Hydrogen Energy* **2018**, *43*, 19902–19917. [CrossRef]

99. Wang, S.; Cao, B.; Liu, X.; Xu, L.; Hu, Y.; Afonaa-Mensah, S.; Abomohra, A.E.; He, Z.; Wang, Q.; Xu, S. A comparative study on the quality of bio-oil derived from green macroalga Enteromorpha clathrata over metal modified ZSM-5 catalysts. *Bioresour. Technol.* **2018**, *256*, 446–455. [CrossRef]

100. Zainan, N.H.; Srivatsa, S.C.; Bhattacharya, S. Catalytic pyrolysis of microalgae Tetraselmis suecica and characterization study using in situ Synchrotron-based Infrared Microscopy. *Fuel* **2015**, *161*, 345–354. [CrossRef]

101. Chagas, B.M.E.; Dorado, C.; Serapiglia, M.J.; Mullen, C.A.; Boateng, A.A.; Melo, M.A.F.; Ataíde, C.H. Catalytic pyrolysis-GC/MS of Spirulina: Evaluation of a highly proteinaceous biomass source for production of fuels and chemicals. *Fuel* **2016**, *179*, 124–134. [CrossRef]

102. Ansah, E.; Wang, L.; Zhang, B.; Shahbazi, A. Catalytic pyrolysis of raw and hydrothermally carbonized Chlamydomonas debaryana microalgae for denitrogenation and production of aromatic hydrocarbons. *Fuel* **2018**, *228*, 234–242. [CrossRef]

103. Rahman, N.A.A.; Fermoso, J.; Sanna, A. Effect of Li-LSX-zeolite on the in-situ catalytic deoxygenation and denitrogenation of Isochrysis sp. microalgae pyrolysis vapours. *Fuel Process. Technol.* **2018**, *173*, 253–261. [CrossRef]

104. Jia, X.; Zhang, X.; Rui, N.; Hu, X.; Liu, C.-j. Structural effect of Ni/ZrO$_2$ catalyst on CO$_2$ methanation with enhanced activity. *Appl. Catal. B Environ.* **2019**, *244*, 159–169. [CrossRef]

105. Aysu, T.; Abd Rahman, N.A.; Sanna, A. Catalytic pyrolysis of Tetraselmis and Isochrysis microalgae by nickel ceria based catalysts for hydrocarbon production. *Energy* **2016**, *103*, 205–214. [CrossRef]

106. Aysu, T.; Fermoso, J.; Sanna, A. Ceria on alumina support for catalytic pyrolysis of Pavlova sp. microalgae to high-quality bio-oils. *J. Energy Chem.* **2018**, *27*, 874–882. [CrossRef]

107. Koley, S.; Khadase, M.S.; Mathimani, T.; Raheman, H.; Mallick, N. Catalytic and non-catalytic hydrothermal processing of Scenedesmus obliquus biomass for bio-crude production—A sustainable energy perspective. *Energy Convers. Manag.* **2018**, *163*, 111–121. [CrossRef]

108. Zhang, B.; He, Z.; Chen, H.; Kandasamy, S.; Xu, Z.; Hu, X.; Guo, H. Effect of acidic, neutral and alkaline conditions on product distribution and biocrude oil chemistry from hydrothermal liquefaction of microalgae. *Bioresour. Technol.* **2018**, *270*, 129–137. [CrossRef]

109. Shakya, R.; Whelen, J.; Adhikari, S.; Mahadevan, R.; Neupane, S. Effect of temperature and Na$_2$CO$_3$ catalyst on hydrothermal liquefaction of algae. *Algal Res.* **2015**, *12*, 80–90. [CrossRef]

110. Bi, Z.; Zhang, J.; Zhu, Z.; Liang, Y.; Wiltowski, T. Generating biocrude from partially defatted Cryptococcus curvatus yeast residues through catalytic hydrothermal liquefaction. *Appl. Energy* **2018**, *209*, 435–444. [CrossRef]

111. Muppaneni, T.; Reddy, H.K.; Selvaratnam, T.; Dandamudi, K.P.R.; Dungan, B.; Nirmalakhandan, N.; Schaub, T.; Omar Holguin, F.; Voorhies, W.; Lammers, P.; et al. Hydrothermal liquefaction of Cyanidioschyzon merolae and the influence of catalysts on products. *Bioresour. Technol.* **2017**, *223*, 91–97. [CrossRef]

112. Yan, L.; Wang, Y.; Li, J.; Zhang, Y.; Ma, L.; Fu, F.; Chen, B.; Liu, H. Hydrothermal liquefaction of Ulva prolifera macroalgae and the influence of base catalysts on products. *Bioresour. Technol.* **2019**, *292*, 121286. [CrossRef] [PubMed]

113. Kumar, V.; Kumar, S.; Chauhan, P.K.; Verma, M.; Bahuguna, V.; Joshi, H.C.; Ahmad, W.; Negi, P.; Sharma, N.; Ramola, B.; et al. Low-temperature catalyst based Hydrothermal liquefaction of harmful Macroalgal blooms, and aqueous phase nutrient recycling by microalgae. *Sci. Rep.* **2019**, *9*, 11384. [CrossRef] [PubMed]

114. Nørskov, J.K.; Bligaard, T.; Logadottir, A.; Bahn, S.; Hansen, L.B.; Bollinger, M.; Bengaard, H.; Hammer, B.; Sljivancanin, Z.; Mavrikakis, M.; et al. Universality in Heterogeneous Catalysis. *J. Catal.* **2002**, *209*, 275–278. [CrossRef]

115. Egesa, D.; Chuck, C.J.; Plucinski, P. Multifunctional Role of Magnetic Nanoparticles in Efficient Microalgae Separation and Catalytic Hydrothermal Liquefaction. *ACS Sustain. Chem. Eng.* **2017**, *6*, 991–999. [CrossRef]

116. Liu, C.; Kong, L.; Wang, Y.; Dai, L. Catalytic hydrothermal liquefaction of spirulina to bio-oil in the presence of formic acid over palladium-based catalysts. *Algal Res.* **2018**, *33*, 156–164. [CrossRef]

117. Xu, D.; Guo, S.; Liu, L.; Lin, G.; Wu, Z.; Guo, Y.; Wang, S. Heterogeneous catalytic effects on the characteristics of water-soluble and water-insoluble biocrudes in chlorella hydrothermal liquefaction. *Appl. Energy* **2019**, *243*, 165–174. [CrossRef]

118. Ma, C.; Geng, J.; Zhang, D.; Ning, X. Hydrothermal liquefaction of macroalgae: Influence of zeolites based catalyst on products. *J. Energy Inst.* **2019**. [CrossRef]

119. Wang, W.; Xu, Y.; Wang, X.; Zhang, B.; Tian, W.; Zhang, J. Hydrothermal liquefaction of microalgae over transition metal supported TiO$_2$ catalyst. *Bioresour. Technol.* **2018**, *250*, 474–480. [CrossRef]

120. Liu, C.; Wufuer, A.; Kong, L.; Wang, Y.; Dai, L. Organic solvent extraction-assisted catalytic hydrothermal liquefaction of algae to bio-oil. *RSC Adv.* **2018**, *8*, 31717–31724. [CrossRef]

121. Kohansal, K.; Tavasoli, A.; Bozorg, A. Using a hybrid-like supported catalyst to improve green fuel production through hydrothermal liquefaction of Scenedesmus obliquus microalgae. *Bioresour. Technol.* **2019**, *277*, 136–147. [CrossRef]

122. Chen, Y.; Mu, R.; Yang, M.; Fang, L.; Wu, Y.; Wu, K.; Liu, Y.; Gong, J. Catalytic hydrothermal liquefaction for bio-oil production over CNTs supported metal catalysts. *Chem. Eng. Sci.* **2017**, *161*, 299–307. [CrossRef]

123. Saber, M.; Golzary, A.; Hosseinpour, M.; Takahashi, F.; Yoshikawa, K. Catalytic hydrothermal liquefaction of microalgae using nanocatalyst. *Appl. Energy* **2016**, *183*, 566–576. [CrossRef]

124. Tian, W.; Liu, R.; Wang, W.; Yin, Z.; Yi, X. Effect of operating conditions on hydrothermal liquefaction of Spirulina over Ni/TiO$_2$ catalyst. *Bioresour. Technol.* **2018**, *263*, 569–575. [CrossRef] [PubMed]

125. Xu, D.; Savage, P.E. Effect of temperature, water loading, and Ru/C catalyst on water-insoluble and water-soluble biocrude fractions from hydrothermal liquefaction of algae. *Bioresour. Technol.* **2017**, *239*, 1–6. [CrossRef] [PubMed]

126. Liu, Z.; Li, H.; Zeng, J.; Liu, M.; Zhang, Y.; Liu, Z. Influence of Fe/HZSM-5 catalyst on elemental distribution and product properties during hydrothermal liquefaction of Nannochloropsis sp. *Algal Res.* **2018**, *35*, 1–9. [CrossRef]

127. Schlagermann, P.; Göttlicher, G.; Dillschneider, R.; Rosello-Sastre, R.; Posten, C. Composition of Algal Oil and Its Potential as Biofuel. *J. Combust.* **2012**, *2012*, 285185. [CrossRef]

128. Liu, Y.; Murata, K.; Inaba, M. Hydrocracking of algae oil to aviation fuel-ranged hydrocarbons over NiMo-supported catalysts. *Catal. Today* **2019**, *332*, 115–121. [CrossRef]

129. Shi, Z.; Zhao, B.; Tang, S.; Yang, X. Hydrotreating lipids for aviation biofuels derived from extraction of wet and dry algae. *J. Clean. Prod.* **2018**, *204*, 906–915. [CrossRef]

130. Ullrich, J.; Breit, B. Selective Hydrogenation of Carboxylic Acids to Alcohols or Alkanes Employing a Heterogeneous Catalyst. *ACS Catal.* **2017**, *8*, 785–789. [CrossRef]

131. Di, L.; Yao, S.; Song, S.; Wu, G.; Dai, W.; Guan, N.; Li, L. Robust ruthenium catalysts for the selective conversion of stearic acid to diesel-range alkanes. *Appl. Catal. B Environ.* **2017**, *201*, 137–149. [CrossRef]

132. Ju, C.; Wang, F.; Huang, Y.; Fang, Y. Selective extraction of neutral lipid from wet algae paste and subsequently hydroconversion into renewable jet fuel. *Renew. Energy* **2018**, *118*, 521–526. [CrossRef]

133. Peng, B.; Yao, Y.; Zhao, C.; Lercher, J.A. Towards quantitative conversion of microalgae oil to diesel-range alkanes with bifunctional catalysts. *Angew. Chem. Int. Ed.* **2012**, *51*, 2072–2075. [CrossRef] [PubMed]

134. Wu, L.; Li, L.; Li, B.; Zhao, C. Selective conversion of coconut oil to fatty alcohols in methanol over a hydrothermally prepared Cu/SiO$_2$ catalyst without extraneous hydrogen. *Chem. Commun.* **2017**, *53*, 6152–6155. [CrossRef] [PubMed]

135. Peng, B.; Yuan, X.; Zhao, C.; Lercher, J.A. Stabilizing catalytic pathways via redundancy: Selective reduction of microalgae oil to alkanes. *J. Am. Chem. Soc.* **2012**, *134*, 9400–9405. [CrossRef]

136. Santillan-Jimenez, E.; Morgan, T.; Loe, R.; Crocker, M. Continuous catalytic deoxygenation of model and algal lipids to fuel-like hydrocarbons over Ni–Al layered double hydroxide. *Catal. Today* **2015**, *258*, 284–293. [CrossRef]

137. Cheng, J.; Zhang, Z.; Zhang, X.; Fan, Z.; Liu, J.; Zhou, J. Continuous hydroprocessing of microalgae biodiesel to jet fuel range hydrocarbons promoted by Ni/hierarchical mesoporous Y zeolite catalyst. *Int. J. Hydrogen Energy* **2019**, *44*, 11765–11773. [CrossRef]

138. Loe, R.; Santillan-Jimenez, E.; Morgan, T.; Sewell, L.; Ji, Y.; Jones, S.; Isaacs, M.A.; Lee, A.F.; Crocker, M. Effect of Cu and Sn promotion on the catalytic deoxygenation of model and algal lipids to fuel-like hydrocarbons over supported Ni catalysts. *Appl. Catal. B Environ.* **2016**, *191*, 147–156. [CrossRef]

139. Santillan-Jimenez, E.; Pace, R.; Marques, S.; Morgan, T.; McKelphin, C.; Mobley, J.; Crocker, M. Extraction, characterization, purification and catalytic upgrading of algae lipids to fuel-like hydrocarbons. *Fuel* **2016**, *180*, 668–678. [CrossRef]

140. Bala, D.D.; Chidambaram, D. Production of renewable aviation fuel range alkanes from algae oil. *RSC Adv.* **2016**, *6*, 14626–14634. [CrossRef]

141. Saber, M.; Nakhshiniev, B.; Yoshikawa, K. A review of production and upgrading of algal bio-oil. *Renew. Sustain. Energy Rev.* **2016**, *58*, 918–930. [CrossRef]

142. Elkasabi, Y.; Chagas, B.M.E.; Mullen, C.A.; Boateng, A.A. Hydrocarbons from Spirulina Pyrolysis Bio-oil Using One-Step Hydrotreating and Aqueous Extraction of Heteroatom Compounds. *Energy Fuels* **2016**, *30*, 4925–4932. [CrossRef]

143. Guo, Q.; Wu, M.; Wang, K.; Zhang, L.; Xu, X. Catalytic Hydrodeoxygenation of Algae Bio-oil over Bimetallic Ni–Cu/ZrO$_2$ Catalysts. *Ind. Eng. Chem. Res.* **2015**, *54*, 890–899. [CrossRef]

144. Patel, B.; Arcelus-Arrillaga, P.; Izadpanah, A.; Hellgardt, K. Catalytic Hydrotreatment of algal biocrude from fast Hydrothermal Liquefaction. *Renew. Energy* **2017**, *101*, 1094–1101. [CrossRef]

145. Shakya, R.; Adhikari, S.; Mahadevan, R.; Hassan, E.B.; Dempster, T.A. Catalytic upgrading of bio-oil produced from hydrothermal liquefaction of Nannochloropsis sp. *Bioresour. Technol.* **2018**, *252*, 28–36. [CrossRef] [PubMed]

146. Biller, P.; Sharma, B.K.; Kunwar, B.; Ross, A.B. Hydroprocessing of bio-crude from continuous hydrothermal liquefaction of microalgae. *Fuel* **2015**, *159*, 197–205. [CrossRef]

147. Duan, P.; Xu, Y.; Wang, F.; Wang, B.; Yan, W. Catalytic upgrading of pretreated algal bio-oil over zeolite catalysts in supercritical water. *Biochem. Eng. J.* **2016**, *116*, 105–112. [CrossRef]

148. Castello, D.; Haider, M.S.; Rosendahl, L.A. Catalytic upgrading of hydrothermal liquefaction biocrudes: Different challenges for different feedstocks. *Renew. Energy* **2019**, *141*, 420–430. [CrossRef]

149. Xu, Y.; Duan, P.; Wang, B. Catalytic upgrading of pretreated algal oil with a two-component catalyst mixture in supercritical water. *Algal Res.* **2015**, *9*, 186–193. [CrossRef]

150. Zan, Y.; Sun, Y.; Kong, L.; Miao, G.; Bao, L.; Wang, H.; Li, S.; Sun, Y. Formic Acid-Induced Controlled-Release Hydrolysis of Microalgae (Scenedesmus) to Lactic Acid over Sn-Beta Catalyst. *ChemSusChem* **2018**, *11*, 2492–2496. [CrossRef]

151. Wang, J.J.; Tan, Z.C.; Zhu, C.C.; Miao, G.; Kong, L.Z.; Sun, Y.H. One-pot catalytic conversion of microalgae (Chlorococcum sp.) into 5-hydroxymethylfurfural over the commercial H-ZSM-5 zeolite. *Green Chem.* **2016**, *18*, 452–460. [CrossRef]

152. Francavilla, M.; Intini, S.; Luchetti, L.; Luque, R. Tunable microwave-assisted aqueous conversion of seaweed-derived agarose for the selective production of 5-hydroxymethyl furfural/levulinic acid. *Green Chem.* **2016**, *18*, 5971–5977. [CrossRef]

153. Chen, Y.; Zhou, Y.; Zhang, R.; Hu, C. Conversion of saccharides in enteromorpha prolifera to furfurals in the presence of FeCl$_3$. *Mol. Catal.* **2019**. [CrossRef]

154. Zhang, R.; Chen, Y.; Zhou, Y.; Tong, D.; Hu, C. Selective Conversion of Hemicellulose in Macroalgae Enteromorpha prolifera to Rhamnose. *ACS Omega* **2019**, *4*, 7023–7028. [CrossRef] [PubMed]

MDPI

St. Alban-Anlage 66

4052 Basel

Switzerland

Tel. +41 61 683 77 34

Fax +41 61 302 89 18

www.mdpi.com

Catalysts Editorial Office

E-mail: catalysts@mdpi.com

www.mdpi.com/journal/catalysts